Published for
**OXFORD INTERNATIONAL
AQA EXAMINATIONS**

# International A2 Level
# MATHEMATICS
## Pure and Mechanics

T0177775

Sue Chandler

**OXFORD**
UNIVERSITY PRESS

Great Clarendon Street, Oxford, OX2 6DP, United Kingdom

Oxford University Press is a department of the University of Oxford. It furthers the University's objective of excellence in research, scholarship, and education by publishing worldwide. Oxford is a registered trade mark of Oxford University Press in the UK and in certain other countries

British Library Cataloguing in Publication Data
Data available

978-0-19-837598-2

13

Paper used in the production of this book is a natural, recyclable product made from wood grown in sustainable forests. The manufacturing process conforms to the environmental regulations of the country of origin.

Printed and bound by CPI Group (UK) Ltd, Croydon, CR0 4YY

**Acknowledgements**
The publishers would like to thank the following for permissions to use their photographs:

**Cover:** Colin Anderson/Getty Images.

**Header:** Shutterstock.

Although we have made every effort to trace and contact all copyright holders before publication this has not been possible in all cases. If notified, the publisher will rectify any errors or omissions at the earliest opportunity.

Links to third party websites are provided by Oxford in good faith and for information only. Oxford disclaims any responsibility for the materials contained in any third party website referenced in this work.

AQA material is reproduced by permission of AQA.

# Contents

# About this book

This book has been specially created for the Oxford AQA International A2 Level Mathematics examination (9660).

It has been written by an experienced team of teachers, consultants and examiners and is designed to help you obtain the best possible grade in your maths qualification.

In each chapter the lessons are organised in a logical order to help you to progress through each topic. At the start of each chapter you can see an Introduction, to show you how you will use the knowledge in this chapter, a Recap of what prior knowledge you will need to recall and a clear list of the Objectives that you will fulfil by the end of the chapter.

The Note boxes give you help and support as you work through the examples and exercises.

Clear, worked examples show you how to tackle each question and the steps needed to reach the answer.

Key points are in bold and the chapter colour to make it clear that this information is important.

Exercises allow you to apply the skills that you have learned, and give the opportunity to practise your reasoning and problem solving abilities.

At the end of a chapter you will find a summary of what you have learned, together with a review section that allows you to test your fluency in the basic skills. Finally there is an Assessment section where you can practise exam-style questions.

At the end of the book you will find a comprehensive glossary of key phrases and terms and a full set of answers to all of the exercises.

We wish you well with your studies and hope that you enjoy this course and achieve exam success.

# 1 Functions

## Introduction

This chapter extends the work on functions introduced at AS-level and gives various methods for expressing algebraic fractions in simpler forms. These methods are needed later in the course for integrating and differentiating fractions.

## Recap

You will need to remember...

- ▶ The properties and the shapes of the graphs of linear, quadratic, exponential and trigonometric functions.
- ▶ The effect of simple transformations on a graph, including translations, one-way stretches and reflections in the $x$- and $y$-axes.
- ▶ The Cartesian equation of a curve gives the relationship between the $x$- and $y$-coordinates of points on the curve.
- ▶ How to complete the square for a quadratic function.
- ▶ How to factorise quadratic expressions.
- ▶ The remainder theorem.

## Objectives

By the end of this chapter, you should know how to...

- ▶ Define a function, range of a function and domain of a function.
- ▶ Introduce inverse functions, composite functions and modulus functions.
- ▶ Use combinations of transformations to help to sketch graphs.
- ▶ Simplify an algebraic fraction by dividing by common factors.
- ▶ Decompose algebraic fractions into simpler fractions.

## 1.1 Functions

When you substitute any number for $x$ in the expression $x^2 - 2x$, you get a single answer.

For example when $x = 3$, $x^2 - 2x = 3$.

However, when you substitute a positive number for $x$ in the expression $\pm\sqrt{x}$, you have two possible answers.

For example when $x = 4$, $\pm\sqrt{x} = -2$ or $2$.

A *function* of one variable is such that when a number is substituted for the variable, there is only one answer.

Therefore $x^2 - 2x$ is an example of a function f and can be written as $f(x) = x^2 - 2x$.

However, $\pm\sqrt{x}$ is not a function of $x$ because any positive value of $x$ gives two answers.

## Domain and range

The set of values which the variable in a function can take is called the *domain* of the function.

The domain does not have to contain all possible values of the variable; it can be as wide, or as restricted, as needed. Therefore to define a function fully, the domain must be stated.

If the domain is not stated, assume that it is the set of all **real numbers** (the set of real numbers is denoted by $\mathbb{R}$).

> For each domain, there is a corresponding set of values of f($x$).
> These are values which the function can take for values of $x$ in that
> particular domain. This set is called the *range* of the function.

Look at the expression $x^2 + 3$.

A function f for this expression can be defined over any domain.
Some examples, with their graphs are given.

**1** $f(x) = x^2 + 3$  for  $x \in \mathbb{R}$
(the symbol $\in$ means 'is a member of').
The range is $f(x) \geq 3$.

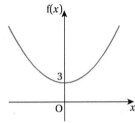

**2** $f(x) = x^2 + 3$  for  $x \geq 0$.
The range is also $f(x) \geq 3$.

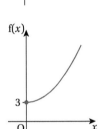

> **Note**
>
> The point on the curve where
> $x = 0$ is included and this is
> denoted this by a solid dot.
> If the domain were $x > 0$,
> then the point would not be
> part of the curve and this is
> indicated by a hollow dot.

**3** $f(x) = x^2 + 3$  for  $x = 1, 2, 3, 4, 5$.
The range is the set of numbers 4, 7, 12, 19, 28.

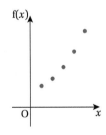

> **Note**
>
> This time the graphical
> representation consists of just
> five separate points.

## Example 1

The function, f, is defined by $f(x) = x^2$  for  $x \leq 0$
and $f(x) = x$  for  $x > 0$.

**a** Find f(4) and f(−4).

**b** Sketch the graph of f($x$).

**c** Give the range of f.

**a** For $x > 0$, $f(x) = x$,
therefore f(4) = 4.
For $x \leq 0$, $f(x) = x^2$,
therefore $f(-4) = (-4)^2 = 16$.

*(continued)*

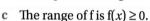

**b** To sketch the graph of a function, use what you know about lines and curves in the *xy*-plane.

So $f(x) = x$ for $x > 0$ is the part of the line $y = x$ which corresponds to positive values of $x$, and $f(x) = x^2$ for $x \leq 0$ is the part of the parabola $y = x^2$ that corresponds to negative values of $x$.

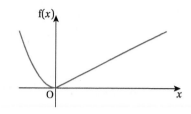

**c** The range of f is $f(x) \geq 0$.

## Exercise 1

**1** Find the range of f in each of the following cases.

   **a** $f(x) = 2x - 3$    for    $x \geq 0$

   **b** $f(x) = x^2 - 5$    for    $x \leq 0$

   **c** $f(x) = 1 - x$    for    $x \leq 1$

   **d** $f(x) = \dfrac{1}{x}$    for    $x \geq 2$

**2** Sketch the graph of each function given in question 1.

**3** The function f is such that $f(x) = -x$    for    $x < 0$

                      and $f(x) = x$    for    $x \geq 0$.

   **a** Find the value of f(5), f(–4), f(–2) and f(0).

   **b** Sketch the graph of the function.

**4** The function f is such that $f(x) = x$    for    $0 \leq x \leq 5$

                      and $f(x) = 5$    for    $x > 5$.

   **a** Find the value of f(0), f(2), f(4), f(5) and f(7).

   **b** Sketch the graph of the function.

   **c** Give the range of the function.

## 1.2 Composite functions

Look at the two functions f and g given by $f(x) = x^2$ and $g(x) = \dfrac{1}{x}$ for $x \neq 0$.

When $g(x)$ replaces $x$ in $f(x)$ this gives the **composite function**

$$f[g(x)] = f\left(\frac{1}{x}\right) = \frac{1}{x^2} \text{ for } x \neq 0$$

A composite function formed this way is also called a **function of a function** and it is denoted by fg.

For example, if $f(x) = 3^x$ and $g(x) = 1 - x$ then $gf(x)$ means the function g of $f(x)$.

$\Rightarrow$       $gf(x) = g(3^x) = 1 - 3^x$

Also     $fg(x) = f(1 - x) = 3^{(1-x)}$

This example shows that $gf(x)$ is *not* always the same as $fg(x)$.

## Exercise 2

**1** The functions f, g and h are defined by $f(x)=x^2$, $g(x)=\dfrac{1}{x}$ for $x\neq 0$ and $h(x)=1-x$.
Find

  **a** $fg(x)$    **b** $fh(x)$    **c** $hg(x)$    **d** $hf(x)$    **e** $gf(x)$

**2** When $f(x)=2x-1$ and $g(x)=x^3$ find the value of

  **a** $gf(3)$    **b** $fg(2)$    **c** $fg(0)$    **d** $gf(0)$

**3** Given that $f(x)=2x$, $g(x)=1+x$ and $h(x)=x^2$, find

  **a** $hg(x)$    **b** $gh(x)$    **c** $gf(x)$

**4** When $f(x)=\sin x$ and $g(x)=3x-4$ find

  **a** $fg(x)$    **b** $gf(x)$

## 1.3 Inverse functions

Look at the function f where $f(x)=2x$ for $x=2,3,4$.

The domain of f is $\{2,3,4\}$ and the range of f is $\{4,6,8\}$. The relationship between the domain and range is shown in the arrow diagram.

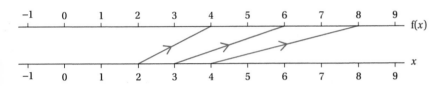

It is possible to reverse this process, so that each member of the range can be mapped back to the corresponding member of the domain by halving each member of the range.

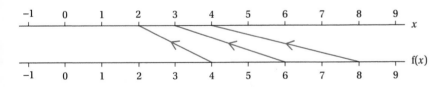

This process can be expressed algebraically.

When $x=4,6,8$, then $x\to\dfrac{1}{2}x$ maps 4 to 2, 6 to 3 and 8 to 4.

This reverse mapping is a function in its own right and it is called the **inverse function** of f where $f(x)=2x$.

Denoting this inverse function by $f^{-1}$ we can write $f^{-1}(x)=\dfrac{1}{2}x$ for $x=4,6,8$.

The function $f(x)=2x$ for $x\in\mathbb{R}$ also has an inverse function, given by $f^{-1}(x)=\dfrac{1}{2}x$ which also has domain $x\in\mathbb{R}$.

> If a function g exists that maps the range of f back to its domain, then g is called the inverse of f and it is denoted by $f^{-1}$.

# The graph of a function and its inverse

Consider the curve g($x$) that is obtained by reflecting $y = f(x)$ in the line $y = x$ (see graph).

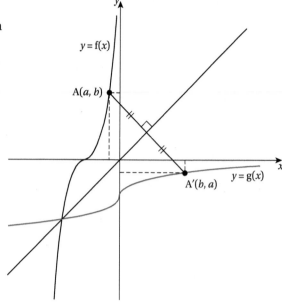

A point A($a$, $b$) on the curve $y = f(x)$ is reflected onto a point A′ on the curve $y = g(x)$, whose coordinates are ($b$, $a$). Hence, interchanging the $x$- and $y$-coordinates of A gives the coordinates of A′.

> The equation of $y = g(x)$ is found by interchanging $x$ and $y$ in the equation $y = f(x)$.

The coordinates of A on $y = f(x)$ are [$a$, f($a$)].

Therefore the coordinates of A′ on $y = g(x)$ are [f($a$), $a$].

So the range of $y = f(x)$ becomes the domain of $y = g(x)$.

> When the equation of the reflected curve is $y = g(x)$ then g is the inverse of f, so $g = f^{-1}$.

Any curve whose equation can be written in the form $y = f(x)$ can be reflected in the line $y = x$. However this reflected curve may not have an equation that can be written in the form $y = f^{-1}(x)$.

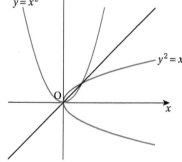

For example, look at the curve $y = x^2$ and its reflection in the line $y = x$ (see graph).

The equation of the reflected curve is $x = y^2$, giving $y = \pm \sqrt{x}$ and $\pm \sqrt{x}$ is not a function.

Therefore the function f where $f(x) = x^2$ does not have an inverse. You can also see this from the diagram, because on the reflected curve, one value of $x$ maps to two values of $y$. So in this case $y$ cannot be written as a function of $x$.

> Not every function has an inverse.

However, by changing the definition of f to $f(x) = x^2$ for $x \geq 0$, then the reflected curve is $y = \sqrt{x}$ for $x \geq 0$, and $\sqrt{x}$ is a function for positive real numbers. You can see this in the graph. Therefore $f^{-1}(x) = \sqrt{x}$ for $x \geq 0$.

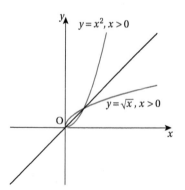

**To summarise:**

▶ The inverse of a function undoes the function, i.e. it maps the range of a function to its domain.

▶ The inverse of the function f is written $f^{-1}$.

▶ Not all functions have an inverse.

▶ When the curve whose equation is $y = f(x)$ is reflected in the line $y = x$, the equation of the reflected curve is $x = f(y)$.

▶ If this equation can be written in the form $y = g(x)$ then g is the inverse of f, so $g(x) = f^{-1}(x)$, and the domain of g is the range of f.

## Example 2

Determine whether there is an inverse of the function f given by $f(x) = 2 + \dfrac{1}{x}$, $x \neq 0$

If $f^{-1}$ exists, express it as a function of $x$ and give its domain.

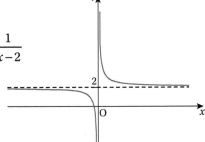

The sketch of $f(x) = 2 + \dfrac{1}{x}$ shows that one value of $f(x)$ maps to one value of $x$, therefore the reverse mapping is a function.

The equation of the reflection of $y = 2 + \dfrac{1}{x}$ can be written as $x = 2 + \dfrac{1}{y} \;\Rightarrow\; y = \dfrac{1}{x-2}$

Therefore when $f(x) = 2 + \dfrac{1}{x}$, $f^{-1}(x) = \dfrac{1}{x-2}$ for $x \in \mathbb{R}$, provided that $x \neq 2$.

## Example 3

The function f is given by $f(x) = 5x - 1$.

a   Find $f^{-1}(4)$.

b   Solve the equation $f^{-1}(x) = x$.

a   Let $y = f(x)$,   that is   $y = 5x - 1$.

The equation of the reflected line is

$$x = 5y - 1 \;\Rightarrow\; y = \frac{1}{5}(x+1)$$

So   $f^{-1}(x) = \dfrac{1}{5}(x+1)$

Therefore   $f^{-1}(4) = \dfrac{1}{5}(4+1) = 1$

b    $f^{-1}(x) = x \;\Rightarrow\; \dfrac{1}{5}(x+1) = x$

$$\Rightarrow \quad x + 1 = 5x$$

Therefore $x = \dfrac{1}{4}$.

## Exercise 3

1   Sketch the graphs of $y = f(x)$ and $y = f^{-1}(x)$ on the same axes.

a   $f(x) = 3x - 1$

b   $f(x) = (x - 1)^3$

c   $f(x) = 2 - x$

d   $f(x) = \dfrac{1}{x-3}$

e   $f(x) = \dfrac{1}{x}$

2   Determine whether f has an inverse function and, if it does, find it.

a   $f(x) = x + 1$

b   $f(x) = x^2 + 1$

c   $f(x) = x^3 + 1$

**d**  $f(x) = x^2 - 4,\ x \geq 0$

**e**  $f(x) = (x+1)^4,\ x \geq -1$

**3**  The function f is given by $f(x) = 1 - \dfrac{1}{x}$ for $x \neq 0$.

    **a**  Find $f^{-1}(4)$.

    **b**  Solve the equation $f^{-1}(x) = 2$.

**4**  The function f is given by $f(x) = \sqrt{3x-2}$ for $x \geq \dfrac{2}{3}$.

    **a**  Find $f^{-1}(x)$.

    **b**  Solve the equation $f^{-1}(x) = 4$.

    **c**  Solve the equation $f^{-1}(x) = x$.

**5**  The functions f and g are defined as $f(x) = x^2$ and $g(x) = \sqrt{1-x}$ for all values of $x < 1$.

    **a**  Find $fg(x)$ and show that the inverse of fg exists.

    **b**  Find $gf(x)$ and explain why gf does not have an inverse.

## 1.4 Modulus functions

The *modulus* of $x$ is written as $|x|$ and it equals the positive value of $x$, whether $x$ itself is positive or negative.

For example, $|2| = 2$ and $|-2| = 2$.

Therefore the graph of $y = |x|$ can be found from the graph of $y = x$ by reflecting the part of the graph where $y$ is negative in the $x$-axis.

In general, the curve $C_1$ whose equation is $y = |f(x)|$ is obtained from the curve $C_2$ with equation $y = f(x)$, by reflecting in the $x$-axis the parts of $C_2$ for which $f(x)$ is negative. The remaining sections of $C_1$ are not changed.

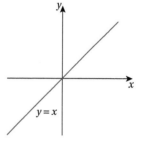

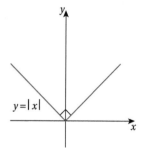

### Example 4

Sketch the curve $y = |(x-1)(x-2)|$.

Start by sketching the curve $y = (x-1)(x-2)$.

Then reflect in the $x$-axis the part of this curve which is below the $x$-axis.

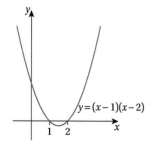

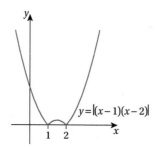

## Exercise 4

Sketch the following graphs.

**1** $y = |2x - 1|$

**2** $y = |x(x - 1)(x - 2)|$

**3** $y = |x^2 - 1|$

**4** $y = |x^2 + 1|$

**5** $y = |\sin x|$ for $0 < x < 2\pi$

**6** $y = |3 - 6x|$

**7** $y = |\cos x|$ for $0 < x < 2\pi$

**8** $y = |1 - x^2|$

**9** $y = |x^3|$

**10** $y = |x^2 - x - 20|$

**11** $y = |2 - x^2|$

**12** $y = |\tan x|$ for $0 < x < \pi$

## The Cartesian equation of a modulus function

When a section of the curve $y = f(x)$ is reflected in the $y$-axis,
the equation of that part of the curve becomes $y = -f(x)$

For example, when $y = |x|$ we can write this equation as

$$\begin{cases} y = x & \text{for } x \geq 0 \\ y = -x & \text{for } x < 0 \end{cases}$$

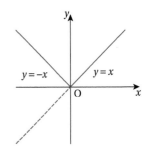

## Intersection and solving equations

To find the points of intersection between two graphs whose equations involve
a modulus, first sketch the graphs to locate the points roughly. Then identify the
equations in Cartesian form for each part of the graph. Write these equations
on the sketch to identify the correct pair of equations to solve simultaneously.

## Example 5

**Question**

Find the exact values of the coordinates of the points of intersection of $y = x - 1$
and $y = |x^2 - 3|$.

**Answer**

The sketch shows the points common to $y = x - 1$ and $y = |x^2 - 3|$.

The sketch also shows that the coordinates of A satisfy the equations

$\quad y = x - 1$ and $y = x^2 - 3$            [1]

and the coordinates of B satisfy the equations

$\quad y = x - 1$ and $y = 3 - x^2$            [2]

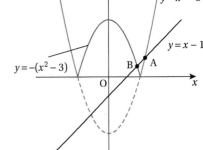

Solving equations [1] gives $x^2 - x - 2 = 0 \implies x = -1$ or 2

The sketch shows that $x \neq -1$, so A is the point (2, 1).

Similarly, solving equations [2] gives $x^2 + x - 4 = 0$

$$\implies x = \frac{1}{2}\left(-1 \pm \sqrt{17}\right)$$

The sketch also shows that the $x$-coordinate of B is positive, so at B,

$$x = \frac{1}{2}\left(-1 + \sqrt{17}\right)$$

*(continued)*

*(continued)*

Then using $y = x - 1$ gives $y = \frac{1}{2}\left(-3 + \sqrt{17}\right)$.

Therefore the coordinates of B are $\left(\frac{1}{2}\left(-1 + \sqrt{17}\right), \frac{1}{2}\left(-3 + \sqrt{17}\right)\right)$

This example again shows how important it is to check solutions to see if they fit the given problem.

An equation such $|2x - 1| = 3x$ can be solved by sketching the graphs of $y = |2x - 1|$ and $y = 3x$ and finding the values of $x$ at their points of intersection.

## Example 6

Solve $|2x - 1| = 3x$.

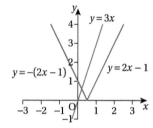

First sketch the graphs of $y = |2x - 1|$ and $y = 3x$.

This shows that there is just one point of intersection of the two graphs, and the value of $x$ at this point is the solution of the equation $|2x - 1| = 3x$.

The graphs $y = |2x - 1|$ and $y = 3x$ intersect where

$$-(2x - 1) = 3x$$

$$\Rightarrow \quad x = \frac{1}{5}$$

## Exercise 5

Find exact values of the co-ordinates of the points of intersection of these graphs.

**1** $y = |x|$ and $y = 1 - x$

**2** $y = x$ and $y = |x^2 - 2x|$

**3** $y = 2|x|$ and $y = 3 + 2x - x^2$

**4** $y = \left|\frac{1}{x}\right|$ and $y = |x|$

**5** $y = |x^2 - 4|$ and $y = 2x + 1$

Solve these equations. Give your answers as exact values.

**6** $|x^2 - 1| = 3x - 1$

**7** $|x + 1| = |4x - 3|$

**8** $2|1 - x| = x$

**9** $|2 - x^2| + 2x + 1 = 0$

**10** $|x + 2| = |x|$

**11** $|x^2 - 1| + 2x = 0$

## Inequalities involving a modulus

Inequalities that involve a modulus function can also be solved by first sketching a graph. Then the points where the graphs intersect can be found.

## Example 7

Find the set of values of $x$ for which $|x + 3| < |x|$.

Start by sketching the graphs $y = |x + 3|$ and $y = |x|$ on the same set of axes.

The sketch shows that the lines intersect where

$$(x + 3) = -x$$

$$\Rightarrow \quad x = -\frac{3}{2}$$

The sketch also shows that $|x + 3| < |x|$ for $x < -\frac{3}{2}$.

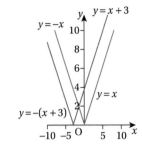

## Exercise 6

In questions 1–7 solve the given inequality. Give your answers in exact form.

**①** $|x - 1| < |x + 2|$

**②** $|x| < |1 - x|$

**③** $|x + 1| < 2x$

**④** $3x - 1 < 1 + |x|$

**⑤** $|3x + 2| > 2 - |x + 1|$

**⑥** $1 + x^2 > |2x + 1|$

**⑦** $|1 - x^2| < 2x + 1$

**⑧** Find the set of values of $x$ between 0 and $2\pi$ for which $|\sin x| < |\cos x|$.

## 1.5 Combinations of transformations

The graphs of many functions can be obtained from transformations of the curves representing basic functions.

The basic transformations are:

The curve $y = f(x - a) + b$ is a translation of the curve $y = f(x)$ by the vector $\begin{bmatrix} a \\ b \end{bmatrix}$.

The curve $y = -f(x)$ is the reflection of the curve $y = f(x)$ in the $x$-axis.

The curve $y = f(-x)$ is the reflection of the curve $y = f(x)$ in the $y$-axis.

The curve $y = af(x)$ is a one-way stretch of the curve $y = f(x)$ by a scale factor $a$ parallel to the $y$-axis.

The curve $y = f(ax)$ is a one-way stretch of the curve $y = f(x)$ by a factor $\frac{1}{a}$ parallel to the $x$-axis.

## Example 8

Sketch the graph of $y = 3 - 2^x$

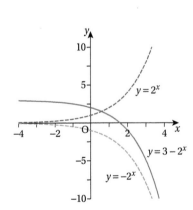

The diagram shows how the shape of the curve is found by combining transformations.

Start with the graph of $y = 2^x$.

Then $y = -2^x$ is the reflection of $y = 2^x$ in the $x$-axis

$y = 3 - 2^x$ is the translation of $y = -2^x$ by $\begin{bmatrix} 0 \\ 3 \end{bmatrix}$.

## Example 9

Describe a sequence of two transformations that map the graph of $y = \sin x$ to the graph of $y = 3 \sin (x - \pi)$.

A translation by the vector $\begin{bmatrix} \pi \\ 0 \end{bmatrix}$ maps $y = \sin x$ to $y = \sin (x - \pi)$.

A one-way stretch by a factor of 3 parallel to the $y$-axis maps $y = \sin (x - \pi)$ to $y = 3 \sin (x - \pi)$.

The sequence of transformations is a translation by the vector $\begin{bmatrix} \pi \\ 0 \end{bmatrix}$ followed by a one-way stretch by a factor of 3 parallel to the $y$-axis.

## Example 10

Describe a sequence of two transformations that maps the graph of $y = x^2$ to the graph of $y = (2x - 3)^2$.

There are two possible combinations:

either $y = x^2$ is mapped to $y = (x - 3)^2$ by the translation $\begin{bmatrix} 3 \\ 0 \end{bmatrix}$, then $y = (x - 3)^2$ is mapped to $y = (2x - 3)^2$ by a one-way stretch parallel to the $x$-axis by a factor $\dfrac{1}{2}$.

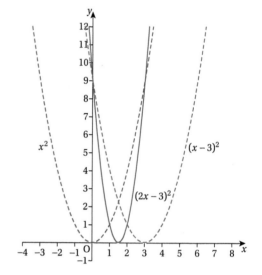

or $y = x^2$ is mapped to $y = (2x)^2 (= 4x^2)$ by a one-way stretch parallel to the $y$-axis by a factor 4, then $y = (2x)^2$ is mapped to $y = (2x - 3)^2 \left( = 4\left(x - \dfrac{3}{2}\right)^2 \right)$ by the translation $\begin{bmatrix} \frac{3}{2} \\ 0 \end{bmatrix}$.

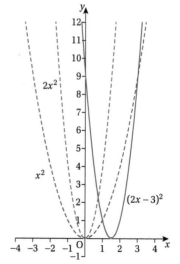

## Exercise 7

In questions 1 to 5, sketch the graph of the equation.

**1** $y = 3 - |x|$ 　　　　 **2** $y = 2 + (x-1)^2$ 　　　　 **3** $y = 2\cos\left(x + \frac{1}{2}\pi\right)$

**4** $y = 1 - 2^{(x-1)}$ 　　　　 **5** $y = 3(2^{-x})$

In questions 6 to 11, describe a sequence of transformations that maps the graph of the first equation to the graph of the second equation.

**6** $y = \sin x$ to $y = \sin(2x + \pi)$ 　　　　 **7** $y = |x|$ to $y = |2x - 1|$

**8** $y = x^2$ to $y = (1 - x^2)$ 　　　　 **9** $y = \cos x$ to $y = \cos(\pi - x)$

**10** $y = \sqrt{x}$ to $y = 2\sqrt{x - 3}$ 　　　　 **11** $y = 4|x|$ to $y = \left|\frac{1}{2}x\right|$

**12** **a** Express $x^2 - 4x - 6$ in the form $(x - a)^2 + b$

　　 **b** Hence describe a sequence of transformations that maps
　　　 $y = x^2$ to $y = 2(x^2 - 4x - 6)$.

# 1.6 Simplification of rational functions

A *rational function* is a fraction where both numerator and denominator are polynomials.

The value of a fraction is unaltered when both the numerator *and* the denominator are multiplied or divided by the same number.

For example　　$\dfrac{3}{6} = \dfrac{1}{2} = \dfrac{2}{4} = \dfrac{7}{14} = \ldots$　　and　　$\dfrac{ax}{ay} = \dfrac{x}{y} = \dfrac{3x}{3y} = \dfrac{x(a+b)}{y(a+b)} = \ldots$

A fraction can be simplified by multiplying or dividing both numerator and denominator by the same factor.

## Example 11

**Question**

Simplify $\dfrac{2a^2 - 2ab}{6ab - 6b^2}$

**Answer**

First factorise the numerator and the denominator, then divide them by any common factors.

$$\frac{2a^2 - 2ab}{6ab - 6b^2} = \frac{\cancel{2}a\cancel{(a-b)}}{{}_3\cancel{6}b\cancel{(a-b)}}$$

$$= \frac{a}{3b}$$

## Example 12

**Question**

Simplify $\dfrac{\frac{1}{2}x^2 - 2}{\frac{1}{4}y^2 + 3}$

Answer

Remove the fractions in the numerator and the denominator by multiplying them both by 4.

$$\frac{\frac{1}{2}x^2-2}{\frac{1}{4}y^2+3}=\frac{2x^2-8}{y^2+12}$$

$$=\frac{2\left(x^2-4\right)}{y^2+12}$$

$$=\frac{2(x-2)(x+2)}{y^2+12}$$

> **Note**
>
> Factorise where possible because there might be a factor that will cancel.

There are no common factors between numerator and denominator which might allow you to simplify further, so the simplest form of the fraction is $\frac{2\left(x^2-4\right)}{y^2+12}$.

## Exercise 8

Simplify, where possible.

**1** $\dfrac{x-2}{4x-8}$

**2** $\dfrac{6x+12}{3x-6}$

**3** $\dfrac{2a+8}{3a+12}$

**4** $\dfrac{3p-3q}{5p-5q}$

**5** $\dfrac{x^2+xy}{xy+y^2}$

**6** $\dfrac{x+2}{x^2-4}$

**7** $\dfrac{a^2-4}{a-2}$

**8** $\dfrac{x^2y+xy^2}{y^2+\frac{2}{5}xy}$

**9** $\dfrac{\frac{1}{3}a-b}{a+\frac{1}{6}b}$

**10** $\dfrac{2x(b-4)}{6x^2(b+4)}$

**11** $\dfrac{(x-4)(x-3)}{x^2-16}$

**12** $\dfrac{4y^2+3}{y^2-9}$

**13** $\dfrac{\frac{1}{3}(x-3)}{x^2-9}$

**14** $\dfrac{x^2-x-6}{2x^2-5x-3}$

**15** $\dfrac{(x-2)(x+2)}{x^2+x-2}$

**16** $\dfrac{\frac{1}{2}(a+5)}{a^2-25}$

**17** $\dfrac{3p+9q}{p^2+6pq+9q^2}$

**18** $\dfrac{a^2+2a+4}{a^2+7a+10}$

**19** $\dfrac{x^2+2x+1}{3x^2+12x+9}$

**20** $\dfrac{4(x-3)^2}{(x+1)(x^2-2x-3)}$

## 1.7 Algebraic division

A fraction where both the numerator and the denominator are polynomials in the same variable is **proper** if the highest power of $x$ in the numerator is less than the highest power of $x$ in the denominator, for example $\dfrac{x+1}{x^2+x+2}$.

When the fraction is **improper** the numerator can be divided by the denominator using long division. This will give a quotient plus a proper fraction.

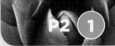
The process of long division is covered at AS-level. This section gives other methods for expressing an improper fraction as the sum of proper fractions.

One method uses the remainder theorem (also covered at AS-level), which states that

> when a polynomial is divided by $(x-a)$ then
> the polynomial is equal to (quotient)$(x-a)$+remainder.

> Hence, the remainder is equal to the value of the polynomial
> when $a$ is substituted for $x$.

Using f$(x)$ to represent the polynomial, the remainder theorem can be expressed as

$$\frac{f(x)}{x-a} = \text{quotient} + \frac{f(a)}{x-a}$$

where $\frac{f(a)}{x-a}$ is a proper fraction and the highest power in the quotient is less than the highest power in f$(x)$.

For example, look at the fraction $\frac{x+5}{x-3}$.

When you divide f$(x) = x+5$ by $x-3$, the remainder is given by f$(3) = 8$ and the quotient is a constant.

Therefore $\frac{x+5}{x-3} = A + \frac{8}{x-3}$ where $A$ is a constant.

You can find the value of $A$ by expressing $A + \frac{8}{x-3}$ as a single fraction:

$$A + \frac{8}{x-3} = \frac{A(x-3)+8}{x-3} = \frac{Ax-3A+8}{x-3}$$

As $\quad \frac{x+5}{x-3} = \frac{Ax-3A+8}{x-3} \quad$ then $\quad x+5 = Ax - 3A + 8.$

Comparing coefficients of $x$ (or the constant) shows that

$$x = Ax \text{ (and } -3A+8=5) \quad \Rightarrow \quad A = 1$$

Therefore $\frac{x+5}{x-3} = 1 + \frac{8}{x-3}.$

Another method uses a rearrangement of the numerator so that part of it is equal to the denominator.

Using $\frac{x+5}{x-3}$ again, subtracting 3 from $x$ and adding 3 to 5 gives

$$\frac{x-3+5+3}{x-3} = \frac{x-3}{x-3} + \frac{8}{x-3} = 1 + \frac{8}{x-3}$$

This method works well with simple fractions.

## Example 13

Question

Express $\frac{2x^3 - x + 5}{x+3}$ as a quadratic expression plus a proper fraction.

$f(x) = 2x^3 - x + 5 \Rightarrow f(-3) = 2(-27) + 3 + 5 = -46$ and the quotient is of the form $Ax^2 + Bx + C$.

Therefore $\dfrac{2x^3 - x + 5}{x + 3} = Ax^2 + Bx + C - \dfrac{46}{x + 3}$

$$= \dfrac{(Ax^2 + Bx + C)(x + 3) - 46}{x + 3}$$

$$= \dfrac{Ax^3 + (B + 3A)x^2 + (3B + C)x + (3C - 46)}{x + 3}$$

$\Rightarrow \qquad 2x^3 - x + 5 = Ax^3 + (B + 3A)x^2 + (3B + C)x + (3C - 46)$

Comparing coefficients of $x^3$, $x^2$ and $x$ gives

$A = 2$

$B + 3A = 0 \quad \Rightarrow \quad B = -6$

$3B + C = -1 \quad \Rightarrow \quad C = 17$

Therefore $\dfrac{2x^3 - x + 5}{x + 3} = 2x^2 - 6x + 17 - \dfrac{46}{x + 3}$.

> **Note**
>
> Check: comparing constants gives $3C - 46 = 5 \Rightarrow C = 17$

## Exercise 9

**1** Express $\dfrac{x - 2}{x + 2}$ in the form $A + \dfrac{B}{x + 2}$.

**2** Express $\dfrac{2x + 5}{x - 2}$ in the form $A + \dfrac{B}{x - 2}$.

**3** Express $\dfrac{2x^2 + 5x - 3}{x + 2}$ in the form $Ax + B + \dfrac{C}{x + 2}$.

**4** Express $\dfrac{x^2 - x + 4}{x + 1}$ in the form $Ax + B + \dfrac{C}{x + 1}$.

**5** Express $\dfrac{4x^3 + x - 1}{x - 1}$ in the form $Ax^2 + Bx + C + \dfrac{D}{x - 1}$.

**6** Express $\dfrac{2x^3 - x^2 + 2}{x - 2}$ in the form $Ax^2 + Bx + C + \dfrac{D}{x - 2}$.

**7** Express these fractions as the sum of a polynomial and a proper fraction.

   **a** $\dfrac{2x}{x - 2}$      **b** $\dfrac{x^2 + 3}{x^2 - 1}$      **c** $\dfrac{x^2}{x - 2}$

## 1.8 Partial fractions

You can add two fractions together to give a single fraction. For example

$$\dfrac{3}{x - 1} + \dfrac{2}{x + 2} = \dfrac{3(x + 2) + 2(x - 1)}{(x - 1)(x + 2)} = \dfrac{5x + 4}{(x - 1)(x + 2)}$$

This sections shows how to reverse this process – that is, how an expression such as $\dfrac{x - 2}{(x + 3)(x - 4)}$ can be expressed as the sum of two separate fractions.

This process is called decomposing an expression into **partial fractions**.

Look again at $\dfrac{x - 2}{(x + 3)(x - 4)}$.

This fraction is a proper fraction because the highest power of $x$ in the numerator (1 in this case) is less than the highest power of $x$ in the denominator (2 in this case, after the brackets have been expanded).

Therefore its separate (or partial) fractions also will be proper.

$\dfrac{x-2}{(x+3)(x-4)}$ can be expressed as $\dfrac{A}{x+3}+\dfrac{B}{x-4}$ where $A$ and $B$ are numbers.

Example 14 shows how you can find the values of $A$ and $B$.

## Example 14

Express $\dfrac{x-2}{x^2-x-12}$ in partial fractions.

First factorise the denominator and express as two separate fractions.

$$\dfrac{x-2}{x^2-x-12}=\dfrac{x-2}{(x+3)(x-4)}=\dfrac{A}{x+3}+\dfrac{B}{x-4}$$

Express the separate fractions on the right-hand side as a single fraction over a common denominator.

$$\dfrac{x-2}{(x+3)(x-4)}=\dfrac{A(x-4)+B(x+3)}{(x+3)(x-4)}$$

The denominators are identical so the numerators also are identical.

$$\Rightarrow\quad x-2\equiv A(x-4)+B(x+3)$$

This is *not* an equation but two ways of writing the same expression, so it follows that the two sides are equal for any value of $x$.

Choosing to substitute 4 for $x$ (to eliminate $A$) gives

$$2=A(0)+B(7)\quad\Rightarrow\quad B=\dfrac{2}{7}$$

Choosing to substitute $-3$ for $x$ (to eliminate $B$) gives

$$-5=A(-7)+B(0)$$
$$\Rightarrow\quad A=\dfrac{5}{7}$$

Therefore $\dfrac{x-2}{(x+3)(x-4)}=\dfrac{\frac{5}{7}}{x+3}+\dfrac{\frac{2}{7}}{x-4}$

$$=\dfrac{5}{7(x+3)}+\dfrac{2}{7(x-4)}$$

## Exercise 10

Decompose these expressions into partial fractions.

1. $\dfrac{x-2}{(x+1)(x-1)}$

2. $\dfrac{2x-1}{(x-1)(x-7)}$

3. $\dfrac{4}{(x+3)(x-2)}$

4. $\dfrac{7x}{(2x-1)(x+4)}$

5. $\dfrac{2}{x(x-2)}$

6. $\dfrac{2x-1}{x^2-3x+2}$

**7** $\dfrac{3}{x^2-9}$    **8** $\dfrac{6x+7}{3x(x+1)}$

**9** $\dfrac{9}{2x^2+x}$    **10** $\dfrac{x+1}{3x^2-x-2}$

## A repeated factor in the denominator

Look at the fraction $\dfrac{2x-1}{(x-2)^2}$.

This is a proper fraction, and can be expressed as two fractions with numerical numerators by rearranging the numerator.

$$\frac{2x-1}{(x-2)^2}=\frac{2(x-2)-1+4}{(x-2)^2}=\frac{2}{x-2}+\frac{3}{(x-2)^2}$$

Any fraction whose denominator is a repeated linear factor can be expressed as separate fractions with numerical numerators.

## Example 15

Express $\dfrac{x-1}{(x+1)(x-2)^2}$ in partial fractions.

$$\frac{x-1}{(x+1)(x-2)^2}=\frac{A}{x+1}+\frac{B}{(x-2)}+\frac{C}{(x-2)^2}$$

$$\Rightarrow \quad x-1=(A)(x-2)^2+B(x+1)(x-2)+C(x+1)$$

$x=2$ gives $C=\dfrac{1}{3}$.

$x=-1$ gives $A=\dfrac{-2}{9}$

Comparing coefficients of $x^2$ gives $0=\dfrac{-2}{9}+B \quad \Rightarrow \quad B=\dfrac{2}{9}$

Therefore $\dfrac{x-1}{(x+1)(x-2)^2}=-\dfrac{2}{9(x+1)}+\dfrac{2}{9(x-2)}+\dfrac{1}{3(x-2)^2}$

To summarise, you can decompose a proper fraction into partial fractions. The form of the partial fractions depends on the form of the factors in the denominator, so that

a linear factor in the denominator gives a partial fraction of the form $\dfrac{A}{ax+b}$

a repeated factor in the denominator gives two partial fractions of the form $\dfrac{A}{ax+b}+\dfrac{B}{(ax+b)^2}$

To decompose an improper fraction into partial fractions, you must first express the improper fraction as the sum of a polynomial and a proper fraction.

## Example 16

Express $\dfrac{x^3}{(x+1)(x-3)}$ in partial fractions.

**Answer**

This fraction is improper and it must be rearranged or divided out to obtain a mixed fraction before it can be expressed in partial fractions. The division method is used here but the method given in section 1.7 can also be used.

$$
\begin{array}{r}
x+2 \\
x^2-2x-3\overline{)x^3+0x^2+0x+0} \\
\underline{x^3-2x^2-3x} \\
2x^2+3x \\
\underline{2x^2-4x-6} \\
7x+6
\end{array}
$$

Therefore $\quad \dfrac{x^3}{(x+1)(x-3)}=x+2+\dfrac{7x+6}{(x+1)(x-3)}$

Next express $\dfrac{7x+6}{(x+1)(x-3)}$ in partial fractions.

$$\dfrac{7x+6}{(x+1)(x-3)}=\dfrac{A}{x+1}+\dfrac{B}{x-3}=\dfrac{A(x-3)+B(x+1)}{(x+1)(x-3)} \quad \Rightarrow \quad A=\dfrac{1}{4} \text{ and } B=\dfrac{27}{4}$$

Therefore $\dfrac{x^3}{(x+1)(x-3)}=x+2+\dfrac{1}{4(x+1)}+\dfrac{27}{4(x-3)}.$

## Exercise 11

Express these as partial fractions.

1. $\dfrac{2}{(x-1)(x+1)^2}$
2. $\dfrac{2x^2+x+1}{(x-3)(x+1)^2}$
3. $\dfrac{x}{(x-1)(x-2)^2}$

4. $\dfrac{(x^2-1)}{x^2(2x+1)}$
5. $\dfrac{x^2-2}{(x+3)(x-1)^2}$
6. $\dfrac{(x-1)}{(x+1)(x+2)^2}$

7. $\dfrac{2}{(x-2)(x-1)}$
8. $\dfrac{5}{(x-1)(x+2)^2}$
9. $\dfrac{3x}{(2x-1)(x-3)}$

10. $\dfrac{x}{(x-2)^2(x+1)}$
11. $\dfrac{x}{(x+2)(x-4)}$
12. $\dfrac{5x}{(2x-1)(x-2)^2}$

13. $\dfrac{5}{(x+2)(x-3)}$
14. $\dfrac{3}{(x-1)^2(x-4)}$
15. $\dfrac{x}{(2x+3)(x+1)}$

16. $\dfrac{3}{(3x-1)(x-1)}$
17. $\dfrac{3}{x^2(2x-3)}$
18. $\dfrac{x^2}{(x+1)(x-1)}$

19. $\dfrac{x^3+3}{(x-1)(x+1)}$
20. $\dfrac{x^2-2}{(x+3)(x-1)}$

# Summary

▶ A function f where $f(x)$ is any expression involving one variable which gives a single value of $f(x)$ for each value of $x$.

▶ The set of values which the variable in a function can take is called the domain of the function.

▶ For each domain, there is a corresponding set of values of $f(x)$. These are values which $f(x)$ can take for values of $x$ in that particular domain. This set is called the range of the function.

▶ The composite function fg means that $g(x)$ replaces $x$ in $f(x)$.

▶ If a function g exists that maps the range of f back to its domain, then g is called the inverse of f and it is denoted by $f^{-1}$.

▶ When curve $y = f(x)$ is reflected in the line $y = x$, the equation of the reflected curve is found by interchanging $x$ and $y$ in the equation $y = f(x)$.

▶ When the equation of the reflected curve is $y = g(x)$, g is called the inverse of f, so $g = f^{-1}$.

▶ The modulus of $f(x)$ is written as $|f(x)|$ and it equals the positive value of $f(x)$, whether $f(x)$ itself is positive or negative.

▶ A rational expression can be simplified by factorising the numerator and the denominator and then dividing both by any common factors.

▶ A proper fraction can be decomposed into partial fractions and the form of the partial fractions depends on the form of the factors in the denominator.

▶ A linear factor gives a partial fraction of the form $\dfrac{A}{ax+b}$.

▶ A repeated factor gives two partial fractions of the form $\dfrac{A}{ax+b} + \dfrac{B}{(ax+b)^2}$.

▶ When the fraction is improper it must first be expressed as the sum of a polynomial and a proper fraction, and can then be decomposed into partial fractions.

# Review

**1** The function f is defined by $f(x) = \sqrt{x-1}$ for $x > 1$.

   **a** Find the range of f.

   **b** Find the value of $f(10)$.

**2** The function f is defined by

      $f(x) = \sin x$    for    $0 \le x < \pi$

      $f(x) = \pi - x$    for    $\pi \le x < 2\pi$.

   **a** Sketch the graph of $f(x)$ for $0 \le x < 2\pi$.

   **b** Find the range of f.

**3** The functions f and g are defined by $f(x) = \sin x$ and $g(x) = \sqrt{x}$ both for $x \ge 0$.

   **a** Find $gf(x)$.

   **b** State a domain of $gf(x)$ so that gf has real values.

**4**  **a** Solve the equation $|x+2| = 1 - x$.

   **b** Show that there are no values of $x$ for which $|x| + 1 = x - |x|$.

**5** Describe a sequence of transformations that maps the graph of $y = 2^x$ to the graph of $y = 3 + 2^{-x}$.

**6** Simplify

a $\dfrac{x^2 - 9}{2x - 6}$

b $\dfrac{4x^2 - 25}{4x^2 + 20x + 25}$

**7** Express $\dfrac{x - 3}{x + 6}$ as a number plus a proper fraction.

**8** Express $\dfrac{3x^2 - 5x + 1}{x + 3}$ as a linear polynomial plus a proper fraction.

**9** Express $\dfrac{x^3 - 4x^2 + 5}{x - 1}$ as a quadratic polynomial plus a proper fraction.

**10** Express in partial fractions.

a $\dfrac{4}{(2x + 1)(x - 3)}$

b $\dfrac{(3x - 2)}{(x + 1)(4x - 3)}$

c $\dfrac{2t}{(t^2 - 1)}$

**11** Express in partial fractions.

a $\dfrac{x + 4}{(x + 3)(x - 5)}$

b $\dfrac{(2x - 3)}{(x - 2)(4x - 3)}$

c $\dfrac{4x^2}{4x^2 - 9}$

**12** Express in partial fractions.

a $\dfrac{3x}{2x^2 - 2x - 4}$

b $\dfrac{3x - 1}{x^2(x - 3)}$

## Assessment

**1** The function f is defined by $f(x) = \sqrt{x - 1}$ for $x \geq 1$.

a Find $f^{-1}(x)$ and state its domain and range.

b Solve the equation $f^{-1}(x) = 2x$.

**2** a Express $\dfrac{x^2}{x^2 - 4}$ as a linear function plus a proper fraction.

b Hence express $\dfrac{x^2}{x^2 - 4}$ in partial fractions.

**3** a Describe a sequence of two transformations that maps the graph of $y = |x + 1|$ to the graph of $y = 1 - |1 + x|$.

b Sketch the graph of $y = 1 - |1 + x|$.

c Find the coordinates of the points of intersection of the graphs of $y = |x + 1|$ and $y = 1 - |1 + x|$.

d Hence find the possible values of $x$ for which $|x + 1| > 1 - |1 + x|$.

**4** Express each rational function in partial fractions.

a $\dfrac{4}{x^2 - 7x - 8}$

b $\dfrac{2x - 1}{(2x + 1)(x - 2)^2}$

c $\dfrac{3}{x(2x + 1)}$

**5** a Sketch the graph of $f(x) = \cos x$ for the domain $0 \leq x \leq 2\pi$.

b State the range of f.

c Given that $g(x) = 1 - |\cos x|$, find $fg(x)$.

d Find the value of $fg\left(\dfrac{\pi}{2}\right)$.

**6** The curve with equation $y = \dfrac{63}{4x-1}$ is sketched below for $1 \leq x \leq 16$.

The function f is defined by $f(x) = \dfrac{63}{4x-1}$ for $1 \leq x \leq 16$.

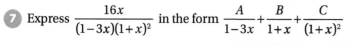

**a** Find the range of f.

**b** The inverse of f is $f^{-1}$.

  **i** Find $f^{-1}$.

  **ii** Solve the equation $f^{-1}(x) = 1$.

**c** The function g is defined by $g(x) = x^2$ for $-4 \leq x \leq -1$

  **i** Write down an expression for $fg(x)$.

  **ii** Solve the equation $fg(x) = 1$.

AQA MPC3 January 2012

**7** Express $\dfrac{16x}{(1-3x)(1+x)^2}$ in the form $\dfrac{A}{1-3x} + \dfrac{B}{1+x} + \dfrac{C}{(1+x)^2}$

AQA MPC4 June 2014 (part question)

**8** **a** Sketch the curve with equation $y = 4 - |2x+1|$, indicating the coordinates where the curve crosses the axes.

**b** Solve the equation $x = 4 - |2x+1|$.

**c** Solve the inequality $x < 4 - |2x+1|$.

**d** Describe a sequence of two geometrical transformations that maps the graph of $y = |2x+1|$ onto the graph of $y = 4 - |2x+1|$.

AQA MPC3 June 2015

# 2 Binomial Series

## Introduction

Many functions can be represented by an infinite polynomial expression. These infinite expressions can be used to find polynomial approximations for the given function, and hence approximate values for such functions at given values of $x$.

## Objectives

By the end of this chapter, you should know how to ...

▶ Give the expansion of a binomial expression to a power that is not a positive integer.

▶ Use this expansion to find approximations.

## Recap

You will need to remember...

▶ The expansion of $(a+b)^n$ where $n$ is a positive integer.

▶ The meaning of a convergent series.

▶ The sum to infinity of a geometric series.

▶ How to express a rational function in partial fractions.

## 2.1 The binomial series for any value of $n$

The expansion of $(1+x)^n$ as a series when $n$ is not a positive integer is very similar to the expansion of $(1+x)^n$ when $n$ is a positive integer. However, there are two important differences:

▶ the series does not terminate but carries on to infinity,

▶ the series converges only for values of $x$ in the range $-1 < x < 1$,

$$\text{so } (1+x)^n = 1 + nx + \frac{n(n-1)}{2!}x^2 + \frac{n(n-1)(n-2)}{3!}x^3 + \cdots$$

for *any* value of $n$ provided that $|x| < 1$.

This series converges only when $x$ has a value within the stated range, therefore this range *must be stated* for every expansion.

There are two particular expansions that illustrate this.

Using the expansion of $(1+x)^n$ with $n = -1$ gives

$$(1+x)^{-1} = 1 + (-1)x + \frac{(-1)(-2)}{2!}x^2 + \frac{(-1)(-2)(-3)}{3!}x^3 + \frac{(-1)(-2)(-3)(-4)}{4!}x^4 + \cdots$$

$$= 1 - x + x^2 - x^3 + x^4 - \cdots$$

The right-hand side is a geometric series with common ratio $-x$, and so has a sum to infinity of $\dfrac{1}{1-(-x)} = \dfrac{1}{1+x} = (1+x)^{-1}$ provided that $|x| < 1$, so

$(1+x)^{-1} = 1 - x + x^2 - x^3 + x^4 - \ldots$ provided that $|x| < 1$.

Replacing $x$ with $-x$ gives

$(1-x)^{-1} = 1 + x + x^2 + x^3 + x^4 + \cdots$ provided that $|x| < 1$.

**Note**

These two expansions are worth memorising.

## Example 1

**Question**

Expand each of the following expressions as a series in ascending powers of $x$ up to and including the term in $x^3$. State the values of $x$ for which each expansion is valid.

**a** $(1+x)^{\frac{1}{2}}$    **b** $(1-2x)^{-3}$    **c** $(2-x)^{-2}$

**Answer**

For $|x| < 1$, $(1+x)^n = 1 + nx + \dfrac{n(n-1)}{2!}x^2 + \dfrac{n(n-1)(n-2)}{3!}x^3 + \ldots$    [1]

**a** Replacing $n$ by $\dfrac{1}{2}$ in [1] gives

$(1+x)^{\frac{1}{2}} = 1 + \dfrac{1}{2}x + \dfrac{\dfrac{1}{2}\left(\dfrac{1}{2}-1\right)}{2!}x^2 + \dfrac{\dfrac{1}{2}\left(\dfrac{1}{2}-1\right)\left(\dfrac{1}{2}-2\right)}{3!}x^3 + \ldots$

$= 1 + \dfrac{1}{2}x + \dfrac{\dfrac{1}{2}\left(-\dfrac{1}{2}\right)}{2!}x^2 + \dfrac{\dfrac{1}{2}\left(-\dfrac{1}{2}\right)\left(-\dfrac{3}{2}\right)}{3!}x^3 + \ldots$

$= 1 + \dfrac{x}{2} - \dfrac{x^2}{8} + \dfrac{x^3}{16} - \cdots$ for $|x| < 1$

**b** Replacing $n$ by $-3$ and $x$ by $-2x$ in [1] gives

$(1-2x)^{-3} = 1 + (-3)(-2x) + \dfrac{(-3)(-4)}{2!}(-2x)^2 + \dfrac{(-3)(-4)(-5)}{3!}(-2x)^3 + \ldots$

$= 1 + 6x + 24x^2 + 80x^3 + \ldots$

provided that $|2x| < 1 \implies -\dfrac{1}{2} < x < \dfrac{1}{2}$.

**c** $(2-x)^{-2} = 2^{-2}\left(1 - \dfrac{1}{2}x\right)^{-2}$

Replacing $n$ by $-2$ and $x$ by $-\dfrac{1}{2}x$ in [1] gives

$(2-x)^{-2} = \dfrac{1}{4}\left[1 + (-2)\left(-\dfrac{1}{2}x\right) + \dfrac{(-2)(-3)}{2!}\left(-\dfrac{1}{2}x\right)^2 + \dfrac{(-2)(-3)(-4)}{3!}\left(-\dfrac{1}{2}x\right)^3 + \ldots\right]$

$= \dfrac{1}{4}\left(1 + x + \dfrac{3}{4}x^2 + \dfrac{1}{2}x^3 + \ldots\right)$

$= \dfrac{1}{4} + \dfrac{1}{4}x + \dfrac{3}{16}x^2 + \dfrac{1}{8}x^3 + \ldots$

The expansion of $\left(1 - \dfrac{1}{2}x\right)^{-2}$ is valid for $\left|\dfrac{1}{2}x\right| < 1 \implies -2 < x < 2$.

Therefore the expansion $(2-x)^{-\frac{1}{2}}$ also is valid for $-2 < x < 2$.

## Exercise 1

Expand each the following expressions as a series in ascending powers of $x$ up to and including the term in $x^3$. In each case give the range of values of $x$ for which the expansion converges.

**1** $(1-2x)^{\frac{1}{2}}$

**2** $(1+5x)^{-2}$

**3** $\left(1-\dfrac{1}{2}x\right)^{-3}$

**4** $(1+x)^{\frac{3}{2}}$

**5** $(3+x)^{-1}$

**6** $(4-x)^{\frac{1}{2}}$

**7** $\left(1+\dfrac{x}{2}\right)^{-\frac{1}{2}}$

**8** $\dfrac{1}{(1-x)^2}$

**9** $\dfrac{1}{(2+x)^3}$

**10** $\sqrt{\dfrac{1}{1+x}}$

**11** $\left(1+\dfrac{x^2}{9}\right)^{-1}$

**12** $(4-3x)^{\frac{1}{2}}$

## Series expansion of rational functions

When a rational function such as $\dfrac{5}{(1+3x)(1-2x)}$ is expressed in partial fractions, each fraction can be expanded as a series. These two series can then be added to give a single series as the binomial expansion of $\dfrac{5}{(1+3x)(1-2x)}$. This is shown in Example 2.

## Example 2

**Question**

Express $\dfrac{5}{(1+3x)(1-2x)}$ in partial fractions.

Hence expand $\dfrac{5}{(1+3x)(1-2x)}$ as a series in ascending powers of $x$ up to and including the term in $x^3$.

State the range of values of $x$ for which the expansion is valid.

**Answer**

Expressing $\dfrac{5}{(1+3x)(1-2x)}$ in partial fractions gives

$$\frac{5}{(1+3x)(1-2x)} = \frac{3}{(1+3x)} + \frac{2}{(1-2x)} = 3(1+3x)^{-1} + 2(1-2x)^{-1}$$

Using $(1+x)^{-1} = 1-x+x^2-x^3+\ldots$ for $-1 < x < 1$ and replacing $x$ by $3x$ gives

$$(1+3x)^{-1} = 1-3x+(3x)^2-(3x)^3+\ldots$$
$$= 1-3x+9x^2-27x^3+\ldots \quad \text{for} \quad -1 < 3x < 1$$

Also using $(1-x)^{-1} = 1+x+x^2+x^3+\ldots$ for $-1 < -x < 1$ and replacing $x$ by $2x$ gives

$$(1-2x)^{-1} = 1+(2x)+(2x)^2+(2x)^3+\ldots$$
$$= 1+2x+4x^2+8x^3+\ldots \quad \text{for} \quad -1 < -2x < 1$$

Hence $\dfrac{5}{(1+3x)(1-2x)} = 3(1+3x)^{-1} + 2(1-2x)^{-1}$

$$= (3+2)+(-9+4)x+(27+8)x^2+(-81+16)x^3+\ldots$$
$$= 5-5x+35x^2-65x^3+\ldots$$

provided that $-\dfrac{1}{3} < x < \dfrac{1}{3}$ and $-\dfrac{1}{2} < x < \dfrac{1}{2}$.

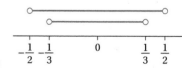

*(continued)*

*(continued)*

Therefore the first four terms of the series are $5 - 5x + 35x^2 - 65x^3$.

The expansion is valid for the range of values of $x$ satisfying both

$-\dfrac{1}{3} < x < \dfrac{1}{3}$ and $-\dfrac{1}{2} < x < \dfrac{1}{2}$. As you can see from the diagram,

the expansion is therefore valid when $-\dfrac{1}{3} < x < \dfrac{1}{3}$.

## Exercise 2

In questions 1 to 8, express the fraction in partial fractions and hence expand the expression as a series in ascending powers of $x$ up to and including the term in $x^3$. In each case give the range of values of $x$ for which the expansion converges.

**1** $\dfrac{1}{(1-x)(1+x)}$

**2** $\dfrac{x+2}{1-x}$

**3** $\dfrac{2}{x(1-x)}$

**4** $\dfrac{1}{(2-x)(1+2x)}$

**5** $\dfrac{2}{9-x^2}$

**6** $\dfrac{x}{(1+x)(1-2x)}$

**7** $\dfrac{3x}{2x^2-2x-4}$

**8** $\dfrac{-x^2}{4-x^2}$

In questions 9 to 11, express the fraction in partial fractions and hence expand the expression as a series in ascending powers of $x$, up to and including the term in $x^2$. Give the range of values of $x$ for which the expansion converges.

**9** $\dfrac{3x-1}{x(3-x)^2}$

**10** $\dfrac{1-x-x^2}{(1-2x)(1-x)^2}$

**11** $\dfrac{1}{(1-3x)(1-x)^2}$

## 2.2 Approximations

A binomial expansion of a function can be used to find a finite polynomial that is approximately equal to the function, for values of $x$ which lie within the interval for which the binomial series converges.

A series can also be used to find an approximate value for an irrational number to a given number of decimal places.

## Example 3

Show that for small values of $x$, $(1-2x)^{\frac{1}{2}} \approx 1 - x - \dfrac{1}{2}x^2$

$$(1-2x)^{\frac{1}{2}} = 1 + \frac{1}{2}(-2x) + \frac{\left(\dfrac{1}{2}\right)\left(-\dfrac{1}{2}\right)}{2!}(-2x)^2 + \ldots$$

$$\approx 1 - x - \frac{1}{2}x^2$$

## Example 4

Use the expansion of $(1-x)^{\frac{1}{2}}$ with $x = 0.02$ to find the decimal value of $\sqrt{2}$ correct to nine decimal places.

$$(1-x)^{\frac{1}{2}} = 1 - \frac{1}{2}x + \frac{\left(\frac{1}{2}\right)\left(-\frac{1}{2}\right)}{2!}(-x)^2 + \frac{\left(\frac{1}{2}\right)\left(-\frac{1}{2}\right)\left(-\frac{3}{2}\right)}{3!}(-x)^3$$

$$+ \frac{\left(\frac{1}{2}\right)\left(-\frac{1}{2}\right)\left(-\frac{3}{2}\right)\left(-\frac{5}{2}\right)}{4!}(-x)^4 + \frac{\left(\frac{1}{2}\right)\left(-\frac{1}{2}\right)\left(-\frac{3}{2}\right)\left(-\frac{5}{2}\right)\left(-\frac{7}{2}\right)}{5!}(-x)^5 - \ldots$$

$$= 1 - \frac{1}{2}x - \frac{1}{8}x^2 - \frac{1}{16}x^3 - \frac{5}{128}x^4 - \frac{7}{256}x^5 - \ldots$$

This series converges when $-1 < x < 1$ and so converges when $x = 0.02$.

Replacing $x$ by 0.02 gives

$$(0.98)^{\frac{1}{2}} = 1 - 0.01 - 0.000\,05 - 0.000\,000\,5 - 0.000\,000\,006\,25 - 0.000\,000\,000\,087\,5 - \ldots$$

The next term in the series is $1.3125 \times 10^{-12}$ and since this does not contribute to the first ten decimal places we do not need it, or any further terms.

So $\sqrt{\dfrac{98}{100}} = 0.989\,949\,493\,7$ to 10 decimal places

$\Rightarrow \dfrac{7}{10}\sqrt{2} = 0.989\,949\,493\,7$ to 10 decimal places

Therefore $\sqrt{2} = 1.414\,213\,562$ correct to 9 decimal places.

## Exercise 3

1. Show that, when $x$ is so small that $x^2$ and higher powers of $x$ may be neglected, $\sqrt{1-x} \approx 1 - \dfrac{1}{2}x$.

2. **a** Express $\dfrac{1}{(1-x)(2+x)}$ in partial fractions.

   **b** Hence show that if $x$ is so small that terms in $x^2$ and higher powers of $x$ may be neglected, then $\dfrac{1}{(1-x)(2+x)} \approx \dfrac{1}{2} + \dfrac{1}{4}x$.

3. **a** Use partial fractions and the binomial series to find a linear expression that is an approximation for $\dfrac{3}{(1-2x)(2-x)}$.

   **b** Give the range of values of $x$ for which the approximation is valid.

4. Find a quadratic expression that approximates to $f(x) = \dfrac{1}{\sqrt[3]{(1-3x)^2}}$ and give the range of values of $x$ for which the approximation is valid.

**5** **a** Show that $\sqrt{121-2x} \approx 11 - \dfrac{x}{11} - \dfrac{x^2}{2662}$ and give the range of values of $x$ for which the approximation is valid.

**b** Hence find $\sqrt{119}$ giving your answer correct to 5 decimal places.

**6** **a** Show that $(125-x)^{\frac{1}{3}} \approx 5 - \dfrac{x}{75} - \dfrac{x^2}{28125}$ and give the range of values of $x$ for which the approximation is valid.

**b** Hence find $\sqrt[3]{124}$ giving your answer correct to 5 decimal places.

**7** **a** Show that $(625-4x)^{\frac{1}{4}} \approx 5 - \dfrac{1}{125}x - \dfrac{3}{156250}x^2$ and give the range of values of $x$ for which the approximation is valid.

**b** Hence find $\sqrt[4]{621}$ giving your answer to 5 decimal places.

**8** **a** Find the binomial expansion of $(169-2x)^{\frac{1}{2}}$ up to and including the term in $x^2$. Give the range of values of $x$ for which the expansion is valid.

**b** Hence find $\sqrt{167}$ giving your answer to 5 decimal places.

**9** **a** Find a quadratic expression that approximates to $(27-2x)^{\frac{1}{3}}$ and give the range of values of $x$ for which the approximation is valid.

**b** Hence find $\sqrt[3]{25}$ giving your answer to 3 decimal places.

**10** By substituting $0.08$ for $x$ in $(1+x)^{\frac{1}{2}}$ and its expansion, find a value for $\sqrt{3}$ correct to four significant figures.

**11** By substituting $\dfrac{1}{10}$ for $x$ in $(1-x)^{-\frac{1}{2}}$ and its expansion, find a value for $\sqrt{10}$ correct to six significant figures.

## Summary

► When $n$ is not a positive integer

$$(1+x)^n = 1 + nx + \frac{n(n-1)}{2!}x^2 + \frac{n(n-1)(n-2)}{3!}x^3 + \ldots$$

for *any* value of $n$ provided that $|x| < 1$.

► In particular $(1+x)^{-1} = 1 - x + x^2 - x^3 + x^4 - \ldots$
and $(1-x)^{-1} = 1 + x + x^2 + x^3 + x^4 - \ldots$ provided that $|x| < 1$.

► When a rational function is expressed in partial fractions, each fraction can be expanded as a binomial series. These two series can then be added to give a single series.

# Review

1. Expand $\dfrac{1}{1+2x}$ as a series in ascending powers of $x$, giving the first three terms.

2. Expand $(1-3x)^{-3}$ as a series in ascending powers of $x$ up to and including the term in $x^2$.

3. a. Express $\dfrac{1}{(1-x)(1-2x)}$ in partial fractions.

   b. Hence expand $\dfrac{1}{(1-x)(1-2x)}$ as a series in ascending powers of $x$ up to and including the term in $x^3$, and state the range of values of $x$ for which the series converges.

4. Find a linear approximation to the curve $y = \dfrac{1}{(1-2x)^2}$ for small values of $x$.

5. a. Show that, when $x$ is small $\dfrac{1}{\sqrt[3]{27-x}} \approx \dfrac{1}{3} + \dfrac{1}{243}x + \dfrac{1}{36561}x^2$.

   b. Use a binomial expansion with $x=1$ to show that $\dfrac{1}{\sqrt[3]{26}} \approx 0.337$.

# Assessment

1. a. Show that $\sqrt[3]{1-2x} \approx 1 - \dfrac{2x}{3} - \dfrac{4x^2}{9} - \dfrac{40x^3}{81}$ and give the values of $x$ for which this approximation is valid.

   b. Show that when $x = \dfrac{1}{10}$, $\sqrt[3]{1-2x} \approx \dfrac{2}{\sqrt[3]{10}}$.

   c. Hence find an approximate value for $\sqrt[3]{10}$, giving your answer in the form $\dfrac{a}{b}$ where $a$ and $b$ are integers.

2. a. Show that when $x = \dfrac{1}{3}$ then $\dfrac{1}{\sqrt{1-x^2}} = \dfrac{3\sqrt{2}}{4}$.

   b. Show that $\dfrac{1}{\sqrt{1-x^2}} \approx 1 + \dfrac{x^2}{2} + \dfrac{ax^4}{8} + \dfrac{bx^6}{16}$ and state the values $a$ and $b$.

   c. Hence find the value of $\sqrt{2}$ giving your answer to 3 decimal places.

3. a. Express $\dfrac{2}{(3-2x)(1-2x)}$ in partial fractions.

   b. Expand $\dfrac{2}{(3-2x)(1-2x)}$ as a series in ascending powers of $x$ up to and including the term in $x^2$. State the range of values of $x$ for which the series converges.

**4** It is given that $f(x) = \dfrac{7x-1}{(1+3x)(3-x)}$.

   **a** Express $f(x)$ in the form $\dfrac{A}{3-x} + \dfrac{B}{1+3x}$,

      where $A$ and $B$ are integers.

   **b**  **i**  Find the first three terms of the binomial expansion
         of $f(x)$ in the form $a + bx + cx^2$,
         where $a$, $b$ and $c$ are rational numbers.

      **ii**  State why the binomial expansion cannot be expected to give a good
         approximation to $f(x)$ at $x = 0.4$.

AQA MPC4 January 2013

**5**  **a**  Find the binomial expansion of $(1+6x)^{-\frac{1}{3}}$ up to and including the
      term in $x^2$.

   **b**  **i**  Find the binomial expansion of $(27+6x)^{-\frac{1}{3}}$ up to and including the
         term in $x^2$, simplifying the coefficients.

      **ii**  Given that $\sqrt[3]{\dfrac{2}{7}} = \dfrac{2}{\sqrt[3]{28}}$, use your binomial expansion from

         part (b)(i) to obtain an approximation to $\sqrt[3]{\dfrac{2}{7}}$ giving your answer to

         six decimal places.

AQA MPC4 June 13

# 3 Trigonometric Functions and Formulae

## Introduction

This chapter extends the work on trigonometric functions started at AS level. It also introduces more trigonometric formulae that can be used to derive further formulae, solve equations and eliminate parameters.

## Recap

You will need to remember...

- ► The definition of a function and an inverse function.
- ► The properties and the graphs of the functions $f(x) = \sin x$, $f(x) = \cos x$ and $f(x) = \tan x$.
- ► The exact values of the sine, cosine and tangent of $\dfrac{\pi}{6}, \dfrac{\pi}{4}, \dfrac{\pi}{3}$.
- ► How to describe a combination of transformations.
- ► The cosine formula and Pythagoras' theorem.
- ► The formulae $\tan \theta = \dfrac{\sin \theta}{\cos \theta}$ and $\cos^2 \theta + \sin^2 \theta = 1$.

## Objectives

By the end of this chapter, you should know how to...

- ► Define the inverse trigonometric functions.
- ► Define the reciprocal trigonometric functions and associated formulae.
- ► Derive and use the compound angle formulae.
- ► Express $a \cos \theta + b \sin \theta$ in the form $r \sin (\theta \pm \alpha)$ or $r \cos (\theta \pm \alpha)$ and use this form to solve equations.
- ► Define and use the double angle formulae.

## 3.1 The inverse trigonometric functions

Look at the function given by $f(x) = \sin x$ for $x \in \mathbb{R}$.

The inverse mapping is given by $\sin x \to x$, but this mapping is not a function because one value of $\sin x$ maps to many values of $x$. As such, $f(x) = \sin x$ does not have an inverse function for the domain $x \in \mathbb{R}$.

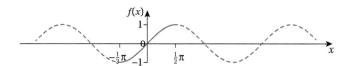

However, when the function $f(x) = \sin x$ is restricted to the domain $-\dfrac{1}{2}\pi \le x \le \dfrac{1}{2}\pi$, one value of $\sin x$ maps to only one value of $x$. Therefore

$f(x) = \sin x$ for the domain $-\dfrac{1}{2}\pi \le x \le \dfrac{1}{2}\pi$ does have an inverse, so $f^{-1}(x)$ exists. The curve $y = f^{-1}(x)$ is found by reflecting $y = \sin x$ in the line $y = x$. Interchanging $x$ and $y$ in the equation $y = \sin x$ gives the equation of the curve $y = f^{-1}(x)$, which in this case is $\sin y = x$, so $y$ is the angle between $-\dfrac{1}{2}\pi$ and $\dfrac{1}{2}\pi$ whose sine is $x$.

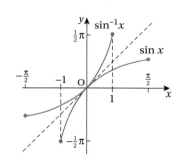

Using $y = \sin^{-1} x$ to mean '$y$ is the angle between $-\dfrac{1}{2}\pi$ and $\dfrac{1}{2}\pi$ whose sine is $x$', where the range of $y = \sin x$ is the domain of the function $y = \sin^{-1} x$, we get the result

**Note**

To remind yourself about inverse functions, look back at Chapter 1, Section 1.3.

when $f(x) = \sin x$ for $-\dfrac{1}{2}\pi \le x \le \dfrac{1}{2}\pi$

then $f^{-1}(x) = \sin^{-1} x$ for $-1 \le x \le 1$

Remember that $\sin^{-1} x$ is an angle, and that angle is between $-\dfrac{1}{2}\pi$ and $\dfrac{1}{2}\pi$.

For example, $\sin^{-1} 0.5$ is the angle between $-\dfrac{1}{2}\pi$ and $\dfrac{1}{2}\pi$ whose sine is 0.5.

Therefore $\sin^{-1} 0.5 = \dfrac{1}{6}\pi$.

Now look at the function $f(x) = \cos x$ for $0 \le x \le \pi$.

The diagram shows that $f^{-1}$ exists and it is denoted by $\cos^{-1}$ where $\cos^{-1} x$ means 'the angle between 0 and $\pi$ whose cosine is $x$'.

The range of $\cos x$ is $-1 \le \cos x \le 1$, and therefore

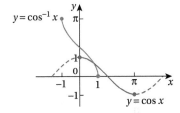

when $f(x) = \cos x$ for $0 \le x \le \pi$

then $f^{-1}(x) = \cos^{-1} x$ for $-1 \le x \le 1$

For example, $\cos^{-1}(-0.5)$ is the angle between 0 and $\pi$ whose cosine is $-0.5$.

Therefore, $\cos^{-1}(-0.5) = \dfrac{2}{3}\pi$.

Similarly, when $f(x) = \tan x$ for $-\dfrac{1}{2}\pi < x < \dfrac{1}{2}\pi$, then $f^{-1}$ exists and is written $\tan^{-1}$ where $\tan^{-1} x$ means 'the angle between $-\dfrac{1}{2}\pi$ and $\dfrac{1}{2}\pi$ whose tangent is $x$'.

The range of $\tan x$ is $\tan x \in \mathbb{R}$, and therefore

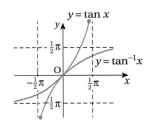

when $f(x) = \tan x$ for $\dfrac{1}{2}\pi < x < \dfrac{1}{2}\pi$

then $f^{-1}(x) = \tan^{-1} x$ for $x \in \mathbb{R}$.

Note that the domain of $\tan^{-1} x$ is all values of $x$.

## Example 1

Find in terms of $\pi$ the exact value of    **a** $\tan^{-1} 1$    **b** $\sin^{-1}\left(-\dfrac{\sqrt{2}}{2}\right)$

**a** $\tan^{-1} 1$ means the angle whose tangent is 1, so if this angle is $\alpha$ then $\tan \alpha = 1$. Hence, $\tan^{-1} 1 = \dfrac{\pi}{4}$.

**b** $\sin^{-1}\left(-\dfrac{\sqrt{2}}{2}\right)$ means the angle whose sine is $-\dfrac{\sqrt{2}}{2}$,

so if this angle is $\beta$ then $\sin \beta = -\dfrac{\sqrt{2}}{2}$. Hence $\sin^{-1}\left(-\dfrac{\sqrt{2}}{2}\right) = -\dfrac{\pi}{4}$.

## Example 2

**a** Sketch the graph of $y = \cos^{-1}(2x - 1)$.

**b** Solve the equation $\cos^{-1}(2x - 1) = \dfrac{\pi}{2}$

**a** The curve $y = \cos^{-1}(2x - 1)$ is a one-way stretch of $y = \cos^{-1} x$ by a factor $\dfrac{1}{2}$ in the direction of the $x$-axis followed by the translation $\begin{bmatrix} 1 \\ 0 \end{bmatrix}$.

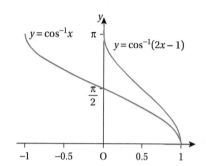

**b** $\cos^{-1}(2x - 1) = \dfrac{\pi}{2}$

$$2x - 1 = \cos\left(\dfrac{\pi}{2}\right)$$

$$2x - 1 = 0$$

$$x = \dfrac{1}{2}$$

## Exercise 1

Find, in terms of $\pi$, the value of

**1** $\tan^{-1}\sqrt{3}$

**2** $\sin^{-1}(-1)$

**3** $\cos^{-1} 0$

**4** $\sin^{-1}\left(-\dfrac{\sqrt{3}}{2}\right)$

**5** $\cos^{-1}\left(-\dfrac{1}{2}\right)$

**6** $\tan^{-1}(-1)$.

Solve these equations.

**7** $\cos^{-1}(2x - 1) = \dfrac{\pi}{3}$

**8** $\sin^{-1}(4x) = \dfrac{\pi}{6}$

**9** $\tan^{-1}(x - 2) = \dfrac{\pi}{6}$

**10** $\tan^{-1}(2 - x) = \dfrac{\pi}{4}$

**11** $2\cos^{-1} x = \pi$

**12** $2\sin^{-1} x = \dfrac{\pi}{2}$

## 3.2 The reciprocal trigonometric functions

The reciprocals of the three main trigonometric functions have their own names.

$$\dfrac{1}{\sin\theta} = \operatorname{cosec}\theta, \quad \dfrac{1}{\cos\theta} = \sec\theta, \quad \dfrac{1}{\tan\theta} = \cot\theta$$

where cosec, sec and cot are abbreviations of **cosecant**, **secant** and **cotangent** respectively.

The graph of $f(\theta) = \operatorname{cosec}\theta$ is shown here.

This graph shows that the cosec function is not continuous as it is undefined when $\theta$ is any integer multiple of $\pi$ (because these are values of $\theta$ where $\sin\theta = 0$, and at these values $\operatorname{cosec}\theta = \dfrac{1}{0}$ which is undefined).

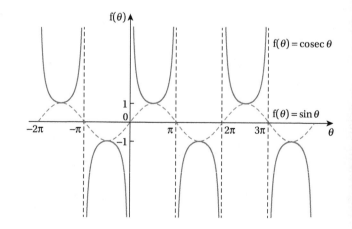

The graph of $f(\theta) = \sec \theta$ is similar to the graph of $y = \operatorname{cosec} \theta$.

However, $\sec(\theta)$ is not defined when the value of $\theta$ is $(2n+1)\dfrac{\pi}{2}$, where $n$ is any integer.

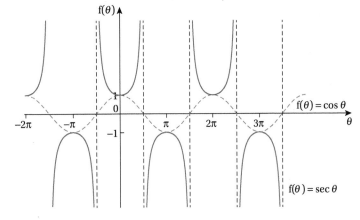

The graph of $f(\theta) = \cot \theta$ is given here. In a similar way to the cosec function, $\cot \theta$ is undefined when $\theta$ is any integer multiple of $\pi$.

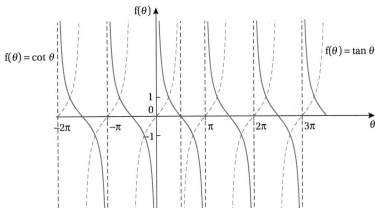

In the right-angled triangle shown in the margin,

$$\tan \alpha = \frac{a}{b} \quad \text{and} \quad \cot \beta = \frac{a}{b} \left( \cot \beta = \frac{1}{\tan \beta} \right)$$

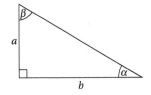

Now $\alpha + \beta = 90°$, so the cotangent of an angle is equal to the tangent of its complement.

This is true for all values of $\theta$ as can be seen from the graph.

Reflecting the curve $y = \tan \theta$ in the vertical axis gives $y = \tan(-\theta)$.

Then translating this curve $\dfrac{1}{2}\pi$ to the left gives $y = \tan\left(\dfrac{1}{2}\pi - \theta\right)$ which is the curve $y = \cot \theta$.

**For *any* angle $\theta$, $\cot \theta = \tan\left(\dfrac{1}{2}\pi - \theta\right)$**

## Example 3

Find the values of $\theta$ for which $\operatorname{cosec} \theta = -8$   for $0 \le \theta \le 360°$

$$\sin \theta = \frac{1}{\operatorname{cosec} \theta} = -\frac{1}{8} = -0.125$$

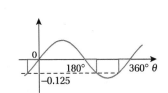

From a calculator $\theta = -7.2°$

From the sketch, the required values of $\theta$ are $187.2°$ and $352.8°$.

## Exercise 2

**1** Find, in the interval $0 \leq \theta \leq 360°$, the values of $\theta$ for which

    **a** $\sec \theta = 2$     **b** $\cot \theta = 0.6$     **c** $\operatorname{cosec} \theta = 1.5$.

**2** Find, in the interval $-180° \leq \theta \leq 180°$, the values of $\theta$ for which

    **a** $\cot \theta = 1.2$     **b** $\sec \theta = -1.5$     **c** $\operatorname{cosec} \theta = -2$.

**3**  **a** Use $\tan \theta = \dfrac{\sin \theta}{\cos \theta}$ to write $\cot \theta$ in terms of $\sin \theta$ and $\cos \theta$.

    **b** Hence show that $\cot \theta - \cos \theta = 0$ can be written in the form
    $\cos \theta (1 - \sin \theta) = 0$, provided that $\sin \theta \neq 0$.

    **c** Find the values in the interval $-\pi \leq \theta \leq \pi$ for which $\cot \theta - \cos \theta = 0$.

**4** Find the values of

    **a** $\cot \dfrac{1}{4}\pi$     **b** $\sec \dfrac{5}{4}\pi$     **c** $\operatorname{cosec} \dfrac{11}{6}\pi$.

**5** Sketch the graph of $f(\theta) = \sec\left(\theta - \dfrac{1}{4}\pi\right)$ for $0 \leq \theta \leq 2\pi$ and give the values of $\theta$ for which $f(\theta) = 1$.

**6** Sketch the graph of $f(\theta) = \cot\left(\theta + \dfrac{1}{3}\pi\right)$ for $-\pi \leq \theta \leq \pi$. Hence give the values of $\theta$ in this interval for which $f(\theta) = 1$.

## 3.3 Trigonometric formulae

When the formula $\cos^2 \theta + \sin^2 \theta = 1$ is divided by $\cos^2 \theta$ this gives

$1 + \dfrac{\sin^2 \theta}{\cos^2 \theta} \equiv \dfrac{1}{\cos^2 \theta}$. Hence

    $1 + \tan^2 \theta = \sec^2 \theta$

When $\cos^2 \theta + \sin^2 \theta = 1$ is divided by $\sin^2 \theta$ this gives  $\dfrac{\cos^2 \theta}{\sin^2 \theta} + 1 = \dfrac{1}{\sin^2 \theta}$. Hence

    $\cot^2 \theta + 1 = \operatorname{cosec}^2 \theta$

These formulae can be used to:

▶ simplify trigonometrical expressions

▶ eliminate trigonometrical terms from pairs of equations

▶ derive a variety of further trigonometrical relationships

▶ solve equations.

## Example 4

**Question**

Simplify $\dfrac{\sin \theta}{1 + \cot^2 \theta}$.

**Answer**

$\dfrac{\sin \theta}{1 + \cot^2 \theta} = \dfrac{\sin \theta}{\operatorname{cosec}^2 \theta}$

    $= \sin^3 \theta$

> **Note**
>
> Using $1 + \cot^2 \theta = \operatorname{cosec}^2 \theta$
> and $\operatorname{cosec} \theta = \dfrac{1}{\sin \theta}$

## Example 5

**Question**

Prove that $(1 - \cos A)(1 + \sec A) = \sin A \tan A$.

**Answer**

Do not start by assuming that the relationship is correct. The left and right hand sides must be isolated throughout the proof, by working on only one of these sides at a time. It often helps to express all ratios in terms of sine and/or cosine as, in general, these are easier to work with.

$\text{LHS} = (1 - \cos A)(1 + \sec A) = 1 + \sec A - \cos A - \cos A \sec A$

$= 1 + \sec A - \cos A - \cos A \left( \dfrac{1}{\cos A} \right)$

$= \sec A - \cos A = \dfrac{1}{\cos A} - \cos A$

$= \dfrac{1 - \cos^2 A}{\cos A} = \dfrac{\sin^2 A}{\cos A}$

$= \sin A \left[ \dfrac{\sin A}{\cos A} \right] = \sin A \tan A = \text{RHS}$

> **Note**
>
> $\cos^2 A + \sin^2 A = 1$

## Example 6

**Question**

Find the solution of the equation $\cot\left( \dfrac{1}{3}\theta - 90° \right) = 1$, for which $0 \le \theta \le 540°$.

**Answer**

Using $\dfrac{1}{3}\theta - 90° = \phi$ gives $\cot\left( \dfrac{1}{3}\theta - 90° \right) = \cot\phi$

The solution of the equation $\cot\phi = 1$ is $\phi = 45°$

But $\phi = \dfrac{1}{3}\theta - 90°$, so $\dfrac{1}{3}\theta - 90° = 45°$

Therefore $\dfrac{1}{3}\theta = 135°$

$\Rightarrow \qquad \theta = 405°$

> **Note**
>
> As you require $\theta$ in the range $0 \le \theta \le 540°$, find $\phi$ in the range
> $\dfrac{1}{3}(0) - 90° \le \phi \le \dfrac{1}{3}(540°) - 90°$
> i.e. $-90° \le \phi \le 90°$.

## Exercise 3

Simplify these expressions given that $\theta$ is an acute angle.

**1** $\dfrac{1 - \sec^2 A}{1 - \operatorname{cosec}^2 A}$

**2** $\dfrac{\sin\theta}{\sqrt{(1 - \cos^2\theta)}}$

**3** $\dfrac{\sin\theta}{\cos\theta} + \dfrac{\cos\theta}{\sin\theta}$

**4** $\dfrac{\sqrt{(1 + \tan^2\theta)}}{\sqrt{(1 - \sin^2\theta)}}$

**5** $\dfrac{1}{\cos\theta\sqrt{(1 + \cot^2\theta)}}$

**6** $\dfrac{\sin\theta}{1 + \cot^2\theta}$

Eliminate $\theta$ from these pairs of equations.

**7** $x = 4\sec\theta$
$y = 4\tan\theta$

**8** $x = a\operatorname{cosec}\theta$
$y = b\cot\theta$

**9** $x = 2\tan\theta$
$y = 3\cos\theta$

**10** $x = 2 + \tan\theta$
$y = 2\cos\theta$

**11** $x = a\sec\theta$
$y = b\sin\theta$

Prove these formulae.

**12** $\cot\theta + \tan\theta = \sec\theta\operatorname{cosec}\theta$

**13** $\dfrac{\cos A}{1 - \tan A} + \dfrac{\sin A}{1 - \cot A} = \sin A + \cos A$

**14** $\tan^2\theta + \cot^2\theta = \sec^2\theta + \operatorname{cosec}^2\theta - 2$

**15** $\dfrac{\sin A}{1 + \cos A} = \dfrac{1 - \cos A}{\sin A}$ ⟵

**16** $\dfrac{\sin A}{1 + \cos A} + \dfrac{1 + \cos A}{\sin A} = \dfrac{2}{\sin A}$

> **Note**
>
> *Hint for question 15:* Multiply top and bottom of LHS by $(1 - \cos A)$.

Solve the equations for values of $\theta$ in the interval $0 \le \theta \le 360°$, giving your answers correct to 1 decimal place.

**17** $\sec^2\theta + \tan^2\theta = 6$

**18** $\cot^2\theta = \operatorname{cosec}\theta$

**19** $\tan\theta + \cot\theta = 2\sec\theta$

**20** $\tan\theta + 3\cot\theta = 5\sec\theta$

Solve the equations for angles in the interval $-\pi \le \theta \le \pi$, giving your answers correct to 3 significant figures.

**21** $4\cot^2\theta + 12\operatorname{cosec}\theta + 1 = 0$

**22** $4\sec^2\theta - 3\tan\theta = 5$

Solve the equations for values of $\theta$ in the interval $0 \le \theta \le \pi$.

**23** $\sec 5\theta = 2$

**24** $\cot\dfrac{1}{2}\theta = -1$

**25** $\cot\left(\theta + \dfrac{1}{4}\pi\right) = 1$

**26** $\sec\left(2\theta - \dfrac{1}{3}\pi\right) = -2$

# 3.4 Compound angle formulae

It is tempting to think that $\cos(A - B)$ can be written as $\cos A - \cos B$ but $\cos(A - B)$ is NOT equal to $\cos A - \cos B$.

The correct formula for $\cos(A - B)$ is found using the diagram below, which shows a circle of radius 1 unit centre O.

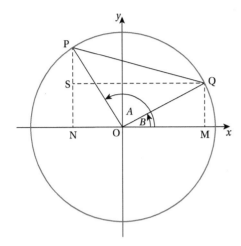

Using the cosine formula in triangle OPQ gives

$$PQ^2 = 1^2 + 1^2 - 2\cos(A - B) = 2 - 2\cos(A - B)$$

$$OM = \cos B \text{ and } ON = -\cos A, \quad \text{so} \quad QS = (-\cos A + \cos B)$$

$$QM = \sin B \text{ and } PN = \sin A \quad \text{so} \quad PS = (\sin A - \sin B)$$

Using Pythagoras' theorem in triangle PQS gives

$$PQ^2 = (-\cos A + \cos B)^2 + (\sin A - \sin B)^2$$
$$= \cos^2 A + \cos^2 B - 2\cos A \cos B + \sin^2 A + \sin^2 B - 2\sin A \sin B$$
$$= 2 - 2(\cos A \cos B + \sin A \sin B)$$

Equating the two expressions for $PQ^2$ gives

$$\cos(A - B) = \cos A \cos B + \sin A \sin A$$

This formula is true for all angles.

The formula derived above is one of the compound angle formulae and we can use it to derive others.

$$\cos(A - B) = \cos A \cos B + \sin A \sin A \qquad [1]$$

Replacing $B$ by $-B$ in [1], $\quad \cos(A + B) = \cos A \cos(-B) + \sin A \sin(-B)$
$$= \cos A \cos B - \sin A \sin B \qquad [2]$$

Replacing $A$ by $\frac{1}{2}\pi - A$ in [1],
$$\cos\left(\left(\frac{1}{2}\pi - A\right) + B\right) = \cos\left(\frac{1}{2}\pi - A\right)\cos B + \sin\left(\frac{1}{2}\pi - A\right)\sin B$$
$$\sin(A + B) = \sin A \cos B + \cos A \sin B \qquad [3]$$

Replacing $B$ by $-B$ in [3], $\quad \sin(A - B) = \sin A \cos(-B) + \cos A \sin(-B)$
$$= \sin A \cos B - \cos A \sin B \qquad [4]$$

Dividing [1] by [3] gives $\dfrac{\sin(A+B)}{\cos(A+B)} = \dfrac{\sin A \cos B + \cos A \sin B}{\cos A \cos B - \sin A \sin B}$

$$\tan(A+B) = \dfrac{\dfrac{\sin A \cos B}{\cos A \cos B} + \dfrac{\cos A \sin B}{\cos A \cos B}}{\dfrac{\cos A \cos B}{\cos A \cos B} - \dfrac{\sin A \sin B}{\cos A \cos B}} = \dfrac{\tan A + \tan B}{1 - \tan A \tan B} \quad [5]$$

Replacing $B$ by $-B$ in [5] gives

$$\tan(A-B) = \dfrac{\tan A - \tan B}{1 + \tan A \tan B} \qquad [6]$$

Collecting these formulae together gives

$\sin (A+B) = \sin A \cos B + \cos A \sin B$

$\sin (A-B) = \sin A \cos B - \cos A \sin B$

$\cos (A+B) = \cos A \cos B - \sin A \sin B$

$\cos (A-B) = \cos A \cos B + \sin A \sin B$

$\tan(A+B) = \dfrac{\tan A + \tan B}{1 - \tan A \tan B}$

$\tan(A-B) = \dfrac{\tan A - \tan B}{1 + \tan A \tan B}$

You do not need to learn these formulae, but you need to be able to recognise the right hand side of these formulae as equivalents to the left hand sides.

You can use these formulae to prove further trigonometrical formulae, as you will see in Examples 7, 8 and 9.

## Example 7

Find exact values for

a   $\sin 75°$

b   $\cos 105°$

To find exact values, express the given angle in terms of angles whose trig ratios are known exact values, that is 30°, 60°, 45°, 90°.

a   $\sin 75° = \sin (45° + 30°) = \sin 45° \cos 30° + \cos 45° \sin 30°$

$$= \left(\dfrac{\sqrt{2}}{2}\right)\left(\dfrac{\sqrt{3}}{2}\right) + \left(\dfrac{\sqrt{2}}{2}\right)\left(\dfrac{1}{2}\right) = \dfrac{\sqrt{2}}{4}(\sqrt{3}+1)$$

b   $\cos 105° = \cos (60° + 45°) = \cos 60° \cos 45° - \sin 60° \sin 45°$

$$= \left(\dfrac{1}{2}\right)\left(\dfrac{\sqrt{2}}{2}\right) - \left(\dfrac{\sqrt{3}}{2}\right)\left(\dfrac{\sqrt{2}}{2}\right) = \dfrac{\sqrt{2}}{4}(1-\sqrt{3})$$

## Example 8

Simplify $\sin \theta \cos \dfrac{1}{3}\pi - \cos \theta \sin \dfrac{1}{3}\pi$ and hence find the smallest positive value of $\theta$ for which the expression has a minimum value.

$$f(\theta) = \sin\theta\cos\frac{1}{3}\pi - \cos\theta\sin\frac{1}{3}\pi = \sin\left(\theta - \frac{1}{3}\pi\right)$$

The graph of $f(\theta) = \sin\left(\theta - \frac{1}{3}\pi\right)$ is a sine wave, but translated $\frac{1}{3}\pi$ in the direction of the positive $\theta$-axis.

Therefore $f(\theta)$ has a minimum value of $-1$ and the smallest positive value of $\theta$ at which this occurs is $\frac{3}{2}\pi + \frac{1}{3}\pi = \frac{11}{6}\pi$.

> **Note**
>
> $\sin\theta\cos\frac{1}{3}\pi - \cos\theta\sin\frac{1}{3}\pi$
> is the expansion of $\sin(A - B)$
> with $A = \theta$ and $B = \frac{1}{3}\pi$.

## Example 9

Prove that $\dfrac{\sin(A - B)}{\cos A\cos B} = \tan A - \tan B$.

Expanding the numerator, the LHS becomes

$$\frac{\sin A\cos B - \cos A\sin B}{\cos A\cos B} = \frac{\sin A\cos B}{\cos A\cos B} - \frac{\cos A\sin B}{\cos A\cos B}$$

$$= \tan A - \tan B$$

## Example 10

Find all the solutions of the equation $2\cos\theta = \sin\left(\theta + \frac{1}{6}\pi\right)$ in the interval $0 \leq \theta \leq 2\pi$.

$$2\cos\theta = \sin\left(\theta + \frac{1}{6}\pi\right) = \sin\theta\cos\frac{1}{6}\pi + \cos\theta\sin\frac{1}{6}\pi = \frac{\sqrt{3}}{2}\sin\theta + \frac{1}{2}\cos\theta$$

Therefore $\dfrac{3}{2}\cos\theta = \dfrac{\sqrt{3}}{2}\sin\theta$

$$\Rightarrow \qquad \frac{3}{\sqrt{3}} = \frac{\sin\theta}{\cos\theta}$$

$$\Rightarrow \qquad \tan\theta = \sqrt{3}$$

Now $\tan\frac{1}{3}\pi = \sqrt{3}$, so the solution is $\theta = \frac{1}{3}\pi, \frac{4}{3}\pi$.

## Exercise 4

Find the exact value of each expression, leaving your answer in surd form where necessary.

**1** $\cos 40° \cos 50° - \sin 40° \sin 50°$

**2** $\sin 37° \cos 7° - \cos 37° \sin 7°$

**3** $\cos 75°$

**4** $\tan 105°$

**5** $\sin 165°$

**6** $\cos 15°$

Simplify each expression.

**7** $\sin\theta\cos 2\theta + \cos\theta\sin 2\theta$

**8** $\cos\alpha\cos(90° - \alpha) - \sin\alpha\sin(90° - \alpha)$

**9** $\dfrac{\tan A + \tan 2A}{1 - \tan A\tan 2A}$

**10** $\dfrac{\tan 3\beta - \tan 2\beta}{1 + \tan 3\beta\tan 2\beta}$

11. Find the greatest value of each expression and the value of $\theta$ between 0 and 360° at which it occurs.

    **a**   $\sin\theta\cos 25° - \cos\theta\sin 25°$      **b**   $\sin\theta\sin 30° + \cos\theta\cos 30°$

    **c**   $\cos\theta\cos 50° - \sin\theta\sin 50°$      **d**   $\sin 60°\cos\theta - \cos 60°\sin\theta$

Prove the formulae.

12. $\cot(A+B) = \dfrac{\cot A\cot B - 1}{\cot A + \cot B}$      13. $\sin(A+B) + \sin(A-B) = 2\sin A\cos B$

14. $\cos(A+B) + \cos(A-B) = 2\cos A\cos B$      15. $\dfrac{\sin(A+B)}{\cos A\cos B} = \tan A + \tan B$

Solve the equations for values of $\theta$ in the interval $0 \le \theta \le 360°$.

16. $\cos(45° - \theta) = \sin\theta$      17. $3\sin\theta = \cos(\theta + 60°)$

18. $\tan(A-\theta) = \dfrac{2}{3}$ and $\tan A = 3$      19. $\sin(\theta + 60°) = \cos\theta$

# 3.5 Expressions of the form $f(\theta) = a\cos\theta + b\sin\theta$

The expression $a\cos\theta + b\sin\theta$ can be expressed in the form $r\sin(\theta + \alpha)$ where $r > 0$.

Starting with $r\sin(\theta + \alpha) = a\cos\theta + b\sin\theta$, expanding the LHS using the compound angle formulae gives

    $r\underline{\sin}\,\theta\cos\alpha + r\underline{\cos}\,\theta\sin\alpha = a\underline{\cos}\,\theta + b\underline{\sin}\,\theta$

Comparing coefficients of $\cos\theta$ and of $\sin\theta$ gives

         $r\sin\alpha = a$                                 [1]

and      $r\cos\alpha = b$                                   [2]

Squaring and adding equations [1] and [2] gives

    $r^2(\sin^2\alpha + \cos^2\alpha) = a^2 + b^2 \quad\Rightarrow\quad r = \sqrt{a^2 + b^2}$

Dividing equation [1] by equation [2] gives

    $\dfrac{r\sin\alpha}{r\cos\alpha} = \dfrac{a}{b} \quad\Rightarrow\quad \tan\alpha = \dfrac{a}{b}$

    Therefore     $r\sin(\theta + \alpha) = a\cos\theta + b\sin\theta$

    where $r = \sqrt{a^2 + b^2}$     and     $\tan\alpha = \dfrac{a}{b}$

Using a similar method it is also possible to express $a\cos\theta + b\sin\theta$ as $r\sin(\theta - \alpha)$ or $r\cos(\theta \pm \alpha)$.

## Example 11

Express    $3\sin\theta - 2\cos\theta$    as    $r\sin(\theta - \alpha)$. Give $\alpha$ to the nearest 0.1°.

$3 \sin \theta - 2 \cos \theta = r \sin (\theta - \alpha) \quad \Rightarrow \quad 3 \underline{\sin} \theta - 2 \underline{\cos} \theta = r \underline{\sin} \theta \cos \alpha - r \underline{\cos} \theta \sin \alpha$

Comparing coefficients of $\sin \theta$ and of $\cos \theta$ gives

$$\left.\begin{array}{r} 3 = r \cos \alpha \\ 2 = r \sin \alpha \end{array}\right\} \quad \Rightarrow \quad \begin{cases} 13 = r^2 \quad \Rightarrow \quad r = \sqrt{13} \\ \tan \alpha = \dfrac{2}{3} \quad \Rightarrow \quad \alpha = 33.7° \end{cases}$$

Therefore $3 \sin \theta - 2 \cos \theta = \sqrt{13} \ \sin (\theta - 33.7°)$.

## Example 12

Find the maximum value of $f(x) = 3 \cos x + 4 \sin x$ and the smallest positive value of $x$ at which the maximum occurs. Give your answer to the nearest $0.1°$.

Expressing $f(x)$ in the form $r \sin (x \pm \alpha)$ or $r \cos (x \pm \alpha)$ means that its maximum value, and the values of $x$ at which this maximum occurs, can be seen from a sketch. In this question you can choose the form in which to express $f(x)$. The solution given here uses $r \cos (x - \alpha)$ because of the plus sign in $f(x)$ it fits better than $r \sin (x + \alpha)$.

$3 \underline{\cos} x + 4 \underline{\sin} x = r \cos (x - \alpha) = r \underline{\cos} x \cos \alpha + r \underline{\sin} x \sin \alpha$

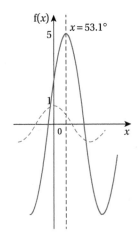

Hence $\left.\begin{array}{r} r \cos \alpha = 3 \\ r \sin \alpha = 4 \end{array}\right\} \quad \Rightarrow \quad \begin{cases} r^2 = 25 \quad \Rightarrow \quad r = 5 \\ \tan \alpha = \dfrac{4}{3} \quad \Rightarrow \quad \alpha = 53.1° \end{cases}$

Therefore $\quad f(x) = 5 \cos (x - 53.1°)$

The graph of $f(x)$ is a cosine wave stretched by a factor of 5 parallel to the $y$-axis and translated by $53.1°$ along the $x$-axis.

Therefore $f(x)$ has a maximum value of 5 and the smallest positive value of $x$ at which it occurs is $53.1°$.

## Example 13

Find, for $-\pi \le x \le \pi$, the solution of the equation $\sqrt{3} \cos x + \sin x = 1$.

First express $\sqrt{3} \cos x + \sin x$ in the form $r \cos (x - \alpha)$.

$\sqrt{3} \underline{\cos} x + \underline{\sin} x = r \cos (x - \alpha)$

$\qquad\qquad\qquad = r \underline{\cos} x \cos \alpha + r \underline{\sin} x \sin \alpha$

Comparing coefficients gives $\begin{cases} r \cos \alpha = \sqrt{3} \\ r \sin \alpha = 1 \end{cases} \quad \Rightarrow \quad \begin{cases} r^2 = 4 \quad \Rightarrow \quad r = 2 \\ \tan \alpha = \dfrac{1}{\sqrt{3}} \quad \Rightarrow \quad \alpha = \dfrac{1}{6}\pi \end{cases}$

Therefore $\quad \sqrt{3} \cos x + \sin x = 2 \cos \left(x - \dfrac{1}{6}\pi\right)$

so $\qquad \sqrt{3} \cos x + \sin x = 1 \quad \Rightarrow \quad 2 \cos \left(x - \dfrac{1}{6}\pi\right) = 1$

$$\Rightarrow \qquad \cos\left(x - \tfrac{1}{6}\pi\right) = \frac{1}{2}$$

$$\Rightarrow \qquad x - \tfrac{1}{6}\pi = \pm\tfrac{1}{3}\pi$$

Therefore $\qquad x = \tfrac{1}{2}\pi, -\tfrac{1}{6}\pi.$

## Exercise 5

**1** Find the values of $r$ and $\alpha$ for which

   **a**   $\sqrt{3}\cos\theta - \sin\theta = r\cos(\theta + \alpha)$

   **b**   $\cos\theta + 3\sin\theta = r\cos(\theta - \alpha)$

   **c**   $4\sin\theta - 3\cos\theta = r\sin(\theta - \alpha).$

   Give angles in degrees to the nearest $0.1°$

**2** Express $\cos 2\theta - \sin 2\theta$ in the form $r\cos(2\theta + \alpha)$. Give $\alpha$ in radians to three significant figures.

**3** Express $2\cos 3\theta + 5\sin 3\theta$ in the form $r\sin(3\theta + \alpha)$. Give $\alpha$ in degrees to the nearest $0.1°$.

**4**  **a**   Express $\cos\theta - \sqrt{3}\sin\theta$ in the form $r\sin(\theta - \alpha)$. Give $\alpha$ in degrees to the nearest $0.1°$. Hence sketch the graph of $f(\theta) = \cos\theta - \sqrt{3}\sin\theta$.

   **b**   Give the maximum and minimum values of $f(\theta)$ and the values of $\theta$ between 0 and $360°$ at which they occur.

**5**  **a**   Express $7\cos\theta - 24\sin\theta$ in the form $r\cos(\theta + \alpha)$. Give $\alpha$ in degrees to the nearest $0.1°$ Hence sketch the graph of $f(\theta) = 7\cos\theta - 24\sin\theta + 3$.

   **b**   Give the maximum and minimum values of $f(\theta)$ and the values of $\theta$ between 0 and $360°$ at which they occur.

**6** Find the solution of these equations in the interval $0 \le x \le 360°$.

   **a**   $\cos x + \sin x = \sqrt{2}$        **b**   $7\cos x + 6\sin x = 2$

   **c**   $\cos x - 3\sin x = 1$        **d**   $2\cos x - \sin x = 2$

## 3.6 Double angle formulae

The compound angle formulae are true for any two angles $A$ and $B$, and can therefore be used for two equal angles, that is, when $B = A$.

Replacing $B$ by $A$ in the trigonometric formulae for $(A + B)$ gives you the set of double angle formulae.

$$\sin 2A = 2\sin A \cos A$$

$$\cos 2A = \cos^2 A - \sin^2 A$$

$$\tan 2A = \frac{2\tan A}{1 - \tan^2 A}$$

Using $\cos^2 A + \sin^2 A = 1$, you can change the right-hand side of the formula for $\cos 2A$ into an expression involving *only* $\sin^2 A$ or $\cos^2 A$.

$$\cos^2 A - \sin^2 A = \begin{cases} (1 - \sin^2 A) - \sin^2 A = 1 - 2\sin^2 A \\ \cos^2 A - (1 - \cos^2 A) = 2\cos^2 A - 1 \end{cases}$$

$$\cos 2A = \begin{cases} \cos^2 A - \sin^2 A \\ 1 - 2\sin^2 A \\ 2\cos^2 A - 1 \end{cases}$$

The last two forms of the cosine double angle formulae can be rearranged to give $\sin^2 A$ or $\cos^2 A$ in terms of $\cos 2A$. These are useful when you need change either $\sin^2 A$ or $\cos^2 A$ into an expression involving $\cos 2A$.

Starting with $\cos 2A = 2\cos^2 A - 1$, and rearranging gives

$$\cos^2 A = \frac{1}{2}(1 + \cos 2A)$$

and starting with $\cos 2A = 1 - 2\sin^2 A$, and rearranging gives

$$\sin^2 A = \frac{1}{2}(1 - \cos 2A).$$

> **Note**
>
> You need to learn all the compound angle and double angle formulae.

## Example 14

When $\tan\theta = \dfrac{3}{4}$, show that $\tan 2\theta = \dfrac{24}{7}$. Hence find the value of $\tan 4\theta$.

Using $\tan 2A = \dfrac{2\tan A}{1 - \tan^2 A}$ with $A = \theta$ and $\tan\theta = \dfrac{3}{4}$ gives

$$\tan 2\theta = \frac{2\left(\dfrac{3}{4}\right)}{1 - \left(\dfrac{3}{4}\right)^2} = \frac{24}{7}$$

Using the formula for $\tan 2A$ again, but this time with $A = 2\theta$, gives

$$\tan 4\theta = \frac{2\tan 2\theta}{1 - \tan^2 2\theta} = \frac{2\left(\dfrac{24}{7}\right)}{1 - \left(\dfrac{24}{7}\right)^2} = -\frac{336}{527}$$

## Example 15

Eliminate $\theta$ from the equations $x = \cos 2\theta$, $y = \sec\theta$.

Using $\cos 2\theta = 2\cos^2\theta - 1$ gives

$$x = 2\cos^2\theta - 1 \text{ and } y = \frac{1}{\cos\theta}$$

therefore $\quad x = 2\left(\dfrac{1}{y}\right)^2 - 1 \quad \Rightarrow \quad (x+1)y^2 = 2$

## Example 16

Prove that $\sin 3A = 3\sin A - 4\sin^3 A$

Answer

$$\sin 3A = \sin(2A + A)$$
$$= \sin 2A \cos A + \cos 2A \sin A$$
$$= (2\sin A \cos A)\cos A + (1 - 2\sin^2 A)\sin A$$
$$= 2\sin A \cos^2 A + \sin A - 2\sin^3 A$$
$$= 2\sin A(1 - \sin^2 A) + \sin A - 2\sin^3 A$$
$$= 3\sin A - 4\sin^3 A$$

## Example 17

Question

Solve the equation $\cos 2x + 3\sin x = 2$, giving values of $x$ in the interval $-\pi \le x \le \pi$.

Answer

When an equation involves a trigonometric functions of different multiples of an angle, it is sensible to express the equation in a form where all the trigonometric functions are of the same angle and, when possible, involving only one trigonometric function.

Using $\cos 2x = 1 - 2\sin^2 x$ gives

$$1 - 2\sin^2 x + 3\sin x = 2$$

$\Rightarrow$ $$2\sin^2 x - 3\sin x + 1 = 0$$

$\Rightarrow$ $$(2\sin x - 1)(\sin x - 1) = 0$$

therefore $\sin x = \frac{1}{2}$ or $\sin x = 1$

When $\sin x = \frac{1}{2}$, $x = \frac{1}{6}\pi, \frac{5}{6}\pi$ and when $\sin x = 1$, $x = \frac{1}{2}\pi$

Therefore the solution is $x = \frac{1}{6}\pi, \frac{5}{6}\pi, \frac{1}{2}\pi$.

## Example 18

Question

Express $4\cos^2 x + 1$ in terms of the angle $2x$.

Answer

Using $\cos^2 x = \frac{1}{2}(1 + \cos 2x)$ gives

$$4\cos^2 x + 1 = 4 \times \frac{1}{2}(1 + \cos 2x) + 1$$
$$= 2(1 + \cos 2x) + 1$$
$$= 3 + 2\cos 2x$$

## Exercise 6

For questions 1 to 8, simplify the expressions, giving an exact value where this is possible.

1. $2\sin 15° \cos 15°$
2. $\cos^2 \frac{1}{8}\pi - \sin^2 \frac{1}{8}\pi$
3. $\sin\theta \cos\theta$
4. $1 - 2\sin^2 4\theta$
5. $\dfrac{2\tan 75°}{1 - \tan^2 75°}$
6. $\dfrac{2\tan 3\theta}{1 - \tan^2 3\theta}$

**7** $2\cos^2\dfrac{3}{8}\pi - 1$

**8** $1 - 2\sin^2\dfrac{1}{8}\pi$

**9** Find the value of $\cos 2\theta$ and $\sin 2\theta$ for an acute angle $\theta$.

   **a** $\cos\theta = \dfrac{3}{5}$
      **b** $\sin\theta = \dfrac{7}{25}$
      **c** $\tan\theta = \dfrac{12}{5}$

**10** Given that $\tan\theta = -\dfrac{7}{24}$ and $\theta$ is obtuse, find

   **a** $\tan 2\theta$
     **b** $\cos 2\theta$
     **c** $\sin 2\theta$
     **d** $\cos 4\theta$.

**11** Eliminate $\theta$ from the pairs of equations.

   **a** $x = \tan 2\theta,\ y = \tan\theta$
      **b** $x = \cos 2\theta,\ y = \cos\theta$

   **c** $x = \cos 2\theta,\ y = \operatorname{cosec}\theta$
      **d** $x = \sin 2\theta,\ y = \sec 4\theta$

**12** Express in terms of $\cos 2x$.

   **a** $2\sin^2 x - 1$
     **b** $4 - 2\cos^2 x$
     **c** $2\cos^2 x + \sin^2 x$

   **d** $2\cos^2 x(1 + \cos^2 x)$
     **e** $\cos^4 x$   (Hint: $\cos^4 x = (\cos^2 x)^2$)
     **f** $\sin^4 x$

**13** Prove that these identities are correct.

   **a** $\dfrac{1 - \cos 2A}{\sin 2A} = \tan A$
      **b** $\sec 2A + \tan 2A = \dfrac{\cos A + \sin A}{\cos A - \sin A}$

   **c** $\cos 4A = 8\cos^4 A - 8\cos^2 A + 1$
     **d** $\sin 2\theta = \dfrac{2\tan\theta}{1 + \tan^2\theta}$

   **e** $\cos 2\theta = \dfrac{1 - \tan^2\theta}{1 + \tan^2\theta}$

**14** Find solutions of the equations for angles from $0$ to $2\pi$.

   **a** $\cos 2x = \sin x$
     **b** $\sin 2x + \cos x = 0$
     **c** $\cos 2x = \cos x$

   **d** $\sin 2x = \cos x$
     **e** $4 - 5\cos\theta = 2\sin^2\theta$
     **f** $\sin 2\theta - 1 = \cos 2\theta$

   **g** $\cos^2\theta = \sin\theta - 1$
     **h** $\cos 2\theta = 1 + \sin\theta$

# Summary

## Inverse trigonometric functions

$\sin^{-1} x$ means the angle whose sine is $x$.

$f(x) = \sin^{-1} x$ has domain $-1 \le x \le 1$ and range $-\dfrac{1}{2}\pi \le f(x) \le \dfrac{1}{2}\pi$.

$\cos^{-1} x$ means the angle whose cosine is $x$.

$f(x) = \cos^{-1} x$ has domain $-1 \le x \le 1$ and range $0 \le f(x) \le \pi$.

$\tan^{-1} x$ means the angle whose tangent is $x$.

$f(x) = \tan^{-1} x$ has domain $x \in \mathbb{R}$ and range $-\dfrac{1}{2}\pi < f(x) < \dfrac{1}{2}\pi$.

## Reciprocal trigonometric functions

$$\operatorname{cosec}\theta = \frac{1}{\sin\theta},\ \sec\theta = \frac{1}{\cos\theta},\ \cot\theta = \frac{1}{\tan\theta}$$

## Trigonometric formulae

$$1 + \tan^2\theta = \sec^2\theta$$

$$\cot^2\theta + 1 = \operatorname{cosec}^2\theta$$

## Compound angle formulae

$$\sin(A+B) = \sin A \cos B + \cos A \sin B$$

$$\sin(A-B) = \sin A \cos B - \cos A \sin B$$

$$\cos(A+B) = \cos A \cos B - \sin A \sin B$$

$$\cos(A-B) = \cos A \cos B + \sin A \sin B$$

$$\tan(A+B) = \frac{\tan A + \tan B}{1 - \tan A \tan B}$$

$$\tan(A-B) = \frac{\tan A - \tan B}{1 + \tan A \tan B}$$

$a \cos\theta \pm b \sin\theta$ can be expressed as $r \sin(\theta \pm \alpha)$ or $r \cos(\theta \pm \alpha)$

where $r = \sqrt{a^2 + b^2}$ and $\tan\alpha = \dfrac{a}{b}$ or $\tan\alpha = \dfrac{b}{a}$

## Double angle formulae

$$\sin 2A = 2 \sin A \cos A$$

$$\tan 2A = \frac{2 \tan A}{1 - \tan^2 A} \qquad\qquad \cos^2 A = \frac{1}{2}(1 + \cos 2A)$$

$$\cos 2A = \begin{cases} \cos^2 A - \sin^2 A & \sin^2 A = \dfrac{1}{2}(1 - \cos 2A) \\ 1 - 2\sin^2 A \\ 2\cos^2 A - 1 \end{cases}$$

## Review

1. Eliminate $\alpha$ from the equations $x = \cos\alpha$, $y = \operatorname{cosec}\alpha$.

2. Find the solution of the equation $\sec\theta + \tan^2\theta = 5$ for values of $\theta$ in the interval $0 \le \theta \le 360°$.

3. Prove that $\left(\cot\theta + \operatorname{cosec}\theta\right)^2 \equiv \dfrac{1 + \cos\theta}{1 - \cos\theta}$.

4. Simplify $\sec^4\theta - \sec^2\theta$.

5. Eliminate $\theta$ from the equations $x = \sec\theta - 3$, $y = 2 - \tan\theta$.

6. State the value of $\cos^{-1}\dfrac{\sqrt{3}}{2}$.

7. Solve the equation $\sin^{-1}(3x - 1) = \dfrac{\pi}{2}$.

8. Eliminate $\theta$ from the equations $x = \sin\theta$ and $y = \cos 2\theta$.

9. Prove the formula $\dfrac{\sin 2\theta}{1 + \cos 2\theta} = \tan\theta$.

10. Prove that $\tan\left(\theta + \dfrac{1}{4}\pi\right)\tan\left(\dfrac{1}{4}\pi - \theta\right) = 1$.

11. When $\cos A = \dfrac{4}{5}$ and $\cos B = \dfrac{5}{13}$ find the possible values of $\cos(A+B)$.

12. Eliminate $\theta$ from the equations $x = \cos 2\theta$ and $y = \cos^2\theta$.

13. Solve the equation $8 \sin\theta \cos\theta = 3$ for values of $\theta$ from $-180°$ to $180°$.

14. Find the solution of the equation $\cos^2\theta - \sin^2\theta = 1$ for $-\pi \le \theta \le \pi$.

15. Show that $\cos^4\theta - \sin^4\theta = \cos 2\theta$.

**16** Simplify the expression $\dfrac{1+\cos 2x}{1-\cos 2x}$.

**17** Find the values of $A$ between 0 and 360° for which
$\sin(60° - A) + \sin(120° - A) = 0$.

**18** **a** Express $2\sin^2\theta + 1$ in terms of $\cos 2\theta$.

 **b** Express $4\cos^2 2A$ in terms of $\cos 4A$ (Hint: Use $2A = x$).

**19** Express $4\sin\theta - 3\cos\theta$ in the form $r\sin(\theta - \alpha)$. Hence find the local

maximum and local minimum values of $\dfrac{7}{4\sin\theta - 3\cos\theta + 2}$.

**20** Express $\cos x + \sin x$ in the form $r\cos(x - \alpha)$. Hence find the smallest

positive value of $x$ for which $\dfrac{1}{(\cos x + \sin x)}$ has a minimum value.

## Assessment

**1** **a** Express $3\cos x - 4\sin x$ in the form $r\cos(x + \alpha)$.

 **b** Hence express $4 + \dfrac{10}{3\cos x - 4\sin x}$ in the form $4 + k\sec(x + \alpha)$.

 **c** Sketch the graph of $y = 4 + \dfrac{10}{3\cos x - 4\sin x}$.

**2** **a** Express $\sin 2\theta - \cos 2\theta$ in the form $r\sin(2\theta - \alpha)$.

 **b** Hence find the smallest positive value of $\theta$ for which $\sin 2\theta - \cos 2\theta$ has
 a maximum value.

**3** **a** Express $4\sin x + 3\cos x$ in the form $r\sin(x + \alpha)$.

 **b** Find the local maximum value of $\dfrac{3}{4\sin x + 3\cos x}$.

**4** **a** Find all the values of $x$ between 0 and 360° for which $\cos x - 2\sin x = 1$.

 **b** Solve the equation $3\cos x - 2\sin x = 1$ for values of $x$ in the interval
 $0 \le x \le 360°$.

**5** **a** Find the values of $x$ in the range $0 \le x \le 360°$ that satisfy the equation
 $3\sec^2 x + 5\tan x = 5$.

 **b** Eliminate $t$ from the equations $x = 4\cos 2t$ and $y = 3\sin t$.

**6** Sketch the curve with equation $y = \sin^{-1} 3x$, where $y$ is in radians.

State the exact values of the coordinates of the end points of the graph.

AQA MPC3 June 2015 part question

**7** **a** Express $2\cos x - 5\sin x$ in the form $R\cos(x + \alpha)$, where $R > 0$ and
 $0 < \alpha < \dfrac{\pi}{2}$, giving your value of $\alpha$, in radians, to three significant figures.

 **b** **i** Hence find the value of $x$ in the interval $0 < x < 2\pi$ for which
 $2\cos x - 5\sin x$ has its maximum value. Give your value of $x$ to three
 significant figures.

 **ii** Use your answer to part (a) to solve the equation $2\cos x - 5\sin x + 1 = 0$
 in the interval $0 < x < 2\pi$, giving your solutions to three significant figures.

AQA MPC4 June 2015

# 4 Exponential and Logarithmic Functions

## Introduction

In this chapter you will use what you learnt about logarithms at AS level to calculate values associated with exponential growth and decay values. This chapter also introduces the number e, which is one of the most important numbers in mathematics, together with the associated exponential and logarithmic functions.

## Objectives

By the end of this chapter, you should know how to...

▶ Use logarithms to solve problems involving growth or decay by a constant factor over equal time intervals.

▶ Define the irrational number e.

▶ Define the exponential and logarithmic functions in terms of e.

## Recap

You will need to remember...

▶ The properties of $f(x) = a^x$ and the shape of its graph, where $a$ is a positive number.
▶ The meaning of $\log_a b$, and that $\log_a b = c \iff b = a^c$.
▶ The laws of logarithms.
▶ How to solve equations of the form $a^x = b$.
▶ How to solve inequalities.
▶ The meaning of an inverse function.
▶ How to use a combination of transformations to sketch curves.

## Applications

Many quantities are said to increase exponentially, such as the infection rate of a disease, the rate of inflation or the growth of capital in a savings account. All these can be calculated when the rate of increase is known.

## 4.1 Exponential growth and decay

There are many examples where a quantity grows or decays by a constant factor over equal time intervals. This is called **exponential growth** or **exponential decay.**

### Exponential growth

Suppose a debt of £100 has 2% interest added each month then, if no repayment is made,

the debt grows to $£100 \times \left(1 + \dfrac{2}{100}\right) = £100 \times 1.02$ after one month

to $£100 \times 1.02 \times 1.02$, that is $£100 \times (1.02)^2$, after two months

and so on, to $£100 \times (1.02)^n$ after $n$ months.

1.02 is called the **growth factor** per month.

Therefore when £$y$ is the debt after $x$ months, $y = 100(1.02)^x$.

This is an **exponential function** that *increases* in value as the exponent $x$ increases, so $y$ grows exponentially.

## Exponential decay

Some business assets (such as vehicles) depreciate over time. Suppose that at the end of each year, the value of a lorry is half its value at the start of the year. In this case, a lorry which was initially worth £$A$ is worth £$A \times (1 - 0.5) = £A \times (0.5)$ after one year, is worth £$A \times (0.5)^2$ after two years, and so on.

If £$y$ is the lorry's value after $n$ years, then $y = A \times (0.5)^x = A\left(\dfrac{1}{2}\right)^n = A \times 2^{-n}$.

The function $2^{-n}$ is an exponential function that *decreases* in value as $n$ increases, so $y$ decays exponentially.

In each example above, the mathematical expression for the relationship is obtained by making certain assumptions and it is not necessarily valid at all times. The first example assumes that interest rates remain constant, and this is not usually the case. Therefore this relationship is valid only for the time during which the interest rate is 2%.

The assumption in the second case is that the rule for writing down the value of assets never changes, but this is not normally true in practice. After a few years the vehicle will have a value small enough to be written off completely.

## Example 1

Ashis invests £1000 in a savings account at a fixed interest rate of 1.5% per year.

a   Find, to the nearest pound, the value of the investment after 8 years.

b   The money is kept invested for $N$ whole years. Find the value of $N$ for which the value of the investment first exceeds £2000.

c   Rajiv invests £800 in a savings account with a different bank at a fixed interest rate of 2% per year. Ashis and Rajiv invest their money at the same time. Find the number of complete years after which Rajiv's investment first exceeds Ashis's investment.

a   The value of Ashis's investment after $n$ years is £$1000(1.015)^n$.

When $n = 8$, $1000(1.015)^8 = 1126.4...$

Therefore the value of the investment after 8 years is £1126 correct to the nearest pound.

b   After $N$ years, the value of the investment is $1000(1.015)^N$.

$$1000(1.015)^N > 2000 \implies (1.015)^N > 2$$

Therefore $N\log(1.015) > \log 2$

$\implies \quad N > \dfrac{\log 2}{\log(1.015)} = 46.5...$

Therefore the value of $N$ for which the value of the investment first exceeds £2000 is 47 (since $N$ is a whole number).

*(continued)*

> **Note**
>
> You can use $\log_{10}$ on your calculator to do this calculation

Answer

c The value of Rajiv's investment after $n$ years is £800$(1.02)^n$.

The value of Rajiv's investment is greater than the value of Ashis's investment when £800$(1.02)^n$ > £1000$(1.015)^n$.

$$800(1.02)^n > 1000(1.015)^n \quad \Rightarrow \quad \left(\frac{1.02}{1.015}\right)^n > 1.25$$

Therefore $\quad n\log\left(\dfrac{1.02}{1.015}\right) > \log 1.25 \quad \Rightarrow \quad n > \dfrac{\log(1.25)}{\log\left(\dfrac{1.02}{1.015}\right)} = 45.4...$

Therefore Rajiv's investment first exceeds Ashis's investment after 46 complete years.

## Exercise 1

1 When a toll bridge opens, the cost for a car to cross the bridge is £3. At the end of the first year, and after each subsequent year, the cost will increase by a fixed rate of 3%. Find the cost for a car to cross the bridge after 5 years.

2 The cost of a new industrial machine is £10 000. Its value depreciates at the fixed rate of 10% per year.

a Find the value of the machine after 4 years.

b Find the number of whole years after which the value of the machine is less than £5000.

3 Mr Brown buys a house for £250 000. The value increases by a constant 5% each year.

Find the number of whole years after which the value of the house is more than £300 000.

4 The cost of a new car is £20 000. The value of the car depreciates by a constant 10% each year.

Find the number of whole years after which the car has lost half its original value.

5 Mr Said takes out a loan. The cost of the loan, including interest, is £5000.

Mr Said pays back a fixed amount of 10% of the outstanding loan each year.

a Find the amount that Mr Said will owe 4 years after he took the loan.

b Find the number of complete years after which the loan is less than £500.

6 Mr Mendes invests £1000 at Bank Avro at a fixed interest rate of 2% per year.

a Find the number of complete years after which the value of the investment first exceeds £1500.

b Mr Nero invests £900 at Bank Bifra at a fixed rate of interest of 2.5% per year.

Find the number of complete years after which the value of Mr Nero's investment is greater than the value of Mr Mendes' investment.

7 Mr Ahmed pays $200 000 for a house. The value of the house increases at the fixed rate of 4% per year.

Mr Daud pays $250 000 for a yacht. The value of the yacht decreases at the fixed rate of 5% per year.

Find the number of complete years after which the value of the yacht is less than the value of the house.

8 Mr Carlos buys a painting for $100 000. The value of the painting is expected to increase by a constant 6% per year.

At the same time Mr Edwards buys a gold goblet for $50 000. The value of the goblet is expected to increase by a constant 10% per year.

Find the number of complete years after which the value of the goblet is expected to be greater than the value of the painting.

9 A culture dish was seeded with 4 mm² of mould. The table shows the area of the mould at six-hourly intervals.

| Time, $x$ hours | 0 | 6 | 12 | 18 | 24 | 30 | 36 |
|---|---|---|---|---|---|---|---|
| Area, $y$ mm² | 4 | 8.1 | 15.9 | 33 | 68 | 118 | 190 |

a Show that, for a time, the growth factor in the area for each interval of 6 hours is roughly 2 (that is, that the area approximately doubles every 6 hours). Hence show that it is reasonable to use an exponential function to model the relationship between time and area, and suggest a possible function.

b Give reasons why

   i the model is approximate

   ii the model ceases to be reasonable after some time.

10 The equation $y = A(1.3)^{-t}$, where $A$ is constant, is used to predict the value $y of a company vehicle when it is $t$ years old.

a What is the meaning of the constant $A$?

b A new car is valued at $15 000. Use the model to predict the value of this car when it is two years old. Give one reason why this value should only be considered as approximate.

## 4.2 The exponential function

The number e is an irrational number (like $\pi$) so it cannot be expressed exactly as a decimal.

The value of e can be calculated, to as many decimal places as needed, from the infinite series

$$1 + \frac{1}{1!} + \frac{1}{2!} + \frac{1}{3!} + \frac{1}{4!} + \frac{1}{5!} + \cdots$$ and is equal to 2.71823 correct to 5 decimal places.

*The* exponential function (as opposed to *an* exponential function) is defined as $f(x) = e^x$ for all real values of $x$.

For any value of $a$ ($a > 0$), $a^x$ is *an* exponential function, however for the base e (e $\approx 2.718$), $e^x$ is *the* exponential function.

The diagrams show sketches of $y = e^x$ and some simple variations of this function.

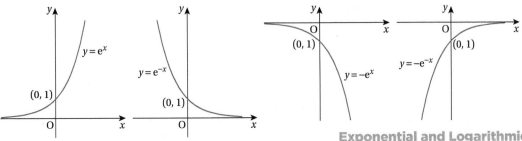

## Exercise 2

**1** Find the value, correct to 3 significant figures, of

    **a** $e^2$          **b** $e^{-1}$          **c** $e^{1.5}$          **d** $e^{-0.3}$.

**2** Sketch the curve whose equation is

    **a** $y = 1 - e^x$          **b** $y = e^x + 1$          **c** $y = e^{(x-1)}$

    **d** $y = 2 - e^x$          **e** $y = 1 + e^{-x}$          **f** $y = e^{2x}$

    **g** $y = 2 - e^{2x}$.

# 4.3 Natural logarithms

Logarithms to the base e are called **natural** (or **Naperian**) **logarithms**. The natural logarithm of $a$, that is $\log_e a$ is written as $\ln a$, so

$$\ln a = b \quad \Leftrightarrow \quad a = e^b$$

Logarithms to the base 10 are called **common logarithms** and are written as log.

The laws used for working with logarithms to a general base also apply to natural logarithms, so

$$\ln a + \ln b = \ln ab$$
$$\ln a - \ln b = \ln \frac{a}{b}$$
$$\ln a^n = n \ln a$$

## Example 2

**Question**

Separate $\ln (\tan x)$ into two terms.

**Answer**

$$\ln (\tan x) = \ln \left( \frac{\sin x}{\cos x} \right)$$
$$= \ln (\sin x) - \ln (\cos x)$$

## Example 3

**Question**

Express $4\ln(x + 1) - \frac{1}{2}\ln x$ as a single logarithm.

**Answer**

$$4\ln(x+1) - \frac{1}{2}\ln x = \ln(x+1)^4 - \ln\sqrt{x}$$

$$= \ln \frac{(x+1)^4}{\sqrt{x}}$$

# 4.4 The logarithmic function

Look at the curve with equation $y = f(x)$ where $f(x) = \ln x$.

When $y = \ln x$ then $x = e^y$, so

   the logarithmic function is the inverse of the exponential function.

It follows that the curve $y = \ln x$ is the reflection of the curve $y = e^x$ in the line $y = x$.

There is no part of the curve $y = \ln x$ for which $x$ is negative.

This is because, when $x = e^y$ (that is when $y = \ln x$), $x$ is positive for all real values of $y$.

Therefore

   ln $x$ does not exist for values of $x$ where $x \leq 0$.

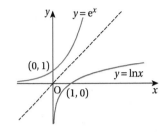

## Example 4

**Question**

Sketch the curve whose equation is $y = \ln(x - 3)$.

**Answer**

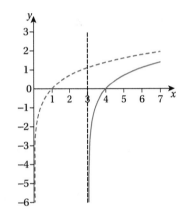

The curve $y = \ln(x - 3)$ is a translation of the curve $y = \ln x$ by the vector $\begin{bmatrix} 3 \\ 0 \end{bmatrix}$.

The curve $y = \ln x$ crosses the $x$-axis where $x = 1$, and the $y$-axis is an asymptote.

Therefore the curve $y = \ln(x - 3)$ crosses the $x$-axis where $x = 4$, and the line $x = 3$ is an asymptote.

## Exercise 3

**1** Write these in logarithmic form.

   **a** $e^x = 4$      **b** $e^2 = y$      **c** $e^{2x} = 3$

   **d** $e^{(x-1)} = 5$      **e** $e = x$

**2** Write these in index form.

   **a** $\ln x = 4$      **b** $\ln 0.5 = x$      **c** $\ln x = y$

   **d** $2 \ln x = 3$      **e** $2 \ln(1 - x) = 1.5$

**3** Find the value of

   **a** $\ln 48$      **b** $\ln e$      **c** $\ln 1$.

**4** Express each as a sum or difference of logarithms.

   **a** $\ln \dfrac{x^2}{(x+1)}$      **b** $\ln(a^2 - b^2)$

   **c** $\ln(\cot x)$      **d** $\ln(\sin^2 x)$

**5** Express each as a single logarithm.

   **a** $\ln(\cos x) - \ln(\sin x)$      **b** $1 + \ln x$      **c** $\dfrac{2}{3}\ln(x-1)$

**6** Solve the equations for $x$.

   **a** $e^x = 8.2$             **b** $e^{2x} + e^x - 2 = 0$ ←

   **c** $e^{2x-1} = 3$           **d** $e^{4x} - e^x = 0$

**Note**

Hint: Use $e^{2x} = (e^x)^2$

**7** Sketch each curve and mark the vertical asymptote on your sketch.

   **a** $y = -\ln x$        **b** $y = \ln(-x)$        **c** $y = 2 + \ln x$

   **d** $y = \ln(x^2)$        **e** $y = \ln(3 - x)$      **f** $y = 3 - \ln x$

## Summary

▶ Exponential growth or decay is where a quantity grows or decays by a constant factor over equal time intervals.

▶ The number $e \approx 2.718...$ is an irrational number.

▶ The function f where $f(x) = e^x$ is the exponential function, and is defined for all real numbers $x$.

▶ The logarithmic function f where $f(x) = \ln x$, is the inverse of the exponential function.

▶ The function f where $f(x) = \ln x$ only exists for positive values of $x$.

## Review

**1** Find the value of

   **a** $e^4$           **b** $e^{-15}$          **c** $e^{\frac{1}{2}}$.

**2** Sketch the curve whose equation is

   **a** $y = e^{3x}$         **b** $y = e^x - 2$.

**3** Write in logarithmic form.

   **a** $e^x = 2$          **b** $e^3 = y$.

**4** Write in index form.

   **a** $\ln x = 2$        **b** $\ln 0.4 = x$      **c** $\ln(x - 1) = y$

**5** Find the value of $\ln 4$.

**6** Express as a sum or difference of logarithms.

   **a** $\ln \dfrac{x}{x^2 + 1}$      **b** $\ln(x^2 - 2x + 1)$

**7** Express as a single logarithm.

   **a** $\ln(x - 1) - \ln x$    **b** $\ln(\sin x) - \ln(\cos x)$

**8** Solve the equation $e^x = 10$.

**9** Sketch the curve whose equation is $y = 2 - \ln(x + 1)$.

## Assessment

**1** Juan buys a vintage car for £2000. The car is expected to increase in value at the fixed rate of 2.5% per year.

   **a** Find the value of the car after 5 years correct to the nearest pound.

   **b** The car is kept for $N$ whole years. Find the value of $N$ for which the value of the car first exceeds £3000.

**c** Mujad invests £1500 in a savings account at a fixed interest rate of 3% per year.

Mujad invests his money at the same time at which Juan buys the car. Find the number of complete years after which Mujad's investment first exceeds the value of Juan's car.

**2** **a** Find the value of $e^{-\frac{1}{3}}$.

**b** Sketch the curve whose equation is $y = 1 - e^{2x}$.

**3** **a** Write $e^{(x-1)} = 3$ in logarithmic form.

**b** Solve the equation $e^{x-1} = 1.5$.

**4** **a** Express $2\ln(x+1) - \ln(x-1)$ as a single logarithm.

**b** Show the equation $2\ln(x+1) - \ln(x-1) = 0$ has no real roots.

**5** The equation $y = A(1.02)^t$, where $A$ is constant, is used to predict the value $\$y$ of a company share $t$ years after it is purchased.

**a** What is the meaning of constant $A$?

**b** Aisha bought 100 of these shares when they cost $1.20 each. Find the cost of these shares.

**c** Use the model to predict the value of these shares five years later. Give one reason why this value should only be considered as approximate.

**6** The functions f is defined by $f(x) = 5 - e^{3x}$ for all real values of $x$

**a** Find the range of f.

**b** The inverse of f is $f^{-1}$.

    **i** Find $f^{-1}(x)$.     **ii** Solve the equation $f^{-1}(x) = 0$.

AQA MPC3 June 2015 (part question)

**7** The curve with equation $y = f(x)$, where $f(x) = \ln(2x - 3)$, $x > \dfrac{3}{2}$, is sketched below.

**a** The inverse of f is $f^{-1}$.

    **i** Find $f^{-1}(x)$.

    **ii** State the range of $f^{-1}$.

    **iii** Sketch, the curve with equation $y = f^{-1}(x)$, indicating the value of the $y$-coordinate of the point where the curve intersects the $y$-axis.

**b** The function g is defined by $g(x) = e^{2x} - 4$, for all real values of $x$

    **i** Find $gf(x)$, giving your answer in the form $(ax - b)^2 - c$, where $a$, $b$ and $c$ are integers.

    **ii** Write down an expression for $fg(x)$, and hence find the exact solution of the equation $fg(x) = \ln 5$.

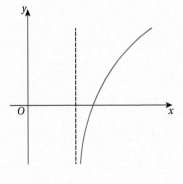

AQA MPC3 June 2013

# 5 Differentiation

## Introduction

This chapter extends differentiation to exponential, logarithmic and trigonometric functions and to combinations of these functions.

Sometimes the equation of a curve giving the direct relationship between $x$ and $y$ is awkward to work with. In this case it helps to express $x$ and $y$ each in terms of a third variable, called a parameter, which makes the equation much easier to work with.

All the differentiation techniques you have seen so far have been used on equations that can be expressed in the form $y = f(x)$. However some curves have equations that cannot easily be written in this way. This chapter shows how gradients of such curves can be found.

## Recap

You will need to remember...

▶ The meaning of the functions $e^x$ and $\ln x$ and the trigonometric functions.
▶ How to differentiate polynomials.
▶ How to find stationary values on a curve.
▶ How to find the equation of a tangent and a normal to a curve at a given point on the curve.
▶ The laws of logarithms.
▶ The properties of the sine and cosine functions.
▶ The relationship $\cos^2\theta + \sin^2\theta = 1$.
▶ How to recognize the equation of a circle.

## Objectives

By the end of this chapter, you should know how to...

▶ Find and use the derivative of the functions $e^x$, $\ln x$, $\sin x$, and $\cos x$.
▶ Find and use formulae for differentiating products and quotients of function for differentiating $\tan x$, and for differentiating a composite function.
▶ Explain the meaning of an implicit function and how to differentiate such a function with respect to $x$.
▶ Understand parameters.
▶ Convert a parametric equation of a curve into the Cartesian equivalent.
▶ Find $\dfrac{dy}{dx}$ from the parametric equation of a curve and the equation of a tangent or a normal to the curve, at a general point on the curve.

## 5.1 The derivative of $e^x$

The general shape of an exponential curve is shown in the AS Book.

This diagram shows three exponential curves.

They all pass through the point $(0, 1)$ because, when $y = a^x$ for any positive value of $a$, then $a^0 = 1$.

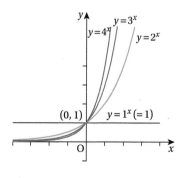

Each exponential function has a unique property: when the gradient at any point on the curve $y = a^x$ is divided by the value of $y$ at that point, the result is always the same number.

The table shows the value of $\dfrac{dy}{dx} \div y$ for some exponential functions.

| $y =$ | $2^x$ | $3^x$ | $4^x$ |
|---|---|---|---|
| $\dfrac{dy}{dx} \div y \approx$ | 0.7 | 1.1 | 1.4 |

The table shows that there is a number, somewhere between 2 and 3, for which

$$\frac{dy}{dx} \div y = 1, \text{ that is } \frac{dy}{dx} = y.$$

This number is e, which you met in Chapter 4.

> When $y = e^x$ then $\dfrac{dy}{dx} = e^x$.

The function f where $f(x) = e^x$ is the only function which is unchanged when differentiated.

## Example 1

**a** Find the gradient at the point on the curve $y = 2e^x - x$ where $x = 2$.

**b** Find the exact value of $x$ at which $f(x) = 2e^x - x$ has a stationary value.

**a** The derivative of $cf(x) =$ is $cf'(x)$ therefore $\dfrac{d}{dx}(2e^x) = 2\dfrac{d}{dx}(e^x)$

$$y = 2e^x - x \implies \frac{dy}{dx} = 2e^x - 1$$

When $x = 2$, $\dfrac{dy}{dx} = 2e^2 - 1$

**b** $f(x)$ has a stationary value where $2e^x - 1 = 0$

$$\implies e^x = \frac{1}{2} \implies x \ln e = \ln\left(\frac{1}{2}\right)$$

$\ln e = 1$   therefore   $x = \ln\left(\dfrac{1}{2}\right) = -\ln 2$

$f(x)$ has a stationary value where $x = -\ln 2$.

> **Note**
>
> $f'(x)$ means the derivative of $f(x)$ with respect to $x$.

## Exercise 1

**1** Write down the derivative of

    **a** $2e^x$         **b** $x^2 - e^x$         **c** $e^x$.

In questions 2 to 4, find the gradient of each curve at the specified value of $x$.

**2** $y = e^x - 2x$   where   $x = 2$

**3** $y = x^2 + 2e^x$   where   $x = 1$

**4** $y = e^x - 3x^3$   where   $x = 0$

**5** Find the value of $x$ at which $f(x) = e^x - x$ has a stationary value.

**6** Find the value of $x$ at which $f(x) = 4x - e^x$ has a stationary value.

**7** Explain why $f(x) = e^x + x$ does not have a stationary value.

## 5.2 The derivative of ln $x$

To find the derivative of ln $x$, a relationship is needed between $\dfrac{dy}{dx}$ and $\dfrac{dx}{dy}$.

The equation $y = \mathrm{f}(x)$, where $\mathrm{f}(x)$ is any function of $x$, is such that

$$\frac{dy}{dx} = \lim_{\delta x \to 0} \frac{\delta y}{\delta x} = \lim_{\delta x \to 0} \left( \frac{1}{\dfrac{\delta x}{\delta y}} \right)$$

Now $\delta y \to 0$ as $\delta x \to 0$, $\displaystyle\lim_{\delta y \to 0} \left( \frac{1}{\dfrac{\delta x}{\delta y}} \right)$, so

$$\frac{dy}{dx} = \frac{1}{\dfrac{dx}{dy}}$$

This relationship can be used to find the derivative of *any* function if the derivative of its inverse is known.

To differentiate $y = \ln x$, remember that

$$y = \ln x \iff x = e^y$$

Differentiating $e^y$ with respect to $y$ gives

$$\frac{dx}{dy} = e^y = x$$

Therefore $\quad \dfrac{dy}{dx} = \dfrac{1}{\left( \dfrac{dx}{dy} \right)} = \dfrac{1}{x}$ so

$$\frac{d}{dx} \ln x = \frac{1}{x}.$$

This result can be used to differentiate many logarithmic functions if they are first simplified by applying the laws of logarithms.

### Example 2

Find the derivative, with respect to $x$, of ln $(2x)$.

$\mathrm{f}(x) = \ln (2x) = \ln 2 + \ln x$

$\dfrac{d}{dx}\{\mathrm{f}(x)\} = 0 + \dfrac{1}{x}$ (as ln 2 is a number)

The derivative of ln $(2x)$ is $\dfrac{1}{x}$.

### Example 3

Find the derivative, with respect to $x$, of

a $\quad \ln\left( \dfrac{1}{x^3} \right)$ 
b $\quad \ln\left( 4\sqrt{x} \right)$.

**a** $f(x) = \ln\left(\dfrac{1}{x^3}\right) = \ln(x^{-3}) = -3\ln x$

$\dfrac{d}{dx}\{f(x)\} = \dfrac{d}{dx}\{-3\ln x\} = \dfrac{-3}{x}$

**b** $f(x) = \ln\left(4\sqrt{x}\right) = \ln 4 + \ln\left(\sqrt{x}\right) = \ln 4 + \dfrac{1}{2}\ln x$

$\dfrac{d}{dx}\{f(x)\} = \dfrac{d}{dx}(\ln 4) + \dfrac{d}{dx}\left(\dfrac{1}{2}\ln x\right) = 0 + \dfrac{\frac{1}{2}}{x} = \dfrac{1}{2x}$

## Exercise 2

**1** Write down the derivative, with respect to $x$, of each of these expressions.

   **a** $\ln x^3$
          **b** $\ln(3x)$
          **c** $\ln(x^{-2})$

   **d** $\ln\left(\dfrac{3}{\sqrt{x}}\right)$
          **e** $\ln\left(\dfrac{1}{x^5}\right)$
          **f** $\ln\left(2x^{\frac{1}{2}}\right)$

   **g** $\ln\left(x^{-\frac{3}{2}}\right)$
          **h** $\ln\left(\dfrac{x^3}{\sqrt{x}}\right)$

**2** Find the stationary points on each curve.

   **a** $y = \ln x - x$
          **b** $y = x^3 - 2\ln x^3$
          **c** $y = \ln x - \sqrt{x}$

## Using $\dfrac{dy}{dx} = \dfrac{1}{\dfrac{dx}{dy}}$ to differentiate an equation of the form $x = f(y)$

When, for example, the equation of a curve is $x = y^3$,
you can find $\dfrac{dy}{dx}$ by first finding $\dfrac{dx}{dy}$.

In this case $\dfrac{dx}{dy} = 3y^2$.

Then using $\dfrac{dy}{dx} = \dfrac{1}{\dfrac{dx}{dy}}$, you can see that $\dfrac{dy}{dx} = \dfrac{1}{3y^2}$.

## Example 4

A curve has equation $x = y + \ln y^2$. Find the value of $\dfrac{dy}{dx}$ when $y = 2$.

$x = y + \ln y^2 \quad \Rightarrow \quad x = y + 2\ln y$

$\dfrac{dx}{dy} = 1 + \dfrac{2}{y} = \dfrac{y+2}{y} \quad$ so $\quad \dfrac{dy}{dx} = \dfrac{1}{\dfrac{(y+2)}{y}} = \dfrac{y}{y+2}$

When $y = 2$, $\dfrac{dy}{dx} = \dfrac{2}{4} = \dfrac{1}{2}$

## Example 5

Question

Find the equation of the tangent to the curve whose equation is $x = 3y^2 - 2y$ at the point where $y = 3$.

Answer

To find the equation of the tangent at the point where $y = 3$, you need to know the coordinates of the point where $y = 3$, and the gradient of the curve at that point.

When $y = 3$, $x = 21$.

$$x = 3y^2 - 2y \quad \Rightarrow \quad \frac{dx}{dy} = 6y - 2$$

Therefore $\dfrac{dy}{dx} = \dfrac{1}{6y - 2}$ so when $y = 3$, the gradient of the tangent is $\dfrac{1}{16}$.

The equation of the tangent is $y - 3 = \dfrac{1}{16}(x - 21) \quad \Rightarrow \quad 16y - x - 27 = 0$.

## Exercise 3

**1** The equation of a curve is $x = 4y^2$. Find $\dfrac{dy}{dx}$ when $y = 1$.

**2** The equation of a curve is $x = y^2 - 3y + 2$. Find $\dfrac{dy}{dx}$ when $y = 3$.

**3** The equation of a curve is $x = 2y^3 + 3y$. Find $\dfrac{dy}{dx}$ when $y = \dfrac{1}{2}$.

**4** The equation of a curve is $x = y^2 + y$. Find $\dfrac{dy}{dx}$ when $y = 1$.

Hence find the equation of the tangent to the curve at the point on the curve where $y = 1$.

**5** The equation of a curve is $x = e^y + y$. Find $\dfrac{dy}{dx}$ when $y = 1$.

Hence find the equation of the tangent to the curve at the point on the curve where $y = 2$.

**6** The equation of a curve is $x = y - \ln y$. Find $\dfrac{dy}{dx}$ when $y = 3$.

Hence find the equation of the tangent to the curve at the point on the curve where $y = 3$.

**7** The equation of a curve is $x = \ln(y^2)$. Find $\dfrac{dy}{dx}$ when $y = 2$.

Hence find the equation of the tangent to the curve at the point on the curve where $y = 2$.

**8** The equation of a curve is $x = 3y + 4e^y$. Find $\dfrac{dy}{dx}$ when $y = 1$.

Hence find the equation of the tangent to the curve at the point on the curve where $y = 1$.

# 5.3 The derivatives of sin *x* and cos *x*

When *x* is measured in radians the gradient function of sin *x* is cos *x*, and the gradient function of cos *x* is −sin *x*.

When $y = \sin x$ then $\dfrac{dy}{dx} = \cos x$.

When $y = \cos x$ then $\dfrac{dy}{dx} = -\sin x$.

These results are found by differentiating from first principles and you can quote them without having to prove them.

*It is important to understand that these results are only valid when x is measured in radians* and, throughout all the work that follows involving the differentiation of trigonometric functions, the angle is measured in radians unless it is stated otherwise.

## Example 6

Find the smallest positive value of *x* for which f(*x*) = *x* + 2 cos *x* has a stationary value.

$f(x) = x + 2\cos x \Rightarrow f'(x) = 1 - 2\sin x$

For stationary values, $f'(x) = 0$

so $1 - 2\sin x = 0 \Rightarrow \sin x = \dfrac{1}{2}$

The smallest positive angle for which $\sin x = \dfrac{1}{2}$ is $\dfrac{\pi}{6}$ radians.

> **Note**
>
> The answer *must* be given in radians because the rule used to differentiate cos *x* is valid only for an angle in radians.

## Example 7

Find the smallest positive value of *θ* for which the curve $y = 2\theta - 3\sin\theta$ has a gradient of $\dfrac{1}{2}$.

$y = 2\theta - 3\sin\theta \Rightarrow \dfrac{dy}{d\theta} = 2 - 3\cos\theta$

When $\dfrac{dy}{d\theta} = \dfrac{1}{2}$, $2 - 3\cos\theta = \dfrac{1}{2} \Rightarrow 3\cos\theta = \dfrac{3}{2} \Rightarrow \cos\theta = \dfrac{1}{2}$

The smallest positive value of *θ* for which $\cos\theta = \dfrac{1}{2}$ is $\dfrac{\pi}{3}$ radians.

## Exercise 4

1. Write down the derivative of each of these functions with respect to the given variable.

   a $\sin x - \cos x$    b $\sin\theta + 4$    c $3\cos\theta$

   d $5\sin\theta - 6$    e $2\cos\theta + 3\sin\theta$    f $4\sin x - 5 - 6\cos x$

**2** Write down the derivative of each of these functions with respect to the given variable.

a $\quad x^2 - \sin x$ 　　　　 b $\quad \cos x + \ln x$ 　　　　 c $\quad 2 \sin x - e^x$

d $\quad 3e^x - \ln x + 4 \cos x$ 　 e $\quad 3 + 2 \ln x + 5 \sin x$ 　 f $\quad x^3 - 3\cos x$

**3** Find the gradient of each curve at the point whose $x$-coordinate is given.

a $\quad y = 2 \cos x; \dfrac{1}{2}\pi$ 　　 b $\quad y = \sin x; 0$ 　　　　 c $\quad y = \cos x + \sin x; \pi$

d $\quad y = x - \sin x; \dfrac{1}{2}\pi$ 　 e $\quad y = 2 \sin x - x^2; -\pi$ 　 f $\quad y = -4 \cos x; \dfrac{1}{2}\pi$

**4** For each of these curves, find the smallest positive value of $\theta$ at which the gradient of the curve has the given value.

a $\quad y = 2 \cos \theta; -1$ 　　　　　　　 b $\quad y = \theta + \cos \theta; \dfrac{1}{2}$

c $\quad y = \sin \theta + \cos \theta; 0$ 　　　　　 d $\quad y = 2\theta + \sin \theta; 1$

**5** Considering only positive values of $x$, find the coordinates of the first two stationary points on each of these curves and determine whether they are maximum or minimum points.

a $\quad y = 2 \sin x - x$ 　　　 b $\quad y = x + 2 \cos x$

**6** Find the equation of the tangent to the curve $y = \cos \theta + 3 \sin \theta$ at the point where $\theta = \dfrac{1}{2}\pi$.

**7** Find the equation of the normal to the curve $y = x^2 + \cos x$ at the point where $x = \pi$.

**8** Find the coordinates of a point on the curve $y = \sin x + \cos x$ at which the tangent is parallel to the line $y = x$.

# 5.4 Differentiating products, quotients and composite functions

## Differentiating a product

When you are given a is curve $y = u(x) \, v(x)$, where $u$ and $v$ are both functions of $x$, for example $y = x^2 \sin x$, you need to know that

$\dfrac{dy}{dx}$ is NOT equal to $\dfrac{d}{dx}(x^2) \times \dfrac{d}{dx}(\sin x)$.

Consider $y = u(x) \, v(x)$. A small increase $(\delta x)$ in $x$, gives corresponding small increases of $\delta u$, $\delta v$ and $\delta y$ in the values of $u$, $v$ and $y$.

Therefore $\quad y + \delta y = (u + \delta u)(v + \delta v)$

$$= uv + u\delta v + v\delta u + \delta u \delta v$$

Now $y = uv$

therefore $\quad\quad \delta y = u\delta v + v\delta u + \delta u \delta v$

$\Rightarrow \quad\quad\quad \dfrac{\delta y}{\delta x} = u\dfrac{\delta v}{\delta x} + v\dfrac{\delta u}{\delta x} + \delta u\dfrac{\delta v}{\delta x}$

As $\delta x \to 0$, $\dfrac{\delta v}{\delta x} \to \dfrac{dv}{dx}$, $\dfrac{\delta u}{\delta x} \to \dfrac{du}{dx}$ and $\delta u \to 0$

Therefore $\dfrac{dy}{dx} = \lim_{\delta x \to 0} \dfrac{\delta y}{\delta x}$

$$= u\dfrac{dv}{dx} + v\dfrac{du}{dx} + 0$$

Therefore $\dfrac{d}{dx}(uv) = v\dfrac{du}{dx} + u\dfrac{dv}{dx}$

> **Note**
>
> Addition can be done in any order, so
> $$u\dfrac{dv}{dx} + v\dfrac{du}{dx} = v\dfrac{du}{dx} + u\dfrac{dv}{dx}$$

Applying this formula to the example $y = (x^2)\sin x$, with $u(x) = x^2$ and $v(x) = \sin x$, gives

$$\dfrac{dy}{dx} = (\sin x)(2x) + (x^2)(\cos x) = 2x \sin x + x^2 \cos x.$$

## Example 8

Differentiate with respect to $x$

a $\quad x^3 \ln x$ 
b $\quad \dfrac{\cos x}{x}$.

a $\quad u = x^3 \quad \Rightarrow \quad \dfrac{du}{dx} = 3x^2$

$\quad v = \ln x \quad \Rightarrow \quad \dfrac{dv}{dx} = \dfrac{1}{x}$

$\quad \dfrac{d}{dx}(uv) = v\dfrac{du}{dx} + u\dfrac{dv}{dx} \quad$ gives $\quad \dfrac{dy}{dx} = (3x^2)\ln x + \left(\dfrac{1}{x}\right)x^3 = x^2 + 3x^2 \ln x$

b $\quad$ Writing $\dfrac{\cos x}{x}$ as $(\cos x)(x^{-1})$

$\quad$ then $\quad u = \cos x \quad$ gives $\quad \dfrac{du}{dx} = -\sin x$

$\quad$ and $\quad v = x^{-1} \quad$ gives $\quad \dfrac{dv}{dx} = -x^{-2}$

$\quad$ Using $\quad \dfrac{d}{dx}(uv) = v\dfrac{du}{dx} + u\dfrac{dv}{dx} \quad$ gives $\quad \dfrac{dy}{dx} = (-x^{-2})(\cos x) + (-\sin x)(x^{-1})$

$$= -\dfrac{\cos x}{x^2} - \dfrac{\sin x}{x} = -\dfrac{\cos x + x \sin x}{x^2}.$$

## Exercise 5

Differentiate each expression with respect to $x$.

**1** $\dfrac{\sin x}{x}$

**2** $e^x \cos x$

**3** $(x^3 - 2)\ln x$

**4** $(x + 1)\sin x$

**5** $\sin x \cos x$

**6** $\dfrac{\ln x}{x^2}$

**7** $(\cos x)\ln x$

**8** $e^x \sin x$

**9** $x^2 \sin x$

**10** $x^3 \ln 2x$

## Differentiating a quotient

To differentiate a function of the form $y = \dfrac{u(x)}{v(x)}$, where $u$ and $v$ are both functions of $x$, the function can be rewritten as $y = u(x)v^{-1}(x)$ and so the formula to differentiate a product can be used. This method was used in part (b) of Example 8 but it is not always the neatest way to differentiate a quotient. The alternative is to apply the formula derived below.

When a function is of the form $y = \dfrac{u(x)}{v(x)}$, where $u$ and $v$ are both functions of $x$, a small increase of $\delta x$ in the value of $x$ gives corresponding small increases of $\delta u$ and $\delta v$ in the values $u$ and $v$. Then, as $\delta x \to 0$, $\delta u$ and $\delta v$ also tend to zero.

When $\qquad y = \dfrac{u}{v}$ then $\quad y + \delta y = \dfrac{(u + \delta u)}{(v + \delta v)}$

so $\qquad \delta y = \dfrac{u + \delta u}{v + \delta v} - \dfrac{u}{v} = \dfrac{v\delta u - u\delta v}{v(v + \delta v)}$

Therefore $\quad \dfrac{\delta y}{\delta x} = \dfrac{\left( v\dfrac{\delta u}{\delta x} - u\dfrac{\delta v}{\delta x} \right)}{v(v + \delta v)}$

$\Rightarrow \qquad \dfrac{dy}{dx} = \lim\limits_{\delta x \to 0} \dfrac{\delta y}{\delta x} = \dfrac{v\dfrac{du}{dx} - u\dfrac{dv}{dx}}{v^2}$

Therefore $\qquad \dfrac{dy}{dx} = \dfrac{v\dfrac{du}{dx} - u\dfrac{dv}{dx}}{v^2}$

## Example 9

Find $\dfrac{dy}{dx}$ when $y = \dfrac{4x - 3}{\sin x}$.

Taking $u = 4x - 3$ gives $\dfrac{du}{dx} = 4$ and $v = \sin x$ gives $\dfrac{dv}{dx} = \cos x$

so $\qquad \dfrac{dy}{dx} = \dfrac{\left( v\dfrac{du}{dx} - u\dfrac{dv}{dx} \right)}{v^2}$

$\qquad = \dfrac{4\sin x - (4x - 3)\cos x}{(\sin x)^2}$

## Example 10

Find $\dfrac{dy}{dx}$ when $y = \dfrac{\sin x}{\cos x}$.

Taking $u = \sin x$ gives $\dfrac{du}{dx} = \cos x$ and taking $v = \cos x$ gives $\dfrac{dv}{dx} = -\sin x$

so $\qquad \dfrac{dy}{dx} = \dfrac{\left( v\dfrac{du}{dx} - u\dfrac{dv}{dx} \right)}{v^2}$

$\qquad = \dfrac{\cos^2 x + \sin^2 x}{\cos^2 x} = \dfrac{1}{\cos^2 x} = \sec^2 x$

> **Note**
>
> $\cos^2 x$ means $(\cos x)^2$. Similarly, $\sin^2 x$ means $(\sin x)^2$, and $\tan^2 x$ means $(\tan x)^2$, etc.

## The derivative of tan $x$

In Example 10, you saw how to differentiate $y = \dfrac{\sin x}{\cos x}$. Since $\dfrac{\sin x}{\cos x} = \tan x$, the result gives the derivative of $\tan x$.

When $y = \tan x$, $\dfrac{dy}{dx} = \sec^2 x$.

## Exercise 6

Use the quotient rule to differentiate each of these expressions with respect to $x$.

**1** $\dfrac{e^x}{x}$

**2** $\dfrac{x^2}{(x+3)}$

**3** $\dfrac{(4-x)}{x^2}$

**4** $\dfrac{\ln x}{x^3}$

**5** $\dfrac{4x}{\sin x + \cos x}$

**6** $\dfrac{2x^2}{(x-2)}$

**7** $\dfrac{x^{\frac{5}{3}}}{(3x-2)}$

**8** $\dfrac{1-\ln x}{x^3}$

**9** $\dfrac{\cos x}{\sin x}$

## Differentiating a composite function (the chain rule)

The function $\sin(x^2)$ is a **composite function**. That is, when $f(x) = \sin x$ and $g(x) = x^2$ then $fg(x) = \sin(x^2)$.

For any equation of the form $y = gf(x)$ you can make the substitution $u = f(x)$, which means $y = gf(x)$ can be expressed in two simple parts: that is $u = f(x)$ and $y = g(u)$.

A small increase of $\delta x$ in the value of $x$ gives a corresponding small increase of $\delta u$ in the value of $u$.

Then if $\delta x \to 0$, it follows that $\delta u \to 0$.

Hence $\dfrac{dy}{dx} = \lim\limits_{\delta x \to 0}\left(\dfrac{\delta y}{\delta x}\right) = \lim\limits_{\delta x \to 0}\left(\dfrac{\delta y}{\delta u} \times \dfrac{\delta u}{\delta x}\right)$

$\Rightarrow \quad \dfrac{dy}{dx} = \lim\limits_{\delta x \to 0}\left(\dfrac{\delta y}{\delta u}\right) \times \lim\limits_{\delta x \to 0}\left(\dfrac{\delta u}{\delta x}\right)$

so $\quad \dfrac{dy}{dx} = \dfrac{dy}{du} \times \dfrac{du}{dx}$

This formula is called the **chain rule**.

## Example 11

**Question**

Find $\dfrac{dy}{dx}$ when $y = \sin(x^2)$.

**Answer**

Letting $u = x^2$ gives $y = \sin u$.

Using $\dfrac{dy}{dx} = \dfrac{dy}{du} \times \dfrac{du}{dx}$ gives $\dfrac{dy}{dx} =$ either $(\cos u) \times 2x$ or $2x \cos u$

But $u = x^2$, therefore $\dfrac{dy}{dx} = 2x\cos(x^2)$.

## Example 12

Given $y = (x^3 + 1)^4$, find $\dfrac{dy}{dx}$.

Letting $u = x^3 + 1$ gives $y = u^4$

Using $\quad \dfrac{dy}{dx} = \dfrac{dy}{du} \times \dfrac{du}{dx}$ gives

$$\dfrac{dy}{dx} = (4u^3)(3x^2) = 12x^2 u^3$$

Replacing $u$ by $x^3 + 1$ gives $\quad \dfrac{dy}{dx} = 12x^2 (x^3 + 1)^3$

## Exercise 7

Use the chain rule to differentiate each expression with respect to $x$.

1. $(3x + 1)^2$
2. $(3 - x)^4$
3. $\sin(3x)$
4. $e^{2x}$
5. $\ln(2x - 1)$
6. $\cos(5x)$
7. $\sin(x^3)$
8. $e^{(3x + 5)}$
9. $\sqrt{3x^3 - 4}$
10. $\ln(x^2 - 2x)$
11. $\cos(3x - 5)$
12. $\ln(3x + x^2)$
13. $(4 - 2x)^5$
14. $e^{(x^2 - x)}$
15. $4 \cos(5x - 6)$
16. $\ln(\sin x)$
17. $e^{(x - x^3)}$
18. $(2 - x^3)^4$
19. $\sqrt[3]{x^2 - x}$
20. $(x^5 - 3)^{-\frac{1}{2}}$
21. $\sin^2 x$

## General results

Using the chain rule on some general composite functions gives some standard results that you can quote without having to prove them.

For example, when $y = \sin f(x)$, taking $u = f(x)$ gives $y = \sin u$.

Then $\quad \dfrac{dy}{dx} = \dfrac{dy}{du} \times \dfrac{du}{dx} \quad \Rightarrow \quad \dfrac{dy}{dx} = (\cos u)\dfrac{du}{dx}.$

**Notice that** when $y = \sin f(x)$ then $\quad \dfrac{dy}{dx} = f'(x) \cos f(x)$

**Similarly** when $y = \cos f(x)$ then $\quad \dfrac{dy}{dx} = -f'(x) \sin f(x)$

**In particular** $\quad \dfrac{d}{dx}(\sin ax) = a \cos ax$

**and** $\quad \dfrac{d}{dx}(\cos ax) = -a \sin ax$

When $y = e^{f(x)}$, using $u = f(x)$ gives $y = e^u$, then $\dfrac{dy}{dx} = \dfrac{dy}{du} \times \dfrac{du}{dx} \quad \Rightarrow \quad \dfrac{dy}{dx} = (e^u)\dfrac{du}{dx}.$

**So when** $y = e^{f(x)}$ then $\dfrac{dy}{dx} = f'(x) e^{f(x)}.$

When $y = \ln f(x)$, using $u = f(x)$ gives $y = \ln u$, then $\dfrac{dy}{dx} = \dfrac{dy}{du} \times \dfrac{du}{dx} \implies \dfrac{dy}{dx} = \left(\dfrac{1}{u}\right)\dfrac{du}{dx}$.

So when $y = \ln f(x)$ then $\dfrac{dy}{dx} = \dfrac{f'(x)}{f(x)}$.

## Example 13

Differentiate $\cos\left(\dfrac{1}{6}\pi - 3x\right)$ with respect to $x$.

$$\frac{d}{dx}\left\{\cos\left(\frac{1}{6}\pi - 3x\right)\right\} = -(-3)\sin\left(\frac{1}{6}\pi - 3x\right)$$
$$= 3\sin\left(\frac{1}{6}\pi - 3x\right)$$

## Example 14

Find $\dfrac{dy}{dx}$ when $y = \ln(2x - 3)$.

$$\frac{dy}{dx} = \frac{2}{2x - 3}$$

## Example 15

Find $\dfrac{dy}{d\theta}$ when $y = \cos^3 \theta$.

$y = \cos^3 \theta = [\cos \theta]^3$

$y = u^3$ where $u = \cos \theta$

$\dfrac{dy}{d\theta} = \dfrac{dy}{du} \times \dfrac{du}{d\theta} = (3u^2)(-\sin\theta) = -3u^2 \sin\theta$

Therefore $u = \cos \theta \implies \dfrac{dy}{d\theta} = -3\cos^2 \theta \sin\theta$

Example 15 includes another example of some standard differentiation results:

When $y = \cos^n x$ then $\dfrac{dy}{dx} = -n\cos^{n-1} x \sin x$.

When $y = \sin^n x$ then $\dfrac{dy}{dx} = n\sin^{n-1} x \cos x$

## Exercise 8

Differentiate each of these expressions with respect to $x$.

**1** $\sin 4x$

**2** $\cos(\pi - 2x)$

**3** $\sin\left(\dfrac{1}{2}x + \pi\right)$

**4** $\cos^2 x$

**5** $e^{\sin x}$

**6** $\ln(\cos x)$

**7** $\sin x^2$

**8** $e^{\cos x}$

**9** $\ln(\sin x)$

**10** $\cos^4 x$

**11** $e^{(x^2 - 2x)}$

**12** $\tan 6x$

**13** $\ln(2x^2 - 3x)$

**14** $\sin(5x - 8)$

# 5.5 Implicit functions

It is difficult to find $y$ in terms of $x$ in the equation $x^2 - y^2 + y = 1$.

A relationship of this type, where $y$ is not given explicitly as a function of $x$, is called an **implicit function**, because it is *implied* in the equation that $y = f(x)$.

## Differentiating an implicit function

An implicit function can be differentiated term by term. To be able to do this, you need a method which will allow you to differentiate terms like $y^2$ with respect to $x$.

When $\quad g(y) = y^2$ and $y = f(x)$

then $\quad g(y) = \{f(x)\}^2 \quad$ is a composite function.

Using the substitution $u = f(x)$ gives $y = u^2$. Differentiating by the chain rule gives

$$\frac{d}{dx}y^2 = \frac{d}{dx}u^2 = \frac{d}{du}(u^2) \times \frac{du}{dx}$$

$$= 2u \times \frac{du}{dx} = 2y\frac{dy}{dx}$$

Therefore $\quad \dfrac{d}{dx}(y^2) = 2y\dfrac{dy}{dx}$.

**In general,** $\quad \dfrac{d}{dx}g(y) = \left(\dfrac{d}{dy}g(y)\right)\left(\dfrac{dy}{dx}\right)$.

To differentiate $f(y)$ with respect to $x$, you must differentiate $f(y)$ with respect to $y$ and then multiply by $\dfrac{dy}{dx}$.

For example, $\quad \dfrac{d}{dx}y^3 = 3y^2\dfrac{dy}{dx} \quad$ and $\quad \dfrac{d}{dx}e^y = e^y\dfrac{dy}{dx}$.

A term that contains a product of both $x$ and $y$, for example $x^2y^3$, can be differentiated using the product rule.

Letting $\quad u = x^2$ and $v = y^3 \quad$ and using $\quad \dfrac{d}{dx}(uv) = v\dfrac{du}{dx} + u\dfrac{dv}{dx}$

gives $\quad \dfrac{d}{dx}(x^2y^3) = y^3\dfrac{d}{dx}(x^2) + x^2\dfrac{d}{dx}(y^3)$.

$$= 2xy^3 + x^2 3y^2\frac{dy}{dx}$$

$$= 2xy^2 + 3x^2y^2\frac{dy}{dx}$$

## Example 16

Find $\dfrac{dy}{dx}$ in terms of $x$ and $y$ when $x^2 - y^2 + x^2 y = 1$.

Differentiating $x^2 - y^2 + x^2 y = 1$ term by term gives

$$\frac{d}{dx}(x^2) - \frac{d}{dx}(y^2) + \frac{d}{dx}(x^2 y) = \frac{d}{dx}(1)$$

$$\Rightarrow \qquad 2x - 2y\frac{dy}{dx} + 2xy + x^2\frac{dy}{dx} = 0$$

Hence $\quad 2x(1+y) = \dfrac{dy}{dx}(2y - x^2) \quad \Rightarrow \quad \dfrac{dy}{dx} = \dfrac{2x(1+y)}{2y - x^2}$.

## Example 17

Differentiate $y = xe^y$ with respect to $x$ and hence find $\dfrac{dy}{dx}$ in terms of $x$ and $y$.

When $\quad y = xe^y \quad$ then $\quad \dfrac{dy}{dx} = \dfrac{d}{dx}(xe^y) = e^y\dfrac{d}{dx}(x) + x\dfrac{d}{dx}(e^y)$

$$\Rightarrow \qquad \frac{dy}{dx} = e^y + xe^y\frac{dy}{dx} \quad \Rightarrow \quad \frac{dy}{dx} - xe^y\frac{dy}{dx} = e^y$$

Hence $\quad \dfrac{dy}{dx} = \dfrac{e^y}{1 - xe^y}$.

## Example 18

Given that $y = \sin^{-1} x$, show that $\dfrac{dy}{dx} = \dfrac{1}{\sqrt{1 - x^2}}$.

$y = \sin^{-1} x \quad \Rightarrow \quad \sin y = x$

Differentiating with respect to $x$ gives

$$\cos y\frac{dy}{dx} = 1 \quad \Rightarrow \quad \frac{dy}{dx} = \frac{1}{\cos y}$$

But $\quad \cos y = \sqrt{1 - \sin^2 y} = \sqrt{1 - x^2} \quad$ so $\quad \dfrac{dy}{dx} = \dfrac{1}{\sqrt{1 - x^2}}$.

## Exercise 9

In questions 1 to 9, differentiate the equation with respect to $x$.

**1** $x^2 + y^2 = 4$

**2** $x^2 + xy + y^2 = 0$

**3** $x(x + y) = y^2$

**4** $\dfrac{1}{x} + \dfrac{1}{y} = e^y$

**5** $\dfrac{1}{x^2} + \dfrac{1}{y^2} = \dfrac{1}{4}$

**6** $\dfrac{x^2}{4} - \dfrac{y^2}{9} = 1$

**7** $\sin x + \sin y = 1$      **8** $\sin x \cos y = 2$      **9** $xe^y = x + 1$

**10** Find $\dfrac{dy}{dx}$ as a function of $x$ when $y^2 = 2x + 1$.

**11** Find the gradient of $x^2 + y^2 = 9$ at the points where $x = 1$.

**12** Find $\dfrac{dy}{dx}$ given that $y = \tan^{-1} x$.

**13** Find the equations of the tangents to $x^2 - 3y^2 = 4y$

     **a**   at the points where $y = 2$

     **b**   at the point $(x_1, y_1)$.

**14** Find the equations of the tangents to $x^2 + xy + y^2 = 3$ at the points where $x = 1$.

**15** Find the equation of the tangent at $\left(1, \dfrac{1}{3}\right)$ to the curve whose equation is $2x^2 + 3y^2 - 3x + 2y = 0$.

# 5.6 Parametric equations

Look at the equations   $x = t^2, y = t - 1$.

A point P$(x, y)$ is on the curve represented by $x = t^2, y = t - 1$ if and only if the coordinates of P are $(t^2, t - 1)$.

The variable $t$ is called a **parameter**.

Substituting some numbers for $t$ gives a pair of corresponding values of $x$ and $y$.

For example, when $t = 3$, $x = 9$ and $y = 2$, therefore $(9, 2)$ is a point on the curve.

The direct relationship between $x$ and $y$ is found by eliminating $t$ from these two **parametric equations**.

In this case $y = t - 1 \quad \Rightarrow \quad t = y + 1$

Substituting $t = y + 1$ in the equation $x = t^2$ gives $x = (y + 1)^2$.

So the Cartesian equation of this curve is $(y + 1)^2 = x$.

## Sketching a curve from parametric equations

Look again at the curve whose parametric equations are

     $x = t^2$ and $y = t - 1$.

The table shows the values of $x$ and $y$ that correspond to some values of $t$.

| $t$ | −2 | −1 | 0 | 1 | 2 |
|-----|----|----|----|----|----|
| $x$ | 4 | 1 | 0 | 1 | 4 |
| $y$ | −3 | −2 | −1 | 0 | 1 |

In order to sketch the curve, begin by plotting these points on a graph.

Afterwards, look at what happens to $x$ and to $y$ as $t$ varies:

     $x \geq 0$ for all values of $t$,

     as $t \to \infty$, $x \to \infty$ and $y \to \infty$,

     as $t \to -\infty$, $x \to \infty$ and $y \to -\infty$,

There are no values of $t$ for which either $x$ or $y$ is undefined so it is reasonable to assume that the curve has no breaks.

There is now enough information to sketch the curve, as shown in the margin.

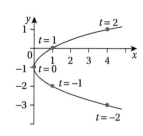

## Example 19

Find the Cartesian equation of the curve whose parametric equations are

**a** $x = t^2$

$y = 2t$

**b** $x = 2\cos\theta$

$y = 2\sin\theta$

**c** $x = 2t$

$y = \dfrac{2}{t}$.

**a** $y = 2t \implies t = \dfrac{1}{2}y$

$x = t^2 \implies x = \left(\dfrac{1}{2}y\right)^2 = \dfrac{1}{4}y^2 \implies y^2 = 4x$

**b** Using $\cos^2\theta + \sin^2\theta = 1$ where $2\cos\theta = x$ and $2\sin\theta = y$ gives

$x^2 + y^2 = 4\cos^2\theta + 4\sin^2\theta = 4$

**c** $y = \dfrac{2}{t} \implies t = \dfrac{2}{y}$

$x = 2t \implies x = \dfrac{4}{y} \implies xy = 4$

## Exercise 10

**1** Find the Cartesian equation of each of these curves.

  **a** $x = 2t^2, y = t$       **b** $x = \cos\theta, y = \sin\theta$       **c** $x = t, y = \dfrac{4}{t}$

**2** Sketch each curve given in question 1.

**3** Find the Cartesian equation of the curve given by the parametric equations

  $x = \dfrac{t}{1-t}$ and $y = \dfrac{t^2}{1-t}$.

**4** Show that the Cartesian equation of the curve given by $x = \dfrac{1}{t}$ and $y = \dfrac{t+1}{t^2}$

  is $y = x + x^2$.

**5** Find the Cartesian equation of the curve given by $x = t$ and $y = t^3 - t$.

**6** Find the Cartesian equation of the curve given by $x = 2\cos\theta$ and $y = 3\sin\theta$.

**7 a** Show that, the parametric equations $x = 4\cos\theta$ and $y = 4\sin\theta$ give the
    equation of a circle.

  **b** State the centre and radius of the circle.

# 5.7 Finding $\dfrac{dy}{dx}$ using parametric equations

When both $x$ and $y$ are given as functions of $t$ then a small increase of $\delta t$ in the value of $t$ results in corresponding small increases of $\delta x$ and $\delta y$ in the values of $x$ and $y$.

Therefore $\dfrac{\delta y}{\delta x} = \dfrac{\delta y}{\delta t} \times \dfrac{\delta t}{\delta x}$

As $\delta t \to 0$, $\delta x$ and $\delta y$ also approach zero, therefore $\dfrac{dy}{dx} = \dfrac{dy}{dt} \times \dfrac{dt}{dx}$

Using the formula $\dfrac{dt}{dx} = \dfrac{1}{\frac{dx}{dt}}$ gives

$$\dfrac{dy}{dx} = \dfrac{\frac{dy}{dt}}{\frac{dx}{dt}}$$

Therefore $\dfrac{dy}{dx}$ is found in terms of $t$ by differentiating both $x$ and $y$ with respect to $t$ and using the formula above.

## Example 20

**Question**

The parametric equations of a curve are $x = t^2$ and $y = t - 1$.

a   Find the gradient of the curve at the point where $t = 1$.

b   Find the coordinates of the point on the curve where the gradient is 2.

c   Show that there are no stationary points on this curve.

**Answer**

a   $x = t^2$ and $y = t - 1$

Using $\dfrac{dy}{dx} = \dfrac{\frac{dy}{dt}}{\frac{dx}{dt}}$ gives

$\dfrac{dy}{dt} = 1$  and  $\dfrac{dx}{dt} = 2t$  $\Rightarrow$  $\dfrac{dy}{dx} = \dfrac{1}{2t}$

Therefore when $t = 1$, $\dfrac{dy}{dx} = \dfrac{1}{2}$.

b   When $\dfrac{dy}{dx} = 2$, $\dfrac{1}{2t} = 2$  $\Rightarrow$  $t = \dfrac{1}{4}$

and when $t = \dfrac{1}{4}$, then $x = \dfrac{1}{16}$  and  $y = -\dfrac{3}{4}$.

The gradient of the curve is 2 at the point $\left( \dfrac{1}{16}, -\dfrac{3}{4} \right)$.

c   There are no values of $t$ for which $\dfrac{dy}{dx} = 0$, so there are no stationary points on this curve.

## Example 21

**Question**

Find the stationary point on the curve whose parametric equations are $x = t^3$, $y = (t + 1)^2$.

**Answer**

$\dfrac{dy}{dt} = 2(t + 1)$   and   $\dfrac{dx}{dt} = 3t^2$   so   $\dfrac{dy}{dx} = \dfrac{\frac{dy}{dt}}{\frac{dx}{dt}} = \dfrac{2(t+1)}{3t^2}$

*(continued)*

*(continued)*

At stationary points $\dfrac{dy}{dx} = 0 \quad \Rightarrow \quad t = -1$

When $t = -1$, then $x = -1$ and $y = 0$.

Therefore the stationary point is $(-1, 0)$.

## Example 22

a  Find the equation of the normal to the curve $x = t^2$, $y = t + \dfrac{2}{t}$, at the point where $t = 1$.

b  Show, without sketching the curve, that this normal does not cross the curve again.

a  $x = t^2$ and $y = t + \dfrac{2}{t}$ give $\dfrac{dy}{dt} = 1 - \dfrac{2}{t^2}$ and $\dfrac{dx}{dt} = 2t$

therefore $\dfrac{dy}{dx} = \dfrac{dy}{dt} \div \dfrac{dx}{dt} = \dfrac{1 - \dfrac{2}{t^2}}{2t} = \dfrac{t^2 - 2}{2t^3}$

When $t = 1$, then $x = 1$, $y = 3$ and $\dfrac{dy}{dx} = -\dfrac{1}{2}$.

Therefore the gradient of the normal at $(1, 3)$ is $\dfrac{-1}{-\dfrac{1}{2}} = 2$.

The equation of this normal is $y - 3 = 2(x - 1)$, that is, $y = 2x + 1$.

b  All points for which $x = t^2$ and $y = t + \dfrac{2}{t}$ are on the given curve.

For any point that is on both the curve and the normal, these coordinates also satisfy the equation of the normal.

At points common to the curve and the normal,

$$t + \dfrac{2}{t} = 2t^2 + 1 \quad \Rightarrow \quad 2t^3 - t^2 + t - 2 = 0 \qquad\qquad [1]$$

When a cubic equation can be factorised, each factor equated to zero gives a root of the equation.

One point where the curve and normal meet is the point where $t = 1$, so $t = 1$ is a root of [1] and $(t - 1)$ is a factor of the LHS.

that is, $(t - 1)(2t^2 + t + 2) = 0$

Therefore, at any other point where the normal meets the curve, the value of $t$ is a root of the equation $2t^2 + t + 2 = 0$

Checking the value of $b^2 - 4ac$ shows that this equation has no real roots so there are no more points where the normal meets the curve.

## Example 23

The parametric equations of a curve are $x = \cos\theta$ and $y = \sin\theta$. Find the equation of the tangent to the curve at the point P $(\cos\alpha, \sin\alpha)$ on the curve.

$\left.\begin{array}{l} x = \cos\theta \\ y = \sin\theta \end{array}\right\}$ are the parametric equations of a circle, centre O and radius 1, as

$x^2 + y^2 = \cos^2\theta + \sin^2\theta = 1$

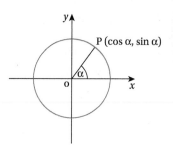

P (cos α, sin α)

Differentiating $x^2 + y^2 = 1$ implicitly gives

$2x + 2y\dfrac{dy}{dx} = 0 \quad \Rightarrow \quad \dfrac{dy}{dx} = -\dfrac{x}{y} = -\dfrac{\cos\theta}{\sin\theta} = -\cot\theta$

Therefore the gradient of the tangent at P is $-\cot\alpha$.

Use $y - y_1 = m(x - x_1)$ to find the equation of the tangent, with $y_1 = \sin\alpha$, $x_1 = \cos\alpha$ and $m = -\cot\alpha$.

The equation of the tangent at P is

$y - \sin\alpha = -\cot\alpha(x - \cos\alpha)$

$\Rightarrow \quad y = -x\cot\alpha + \operatorname{cosec}\alpha$

In the example above, $\alpha$ can be *any* of the possible values of $\theta$, so $\alpha$ can be replaced by $\theta$ in the equation of the tangent, giving

$y = -x\cot\theta + \operatorname{cosec}\theta$

This is the equation of the tangent at *any* point $(\cos\theta, \sin\theta)$ on the curve. It is a *general equation* for a tangent to the curve because, by taking a particular value of $\theta$, you can find the equation of the tangent at that point on the curve. You can also derive this equation directly, without first using $\alpha$ as the parameter.

## Example 24

Find, in terms of $t$, the equation of the tangent to the curve $x = t^2$, $y = t + \dfrac{2}{t}$ at a general point $\left(t^2, t + \dfrac{2}{t}\right)$ on the curve.

From Example 22, the gradient at any point is given by $\dfrac{dy}{dx} = \left(\dfrac{dy}{dt}\right) \div \left(\dfrac{dt}{dx}\right) = \dfrac{t^2 - 2}{2t^3}$.

Hence, the equation of the tangent at any point is $y - \left(t + \dfrac{2}{t}\right) = \left(\dfrac{t^2 - 2}{2t^3}\right)(x - t^2)$.

This simplifies to $2t^3 y - x(t^2 - 2) - (t^4 + 6t^2) = 0$.

## Exercise 11

**1** Find $\dfrac{dy}{dx}$, in terms of the parameter, for each curve.

   **a** $x = 2t^2, y = t$      **b** $x = \cos\theta, y = \sin\theta$      **c** $x = t, y = \dfrac{4}{t}$

**2**  **a** Given that $x = \dfrac{t}{1-t}$ and $y = \dfrac{t^2}{1-t}$, find $\dfrac{dy}{dx}$ in terms of $t$.

   **b** Find the value of $\dfrac{dy}{dx}$ at the point where $x = 1$.

**3** Given that $x = t^2$ and $y = t^3$, find $\dfrac{dy}{dx}$ in terms of $t$.

**4** Find the stationary points of the curve whose parametric equations are $x = t, y = t^3 - t$, and distinguish between them.

**5** A curve has parametric equations $x = \theta - \cos\theta$, $y = \sin\theta$.

Find the smallest positive value of $\theta$ at which the gradient of this curve is zero.

**6** Find the equation of the tangent to the curve $x = t^2$, $y = 4t$ at the point where $t = -1$.

**7 a** Find the equation of the normal to the curve $x = 2\cos\theta$, $y = 3\sin\theta$ at the point where $\theta = \dfrac{1}{4}\pi$.

   **b** Find the coordinates of the point where this normal cuts the curve again.

**8 a** Find the equation of the tangent at a general point to each of the curves in question 1 parts **a** and **c**.

   **b** Find the equation of the normal at a general point to each of the curves in question 1 parts **a** and **c**.

**9 a** Find the equation of the normal at the point $\left(2s, \dfrac{2}{s}\right)$ to the curve whose parametric equations are $x = 2s$, $y = \dfrac{2}{s}$.

   **b** Find, in terms of $s$, the coordinates of the point where this normal cuts the curve again.

**10** The parametric equations of a curve are $x = t$ and $y = \dfrac{1}{t}$.

   **a** Find the general equation of the tangent to this curve, that is the equation of the tangent at the point $\left(t, \dfrac{1}{t}\right)$.

   **b** Find, in terms of $t$, the coordinates of the points at which the tangent cuts the $x$ and $y$ axes.

**11** A curve has parametric equations $x = t^2$, $y = 4t$.

   **a** Find the equation of the normal to this curve at the point $(t^2, 4t)$.

   **b** Find the coordinates of the points where this normal cuts the coordinate axes.

## Summary

**Derivatives of standard functions**

$$\frac{d}{dx}(e^x) = e^x$$

$$\frac{d}{dx}(\ln x) = \frac{1}{x}$$

$$\frac{d}{dx}(\sin x) = \cos x$$

$$\frac{d}{dx}(\cos x) = -\sin x$$

$$\frac{d}{dx}(\tan x) = \sec^2 x$$

## Formulae for differentiating combinations of functions

$$\frac{dy}{dx} = \frac{1}{\dfrac{dx}{dy}}$$

$$\frac{d}{dx}(uv) = v\frac{du}{dx} + u\frac{dv}{dx}$$

$$\frac{dy}{dx} = \frac{v\dfrac{du}{dx} - u\dfrac{dv}{dx}}{v^2}$$

$$\frac{dy}{dx} = \frac{dy}{du} \times \frac{du}{dx}$$

## Derivatives of some composite functions

$$\frac{d}{dx}\big(\sin f(x)\big) = f'(x)\cos f(x)$$

$$\frac{d}{dx}\big(\cos f(x)\big) = -f'(x)\sin f(x)$$

$$\frac{d}{dx}(\sin ax) = a\cos ax$$

$$\frac{d}{dx}(\cos ax) = -a\sin ax$$

$$\frac{d}{dx}e^{f(x)} = f'(x)\,e^{f(x)}$$

$$\frac{d}{dx}\big(\ln f(x)\big) = \frac{f'(x)}{f(x)}$$

▶ To differentiate a function of $y$ with respect to $x$ use $\dfrac{d}{dx}g(y) = \left(\dfrac{d}{dy}g(y)\right)\left(\dfrac{dy}{dx}\right)$.

▶ When a curve is defined in terms of a parameter $t$, the gradient of the curve in terms of $t$ can be found by using $\dfrac{dy}{dx} = \dfrac{\dfrac{dy}{dt}}{\dfrac{dx}{dt}}$.

# Review

In questions 1 to 19, find the derivative of each expression.

**1**   **a**   $-\sin 4\theta$      **b**   $\theta - \cos\theta$      **c**   $\sin^3\theta + \sin 3\theta$

**2**   **a**   $x^3 + e^x$      **b**   $e^{(2x+3)}$      **c**   $e^x \sin x$

**3**   **a**   $\ln\left(\dfrac{1}{3}x^{-3}\right)$      **b**   $\ln\left(\dfrac{2}{x^2}\right)$      **c**   $\ln\left(\dfrac{\sqrt{x}}{4}\right)$

**4**   **a**   $3\sin x - e^{-x}$      **b**   $\ln x^{\frac{1}{2}} - \dfrac{1}{2}\cos x$

     **c**   $x^4 + 4e^x - \ln 4x$      **d**   $\dfrac{1}{2}e^{-x} + x^{-\frac{1}{2}} - \ln\dfrac{1}{2}x$

**5**   $(x+1)\ln x$      **6**   $\sin^2 3x$      **7**   $(4x-1)^{\frac{2}{3}}$

**8**   $(3\sqrt{x} - 2x)^2$      **9**   $\dfrac{(x^4-1)}{(x+1)^3}$      **10**   $\dfrac{\ln x}{\ln(x-1)}$

**11** $\ln(\cot x)$  **12** $x^2 \tan x$  **13** $\dfrac{e^x}{x-1}$

**14** $\dfrac{1+\sin x}{1-\sin x}$  **15** $x^2\sqrt{x-1}$  **16** $(1-x^2)(1-x)^2$

**17** $\ln\sqrt{\dfrac{(x+3)^3}{x^2+2}}$  **18** $\sin x \cos^3 x$  **19** $e^{\cos^2 x}$

**20** Find the value(s) of $x$ for which these expressions have stationary values.

   **a**  $3x - e^x$      **b**  $x^2 - 2\ln x$

In questions 21 to 24, find

**a**  the gradient of the curve at the given point

**b**  the equation of the tangent to the curve at that point

**c**  the equation of the normal to the curve at that point.

**21** $y = \sin x - \cos x;\ x = \dfrac{1}{2}\pi$

**22** $y = x + e^x;\ x = 1$

**23** $y = 1 + x + \sin x;\ x = 0$

**24** $x = \cos y;\ y = \dfrac{\pi}{2}$

**25** Find the coordinates of a point on the curve where the tangent is parallel to the given line.

   **a**  $y = 3x - 2\cos x;\ y = 4x$     **b**  $y = 2\ln x - x;\ y = x$

**26** Differentiate with respect to $x$.

   **a**  $y^4$              **b**  $xy^2$              **c**  $\dfrac{1}{y}$

   **d**  $x\ln y$         **e**  $\sin y$          **f**  $e^y$

   **g**  $y\cos x$       **h**  $y\cos y$

In questions 27 to 30, find $\dfrac{dy}{dx}$ in terms of $x$ and $y$.

**27** $x^2 - 2y^2 = 4$      **28** $\dfrac{1}{x} + \dfrac{1}{y} = 2$

**29** $x^2 y^3 = 9$         **30** $x^2 y^2 = \dfrac{(y+1)}{(x+1)}$

In questions 31 to 36, find $\dfrac{dy}{dx}$ in terms of the parameter.

**31** $x = t^2,\ y = t^3$      **32** $x = (t+1)^2,\ y = t^2 - 1$

**33** $x = \sin^2\theta,\ y = \cos^3\theta$     **34** $x = 4t,\ y = \dfrac{4}{t}$

**35** $x = e^t,\ y = 1 - t$      **36** $x = \dfrac{t}{1-t},\ y = \dfrac{t^2}{1-t}$

**37** Find the equation of the tangent to the curve $x = \cos\theta,\ y = 2\sin\theta$ at the point where $\theta = \dfrac{3}{4}\pi$.

## Assessment

**1** The equation of a curve is $y = 3 - x^2 + \ln x$.

  **a** Find $\dfrac{dy}{dx}$.

  **b** Find the equation of the tangent to the curve at the point where $x = 1$.

  **c** Find the equation of the normal to the curve at the point where $x = 1$.

**2** The equation of a curve is $x = 3y^2 - 2y$.

  **a** Find $\dfrac{dy}{dx}$ in terms of $y$.

  **b** Find the equation of the tangent to the curve at the point where $y = 1$.

  **c** Find the equation of the normal to the curve at the point where $y = 1$.

**3** The equations of a curve are $x = \sin t$ and $y = \cos 2t$.

  **a** Find $\dfrac{dy}{dx}$ in terms of $x$.

  **b** Find the coordinates of the stationary point on the curve.

**4** The equations of a curve are $x = e^t - t$ and $y = e^{2t} - 2t$.

  **a** Show that $\dfrac{dy}{dx} = 2(e^t + 1)$.

  **b** Hence show that there are no stationary point on the curve.

**5** **a** Differentiate $y^2 - 2xy + 3y = 2x$ with respect to $x$.

  **b** Find the coordinates of the points on the curve where $x = 1$.

  **c** Find the equations of the tangent to the curve $y^2 - 2xy + 3y = 2x$ at the points where $x = 1$.

**6** A curve is given by the parametric equations $x = t$, $y = \dfrac{1}{t}$.

  **a** Find the equation of the normal to the curve at the point $\left( a, \dfrac{1}{a} \right)$.

  **b** This normal cuts the curve again at the point with parameter $b$. Find a relationship between $a$ and $b$.

**7** The parametric equations of a curve are

  $x = 2 \cos \theta$, $y = 3 \sin \theta$.

  Find the equation of the tangent to this curve at the point $(2 \cos \theta, 3 \sin \theta)$.

**8** The parametric equations of a curve are $x = \cos \theta$ and $y = 1 + \sin \theta$. Find the equation of the normal to the curve at the point $(\cos \theta, 1 + \sin \theta)$.

**9** The equation of a curve is $y = \ln \dfrac{1}{x} + 4x$.

  Find the coordinate of the stationary point on the curve.

**10** The curve shown in the diagram is called an asteroid.

  The parametric equations of this curve are

  $x = 27 \cos^3 \theta$  and  $y = 27 \sin^3 \theta$  for  $0 \le \theta \le 2\pi$.

  **a** Find the coordinates of the point on the curve where $\theta = \dfrac{\pi}{2}$.

  **b** Show that the Cartesian equation of the curve is $x^{\frac{2}{3}} + y^{\frac{2}{3}} = 9$.

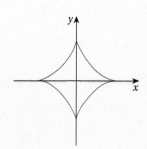

**11** Given that $x = \dfrac{1}{\sin\theta}$, use the quotient rule to show that $\dfrac{dx}{d\theta} = -\text{cosec}\,\theta\,\cot\theta$.

*AQA MPC3 January 2012 (part question)*

**12 a** Find $\dfrac{dy}{dx}$ when $y = e^{3x} + \ln x$

**b i** Given that $u = \dfrac{\sin x}{1 + \cos x}$, show that $\dfrac{du}{dx} = \dfrac{1}{1 + \cos x}$

   **ii** Hence show that if $y = \ln\left(\dfrac{\sin x}{1 + \cos x}\right)$, then $\dfrac{dy}{dx} = \text{cosec}\,x$

*AQA MPC3 January 2013*

**13** A curve is defined by the parametric equations $x = \cos 2t$, $y = \sin t$.

The point P on the curve is where $t = \dfrac{\pi}{6}$,

**a** Find the gradient at P.

**b** Find the equation of the normal to the curve at P in the form $y = mx + c$.

**c** The normal at P intersects the curve again at the point Q $(\cos 2q, \sin q)$.

Use the equation of the normal to form a quadratic equation in $\sin q$ and hence find the $x$-coordinate of Q.

*AQA MPC4 June 2015*

**14** A curve is defined by the equation $9x^2 - 6xy + 4y^2 = 3$.

Find the coordinates of the two stationary points of this curve.

*AQA MPC4 June 2012*

# 6 Integration

## Introduction

In the history of calculus, integration was studied before differentiation as a process of summation – for example, to find areas under curves. This chapter looks at the other interpretation of integration as the reverse of differentiation.

It also demonstrates how you can use products and fractions to integrate some expressions that cannot be integrated by recognition or by substitution. The chapter ends with an application for finding volumes.

## Recap

You will need to remember...

▶ The derivatives of polynomial, exponential, logarithmic and trigonometric functions.

▶ How to integrate polynomial functions.

▶ How to evaluate a definite integral.

▶ How to express a rational function as partial fractions.

▶ How to use the trigonometric double angle formulae.

▶ The product rule and the chain rule for differentiating.

## Objectives

By the end of this chapter, you should know how to...

▶ Find the integrals of functions by recognition of their derivatives.

▶ Use substitution to find the integrals of functions that are not immediately recognised as derivatives of known functions.

▶ Integrate products of functions using integration by parts.

▶ Express a rational function as partial fractions that can be integrated.

▶ Use trigonometric formulae to convert functions involving powers of sin $x$, cos $x$ or tan $x$ to forms that can be integrated.

▶ Find volumes formed when the region between a curve and the $x$- or $y$-axis is rotated about that axis.

## 6.1 Standard integrals

Whenever you recognise a function $f'(x)$ as the derivative of another function $f(x)$, then you can integrate $f'(x)$ 'by sight':

$$\frac{d}{dx}f(x) = f'(x) \quad \Rightarrow \quad \int f'(x)dx = f(x) + K$$

### Integrating exponential functions

You have seen that $\frac{d}{dx}e^x = e^x$, so therefore

$$\int e^x\, dx = e^x + K$$

You have seen that $\frac{d}{dx}(ce^x) = ce^x$, so therefore

$$\int ce^x\, dx = ce^x + K$$

You have seen that $\frac{d}{dx}\left(e^{f(x)}\right) = f'(x)e^{f(x)}$, so therefore,

$$\int f'(x)e^{f(x)}dx = e^{f(x)} + K$$

For example, $\int 2e^{2x-1}\, dx = e^{2x-1} + K$.

## Example 1

Write down the integral of $2e^x$ with respect to $x$.

$$\int 2e^x \, dx = 2\int e^x \, dx = 2e^x + K$$

## Exercise 1

In questions 1 to 10, integrate each expression with respect to $x$.

**1** $e^{4x}$  **2** $4e^{-x}$  **3** $e^{(3x-2)}$

**4** $2e^{(1-5x)}$  **5** $6e^{-2x}$  **6** $5e^{(x-3)}$

**7** $e^{\left(2+\frac{x}{2}\right)}$  **8** $e^{(2+x)}$  **9** $e^{2x} + \dfrac{1}{e^{2x}}$

In questions 10 to 13, evaluate

**10** $\displaystyle\int_0^2 e^{2x} \, dx$  **11** $\displaystyle\int_{-1}^1 2e^{(x+1)} \, dx$  **12** $\displaystyle\int_2^3 e^{(2-x)} \, dx$  **13** $\displaystyle\int_0^2 e^{-x} \, dx.$

## Integrating $\dfrac{1}{x}$

Trying to integrate $\dfrac{1}{x} = x^{-1}$ using the rule $\displaystyle\int x^n \, dx = \dfrac{1}{n+1}x^{(n+1)} + K$ fails because

when $n = -1$, then $\dfrac{1}{n+1} = \dfrac{1}{0}$ which is meaningless.

However $\dfrac{1}{x}$ can be *recognised* as the derivative of $\ln x$, but $\ln x$ is defined only when $x > 0$.

Therefore, provided that $x > 0$  $\dfrac{d}{dx}(\ln x) = \dfrac{1}{x}$ $\Leftrightarrow$ $\displaystyle\int \dfrac{1}{x} \, dx = \ln x + K.$

When $x < 0$ the statement $\displaystyle\int \dfrac{1}{x} \, dx = \ln x$ is not valid because the logarithm of a negative number does not exist.

However, the function $\dfrac{1}{x}$ exists for negative values of $x$, as the graph of $y = \dfrac{1}{x}$ shows.

Also, the definite integral $\displaystyle\int_c^d \dfrac{1}{x} \, dx$, which is represented by the shaded area on the graph region, clearly exists.

So it must be possible to integrate $\dfrac{1}{x}$ when $x$ is negative.

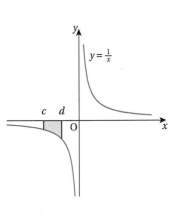

$y = \frac{1}{x}$

When $x < 0$ then $-x > 0$ so  $\displaystyle\int \dfrac{1}{x} \, dx = \int \dfrac{-1}{(-x)} \, dx = \ln(-x) + K$

Therefore when $x < 0$,  $\displaystyle\int \dfrac{1}{x} \, dx = \ln(-x) + K$

and when $x > 0$,  $\displaystyle\int \dfrac{1}{x} \, dx = \ln x + K.$

These two results can be combined using $|x|$ so that, for both positive and negative values of $x$,

$$\int \dfrac{1}{x} \, dx = \ln|x| + K$$

Also $\dfrac{d}{dx}(\ln x^a)=\dfrac{d}{dx}(a\ln x)=\dfrac{a}{x}$.

Therefore, $\displaystyle\int\dfrac{a}{x}\,dx=a\ln|x|+K$

For example, $\displaystyle\int\dfrac{4}{x}\,dx=4\ln|x|+K$.

## Example 2

Find $\displaystyle\int\dfrac{2x-1}{2x}\,dx$.

First express $\dfrac{2x-1}{2x}$ as a proper fraction.

$$\dfrac{2x-1}{2x}=1-\dfrac{1}{2x}=1-\left(\dfrac{1}{2}\right)\left(\dfrac{1}{x}\right)$$

Therefore $\displaystyle\int\dfrac{2x-1}{2x}\,dx=\int\left(1-\dfrac{1}{2}\left(\dfrac{1}{x}\right)\right)dx=x-\dfrac{1}{2}\ln|x|+K$.

## Exercise 2

Integrate with respect to $x$.

**1** $\dfrac{2}{x}$
  **2** $\dfrac{1}{4x}$
  **3** $\dfrac{3}{2x}$

**4** $\dfrac{x+1}{x}$
  **5** $\dfrac{x^2+x-1}{x}$
  **6** $e^x+\dfrac{2}{x}$

Evaluate these integrals.

**7** $\displaystyle\int_1^2\dfrac{1}{3x}\,dx$
  **8** $\displaystyle\int_1^3\left(1-\dfrac{1}{x}\right)dx$
  **9** $\displaystyle\int_1^2\left(\dfrac{1-x}{x}\right)dx$

**10** $\displaystyle\int_2^3\left(e^x-\dfrac{1}{x}\right)dx$
  **11** $\displaystyle\int_4^5\left(\dfrac{2-x}{3x}\right)dx$
  **12** $\displaystyle\int_{-3}^{-2}\left(2x^2-\dfrac{1}{x}\right)dx$

## Integrating $(ax+b)^n$

Look at the function $f(x)=(2x+3)^4$.

To differentiate $f(x)$ using the chain rule, make the substitution

$$u=2x+3\implies f(x)=u^4$$

and using $\dfrac{dy}{dx}=\dfrac{dy}{du}\times\dfrac{du}{dx}$   gives   $\dfrac{d}{dx}(2x+3)^4=4u^3\times2=(4)(2)(2x+3)^3$

Therefore   $\displaystyle\int(4)(2)(2x+3)^3\,dx=(2x+3)^4+K$

Hence   $\displaystyle\int(2x+3)^3\,dx=\dfrac{1}{(2)(4)}(2x+3)^4+K$

Working with $f(x)=(ax+b)^{n+1}$ in a similar way gives the general result

$$\int(ax+b)^n\,dx=\dfrac{1}{(a)(n+1)}(ax+b)^{n+1}+K.$$

## Exercise 3

Integrate these functions with respect to $x$.

**1** $(2x-3)^3$  **2** $(3x+1)^4$  **3** $(5x-2)^4$  **4** $(2-x)^{-2}$

**5** $(x+3)^{-2}$  **6** $(1+x)^{\frac{1}{2}}$  **7** $(1+3x)^5$  **8** $(2-5x)^4$

## Integrating trigonometric functions

As you have seen the derivatives $\sin x$, $\cos x$ and $\tan x$, you will recognise these integrals.

$$\frac{d}{dx}(\sin x)=\cos x \quad \Leftrightarrow \quad \int \cos x\, dx=\sin x+K$$

$$\frac{d}{dx}(\cos x)=-\sin x \quad \Leftrightarrow \quad \int \sin x\, dx=-\cos x+K$$

$$\frac{d}{dx}(\tan x)=\sec^2 x \quad \Leftrightarrow \quad \int \sec^2 x\, dx=\tan x+K$$

Also, when $y=\sin(f(x))$ then $\dfrac{dy}{dx}=f'(x)\cos f(x)$

so $\quad \int f'(x)\cos f(x)\, dx=\sin f(x)+K.$

Similarly, when $y=\cos(f(x))$ then $\dfrac{dy}{dx}=-f'(x)\sin f(x)$

so $\quad \int -f'(x)\sin f(x)\, dx=\cos f(x)+K.$

In particular, $\quad \int \cos(ax+b)\, dx=\dfrac{1}{a}\sin(ax+b)+K$

and $\quad \int \sin(ax+b)\, dx=-\dfrac{1}{a}\cos(ax+b)+K.$

## Example 3

**Question**

Find the integral of $\sec^2(3x-\pi)$ with respect to $x$.

**Answer**

When $y=\tan(f(x))$ then $\dfrac{dy}{dx}=f'(x)\sec^2 f(x)$

So $\quad \int f'(x)\sec^2 f(x)\, dx=\tan f(x)+K$

$\quad \int \sec^2(3x-\pi)\, dx=\dfrac{1}{3}\tan(3x-\pi)+K$

**Note**

Check your answer by differentiation:

$\dfrac{d}{dx}\left(\dfrac{1}{3}\tan(3x-\pi)\right)$

$=\left(\dfrac{1}{3}\right)(3)\sec^2(3x-\pi)$

$=\sec^2(3x-\pi)$

## Exercise 4

In questions 1 to 10, integrate the expression with respect to $x$.

**1** $\sin 2x$  **2** $\cos 7x$  **3** $\sec^2 4x$

**4** $\sin\left(\dfrac{1}{4}\pi+x\right)$  **5** $3\cos\left(4x-\dfrac{1}{2}\pi\right)$  **6** $\sec^2\left(\dfrac{1}{3}\pi+2x\right)$

**7** $2\sin(3x-\alpha)$  **8** $5\cos\left(\alpha-\dfrac{1}{2}x\right)$  **9** $\cos 3x-\cos x$

**10** $\sec^2 2x$

Evaluate these integrals.

**11** $\displaystyle\int_0^{\frac{\pi}{6}} \sin 3x \, dx$

**12** $\displaystyle\int_{\frac{\pi}{6}}^{\frac{\pi}{4}} \cos\left(2x - \frac{1}{2}\pi\right) dx$

**13** $\displaystyle\int_0^{\frac{\pi}{2}} 2\sin\left(2x - \frac{1}{2}\pi\right) dx$

**14** $\displaystyle\int_0^{\frac{\pi}{8}} \sec^2 2x \, dx$

Questions 15 to 35 contain a variety of functions, including some integrals covered in the AS book. Integrate each function with respect to $x$.

You can check your answers are correct by mentally differentiating them. You should get the original function.

**15** $\sin\left(\dfrac{1}{2}\pi - 2x\right)$      **16** $e^{(4x-1)}$      **17** $\sec^2 7x$

**18** $\dfrac{1}{2x-3}$      **19** $\dfrac{1}{\sqrt{2x-3}}$      **20** $\dfrac{1}{(3x-2)^2}$

**21** $e^{5x}$      **22** $\dfrac{x-1}{x}$      **23** $(3x-5)^2$

**24** $e^{(4x-5)}$      **25** $\sqrt{4x-5}$      **26** $\sin\left(5x - \dfrac{1}{3}\pi\right)$

**27** $\dfrac{3}{2(1-x)}$      **28** $\dfrac{4}{(x-6)^4}$      **29** $\cos\left(3x - \dfrac{1}{3}\pi\right)$

**30** $(x-2)(2x+4)$      **31** $x(x-3)^2$      **32** $3-2x(1-3x)$

**33** $\dfrac{2}{3x^3}$      **34** $\dfrac{3}{2x}$      **35** $\dfrac{2}{3}(x-3)^2$

# 6.2 Integrating products by substitution

Section 6.1 demonstrated how you need to be able to recognise some functions as being the derivatives of others, in order to be able to to integrate them. Recognition is equally important to help you avoid errors in integration.

For example, look at the derivative of the product $x^2 \sin x$.

Using the product formula gives $\dfrac{d}{dx}(x^2 \sin x) = 2x \sin x + x^2 \cos x$

The derivative is not a simple product, therefore in general,

$\displaystyle\int uv \, dx$ is NOT $\left(\displaystyle\int u \, dx\right)\left(\displaystyle\int v \, dx\right)$.

However, using the chain rule to differentiate the composite function $(1 + x^2)^3$ gives

$$\frac{d}{dx}(1+x^2)^3 = 6x(1+x^2)^2$$

This time the derivative *is* a product, so clearly the integral of a product *may* be a composite function.

First look at the function $e^u$ where $u$ is a function of $x$.

Differentiating as a composite function gives

$$\frac{d}{dx}(e^u) = \left(\frac{du}{dx}\right)(e^u)$$

Therefore any product of the form $\left(\dfrac{du}{dx}\right)e^u$ can be integrated by recognition,

since $\displaystyle\int\left(\dfrac{du}{dx}\right)e^u\,dx=e^u+K$.

For example $\displaystyle\int 2xe^{x^2}\,dx=e^{x^2}+K$ (recognising that $u=x^2$)

$$\int\cos xe^{\sin x}\,dx=e^{\sin x}+K \quad\text{(recognising that } u=\sin x)$$

$$\int x^2e^{x^3}\,dx=\frac{1}{3}\int 3x^2e^{x^3}\,dx=\frac{1}{3}e^{x^3}+K \quad\text{(recognising that } u=x^3)$$

In these simple cases the substitution of $u$ for f$(x)$ can be done mentally. All the results can be checked by differentiating them mentally.

Similar, but slightly less simple functions, can also be integrated by changing the variable when the substitution is written down.

## Integration by substitution

Look at the general function g$(u)$ where $u$ is a function of $x$.

$$\frac{d}{dx}g(u)=\frac{du}{dx}g'(u)=g'(u)\frac{du}{dx}$$

Therefore $\displaystyle\int g'(u)\frac{du}{dx}\,dx=g(u)+K$      [1]

and $\displaystyle\int g'(u)\,du=g(u)+K$      [2]

Comparing [1] and [2] gives $\displaystyle\int g'(u)\frac{du}{dx}\,dx=\int g'(u)\,du$

so $\displaystyle\cdots\frac{du}{dx}\,dx=\cdots dx$      [3]

Therefore integrating f$(u)\dfrac{du}{dx}$ with respect to $x$ is *equivalent* to integrating f$(u)$ with respect to $u$.

So the relationship in [3] is neither an equation nor an identity, but is a pair of equivalent operations.

For example, to find $\displaystyle\int 2x(x^2+1)^5\,dx$,

making the substitution $u=x^2+1$ gives

$$\int(x^2+1)^5 2x\,dx=\int u^5(2x)\,dx$$

But $\dfrac{du}{dx}=2x$ and as $\cdots\dfrac{du}{dx}\,dx=\cdots du$

then $\cdots 2x\,dx=\cdots du$

so $\displaystyle\int(x^2+1)^5 2x\,dx=\int u^5\,du=\frac{1}{6}u^6+K=\frac{1}{6}(x^2+1)^6+K$

To use this method, the integral needs to be expressed as an integral only in terms of $u$ and $du$.

In practice, you can go direct from $\dfrac{du}{dx}=2x$ to $\cdots 2x\,dx=\cdots du$ by 'separating the variables'.

Products which can be integrated by this method are those in which one factor is basically the derivative of the function in the other factor, so substitute $u$ for this function.

## Example 4

Integrate $x^2\sqrt{x^3+5}$ with respect to $x$.

In this product $x^2$ is basically the derivative of $x^3+5$, so use the substitution $u=x^3+5$.

When $u=x^3+5$, $\dfrac{du}{dx}=3x^2$

$\Rightarrow$ $\quad\quad ...du=...3x^2\,dx$

Therefore $\quad \displaystyle\int x^2\sqrt{x^3+5}\,dx=\frac{1}{3}\int(x^3+5)^{\frac{1}{2}}(3x^2\,dx)=\frac{1}{3}\int u^{\frac{1}{2}}\,du$

$$=\left(\frac{1}{3}\right)\left(\frac{2}{3}\right)u^{\frac{3}{2}}+K=\frac{2}{9}(x^3+5)^{\frac{3}{2}}+K$$

## Example 5

Find $\displaystyle\int\cos x\sin^3 x\,dx$.

Writing the given integral in the form $\cos x\,(\sin x)^3$ suggests substituting $u=\sin x$.

When $u=\sin x$, $\dfrac{du}{dx}=\cos x\Rightarrow ...du=...\cos x\,dx$

$\Rightarrow \quad \displaystyle\int\cos x\sin^3 x\,dx=\int(\sin x)^3\cos x\,dx=\int u^3\,du$

$$=\frac{1}{4}u^4+K=\frac{1}{4}\sin^4 x+K$$

> **Note**
>
> The integrals in Examples 4 and 5 can also be done by inspection.

## Example 6

Find $\displaystyle\int\frac{\ln x}{x}\,dx$.

This looks like a fraction but writing $\dfrac{(\ln x)}{x}$ as $\left(\dfrac{1}{x}\right)(\ln x)$ and since $\dfrac{1}{x}=\dfrac{d}{dx}(\ln x)$ you can make the substitution $u=\ln x$.

When $u=\ln x$, $\dfrac{du}{dx}=\dfrac{1}{x}\Rightarrow ...\,du=...\dfrac{1}{x}\,dx$

Hence $\quad \displaystyle\int\ln x\frac{1}{x}\,dx=\int u\,du=\frac{1}{2}u^2+K$

So $\quad \displaystyle\int\frac{\ln x}{x}\,dx=\frac{1}{2}(\ln x)^2+K$

> **Note**
>
> $(\ln x)^2$ is *not* the same as $\ln x^2$.

## Exercise 5

In questions 1 to 8, integrate the function with respect to $x$.

**1** $4x^3 e^{x^4}$

**2** $\sin x e^{\cos x}$

**3** $(\sec^2 x)e^{\tan x}$

**4** $(2x+1)e^{(x^2+x)}$

**5** $\sec^2 x e^{(1-\tan x)}$

**6** $(1+\cos x)e^{(x+\sin x)}$

**7** $2xe^{(1+x^2)}$

**8** $(3x^2-2)e^{(x^3-2x)}$

In questions 9 to 18, integrate by making the suggested substitution.

**9** $\int x(x^2-3)^4\,dx;\ u=x^2-3$

**10** $\int x\sqrt{1-x^2}\,dx;\ u=1-x^2$

**11** $\int \cos 2x(\sin 2x+3)^2\,dx;\ u=\sin 2x+3$

**12** $\int x^2(1-x^3)dx;\ u=1-x^3$

**13** $\int e^x\sqrt{1+e^x}\,dx;\ u=1+e^x$

**14** $\int \cos x\sin^4 x\,dx;\ u=\sin x$

**15** $\int \sec^2 x\tan^3 x\,dx;\ u=\tan x$

**16** $\int x^n(1+x^{n+1})^2\,dx;\ u=1+x^{n+1};\ n\neq -1$

**17** $\int 3\sin x(1-\cos x)^4\,dx;\ u=1-\cos x$

**18** $\int \sqrt{x}\sqrt{1+x^{\frac{3}{2}}}\,dx;\ u=1+x^{\frac{3}{2}}$

In questions 19 to 24, use a suitable substitution to find the integral or integrate by sight.

**19** $\int x^3(x^4+4)^2\,dx$

**20** $\int e^x(1-e^x)^3\,dx$

**21** $\int \sin\theta\sqrt{1-\cos\theta}\,d\theta$

**22** $\int (x+1)\sqrt{x^2+2x+3}\,dx$

**23** $\int xe^{x^2+1}\,dx$

**24** $\int \sec^2 x(1+\tan x)dx$

## 6.3 Integration by parts

Some functions (which are products of simpler functions) cannot be expressed in the form $f(u)\dfrac{du}{dx}$, so you cannot integrate these functions by using a substitution.

Using the product rule to differentiate a product $u(x)v(x)$ gives

$$\frac{d}{dx}(uv) = v\frac{du}{dx}+u\frac{dv}{dx} \quad\Rightarrow\quad v\frac{du}{dx} = \frac{d}{dx}(uv)-u\frac{dv}{dx}$$

Using $v\dfrac{du}{dx}$ to represent a product which is to be integrated with respect to $x$ gives

$$\int v\frac{du}{dx}dx = \int \frac{d}{dx}(uv)dx - \int u\frac{dv}{dx}dx,\ \text{and hence}$$

$$\int v\frac{du}{dx}dx = uv - \int u\frac{dv}{dx}dx$$

To give the right-hand side of the formula, one factor, $\dfrac{du}{dx}$, has to be integrated to give $u$. The other factor, $v$, has to be differentiated to give $\dfrac{dv}{dx}$. If either factor can be chosen, choose the one that gives the simpler expression when differentiated.

This method for integrating a product is called **integrating by parts**.

## Example 7

Integrate $x\mathrm{e}^x$ with respect to $x$.

Taking $v = x$ and $\dfrac{\mathrm{d}u}{\mathrm{d}x} = \mathrm{e}^x$ gives $\dfrac{\mathrm{d}v}{\mathrm{d}x} = 1$ and $u = \mathrm{e}^x$

Then $\displaystyle\int v\frac{\mathrm{d}u}{\mathrm{d}x}\,\mathrm{d}x = uv - \int u\frac{\mathrm{d}v}{\mathrm{d}x}\,\mathrm{d}x$

gives $\displaystyle\int x\mathrm{e}^x\,\mathrm{d}x = (\mathrm{e}^x)(x) - \int (\mathrm{e}^x)(1)\,\mathrm{d}x$

$$= x\mathrm{e}^x - \mathrm{e}^x + K$$

## Example 8

Find $\displaystyle\int x^4 \ln x \,\mathrm{d}x$.

Because $\ln x$ can be differentiated but *not integrated*, use $v = \ln x$.

Taking $v = \ln x$ and $\dfrac{\mathrm{d}u}{\mathrm{d}x} = x^4$

gives $\dfrac{\mathrm{d}v}{\mathrm{d}x} = \dfrac{1}{x}$ and $u = \dfrac{1}{5}x^5$

Integrating by parts then gives

$$\int x^4 \ln x \,\mathrm{d}x = \left(\frac{1}{5}x^5\right)(\ln x) - \int\left(\frac{1}{5}x^5\right)\left(\frac{1}{x}\right)\mathrm{d}x = \frac{1}{5}x^5 \ln x - \frac{1}{5}\int x^4 \,\mathrm{d}x$$

$$\Rightarrow \qquad \int x^4 \ln x \,\mathrm{d}x = \frac{1}{5}x^5 \ln x - \frac{1}{25}x^5 + K$$

## Example 9

Find $\displaystyle\int \ln x \,\mathrm{d}x$.

$\displaystyle\int \ln x \,\mathrm{d}x$ has not yet been found, but integration by parts can be used when $\ln x$ is regarded as the product of 1 and $\ln x$ as follows.

Taking $v = \ln x$ and $\dfrac{\mathrm{d}u}{\mathrm{d}x} = 1$ gives $\dfrac{\mathrm{d}v}{\mathrm{d}x} = \dfrac{1}{x}$ and $u = x$

Then $\displaystyle\int v\frac{\mathrm{d}u}{\mathrm{d}x}\,\mathrm{d}x = uv - \int u\frac{\mathrm{d}v}{\mathrm{d}x}\,\mathrm{d}x$

becomes $\displaystyle\int \ln x \,\mathrm{d}x = x\ln x - \int x\left(\frac{1}{x}\right)\mathrm{d}x = x\ln x - x + K$

Therefore $\displaystyle\int \ln x \,\mathrm{d}x = x(\ln x - 1) + K$

## Exercise 6

In questions 1 to 16, integrate each expression with respect to $x$.

**1** $x \cos x$

**2** $xe^x$

**3** $x \ln 3x$

**4** $xe^{-x}$

**5** $3x \sin x$

**6** $x \sin 2x$

**7** $xe^{2x}$

**8** $x^2 e^{4x}$

**9** $x\sin x$

**10** $\ln 2x$

**11** $e^x(x+1)$

**12** $x(1+x)^7$

**13** $x\sin\left(x + \dfrac{1}{6}\pi\right)$

**14** $x \cos nx; \; n \neq 0$

**15** $x \ln x$

**16** $3x \cos 2x$

**17** By writing $\cos^3 \theta$ as $(\cos^2 \theta)(\cos \theta)$, use integration by parts to find $\displaystyle\int \cos^3 \theta \, d\theta$.

Each product in questions 18 to 26 can be integrated either:

**a** by immediate recognition, or

**b** by a suitable substitution, or

**c** by parts.

Choose the best method in each case and integrate each function.

**18** $(x-1)e^{(x^2-2x+4)}$

**19** $(x+1)^2 e^x$

**20** $\sin x(4+\cos x)^3$

**21** $\cos x \, e^{\sin x}$

**22** $x^4 \sqrt{1+x^5}$

**23** $e^x(e^x+2)^4$

**24** $(x-1)e^{(2x-1)}$

**25** $x(1-x^2)^9$

**26** $\cos x \sin^5 x$

## 6.4 Integrating fractions

There are several different ways to integrate a fraction.

### Integrating fractions using recognition

Using the chain rule to differentiate $y = \ln f(x)$ with $u = f(x)$ gives

$$\frac{d}{dx}\ln u = \left(\frac{1}{u}\right)\left(\frac{du}{dx}\right) = \frac{\dfrac{du}{dx}}{u}$$

so $\dfrac{d}{dx}\ln f(x) = \dfrac{f'(x)}{f(x)}$

Therefore $\displaystyle\int \frac{f'(x)}{f(x)} \, dx = \ln|f(x)| + K$

As such, all fractions of the form $\dfrac{f'(x)}{f(x)}$ can be integrated *immediately* by recognition.

For example $\displaystyle\int \frac{\cos x}{1+\sin x} \, dx = \ln|1+\sin x| + K$ because $\dfrac{d}{dx}(1+\sin x) = \cos x$

and $\displaystyle\int \frac{e^x}{e^x+4} \, dx = \ln|e^x+4| + K$ because $\dfrac{d}{dx}(e^x+4) = e^x$.

Recognition works only for an integral whose numerator is basically the derivative of the *complete denominator*.

## Example 10

Find $\int \dfrac{x^2}{1+x^3}\,dx$.

$$\int \frac{x^2}{1+x^3}\,dx = \frac{1}{3}\int \frac{3x^2}{1+x^3}\,dx$$

The integral is now in the form $\int \dfrac{f'(x)}{f(x)}\,dx$    so    use $\int \dfrac{f'(x)}{f(x)}\,dx = \ln|f(x)|$

Hence $\int \dfrac{x^2}{1+x^3}\,dx = \dfrac{1}{3}\ln|1+x^3| + K$

## Example 11

By writing $\tan x$ as $\dfrac{\sin x}{\cos x}$,    find $\int \tan x\,dx$.

$\int \tan x\,dx = \int \dfrac{\sin x}{\cos x}\,dx = -\int \dfrac{f'(x)}{f(x)}\,dx$    where    $f(x) = \cos x$

So      $\int \dfrac{\sin x}{\cos x}\,dx = -\ln|\cos x| + K$

Therefore    $\int \tan x\,dx = K - \ln|\cos x|$   or   $K + \ln|\sec x|$

> **Note**
>
> Similarly,
> $$\int \cot x\,dx = \ln|\sin x| + K$$

### Integrating fractions using substitution

An integral whose numerator is the derivative not of the complete denominator, but of an expression *within* the denominator, belongs to the next type of integral. For example

$$\int \frac{2x}{\sqrt{x^2+1}}\,dx.$$

As $2x$ is the derivative of $x^2 + 1$, use the substitution $u = x^2 + 1$.

Then $\dfrac{du}{dx} = 2x$, which gives … $du = …\, 2x\,dx$.

This substitution converts the given integral into the form $\int \dfrac{1}{\sqrt{u}}\,du$.

## Example 12

Find $\int \dfrac{e^x}{(1-e^x)^2}\,dx$.

$e^x$ is basically the derivative of $1 - e^x$, but not of $(1 - e^x)^2$, so make the substitution $u = 1 - e^x$.

When $u = 1 - e^x$ then …$du = …\,{-e^x}\,dx$

So    $\int \dfrac{e^x}{(1-e^x)^2}\,dx = \int \dfrac{-1}{u^2}\,du = \dfrac{1}{u} + K$

$\Rightarrow$    $\int \dfrac{e^x}{(1-e^x)^2}\,dx = \dfrac{1}{1-e^x} + K$

## Example 13

Find $\displaystyle\int \frac{\sec^2 x}{\tan^3 x}\,dx$.

$\sec^2 x$ is the derivative of $\tan x$ but not of $\tan^3 x$.

Taking $u = \tan x$ gives $\dots du = \dots \sec^2 x\, dx$

Then $\displaystyle\int \frac{\sec^2 x}{\tan^3 x}\,dx = \int \frac{1}{u^3}\,du = -\frac{1}{2}u^{-2} + K$

Therefore $\displaystyle\int \frac{\sec^2 x}{\tan^3 x}\,dx = \frac{-1}{2\tan^2 x} + K$

## Exercise 7

In questions 1 to 18 integrate the expression with respect to $x$.

**1** $\displaystyle\frac{\cos x}{4 + \sin x}$

**2** $\displaystyle\frac{e^x}{3e^x - 1}$

**3** $\displaystyle\frac{x}{(1 - x^2)^3}$

**4** $\displaystyle\frac{\sin x}{\cos^3 x}$

**5** $\displaystyle\frac{x^3}{1 + x^4}$

**6** $\displaystyle\frac{2x + 3}{x^2 + 3x - 4}$

**7** $\displaystyle\frac{x^2}{\sqrt{2 + x^3}}$

**8** $\displaystyle\frac{\cos x}{(\sin x - 2)^2}$

**9** $\displaystyle\frac{1}{x\ln x}$ $\left(\text{this is equivalent to } \dfrac{\frac{1}{x}}{\ln x}\right)$

**10** $\displaystyle\frac{\cos x}{\sin^6 x}$

**11** $\displaystyle\frac{2x}{1 - x^2}$

**12** $\displaystyle\frac{e^x}{\sqrt{1 - e^x}}$

**13** $\displaystyle\frac{x - 1}{3x^2 - 6x + 1}$

**14** $\displaystyle\frac{\cos x}{\sin^n x}\ n \neq 0$

**15** $\displaystyle\frac{\sin x}{\cos^n x}\ n \neq 0$

**16** $\displaystyle\frac{\sin x}{4 + \cos x}$

**17** $\displaystyle\frac{x - 1}{x(x - 2)}$

**18** $\displaystyle\frac{e^x - 1}{(e^x - x)^2}$

Evaluate these integrals.

**19** $\displaystyle\int_1^2 \frac{2x + 1}{x^2 + x}\,dx$

**20** $\displaystyle\int_0^1 \frac{x}{x^2 + 1}\,dx$

**21** $\displaystyle\int_2^3 \frac{2x}{(x^2 - 1)^3}\,dx$

**22** $\displaystyle\int_0^1 \frac{e^x}{(1 + e^x)^2}\,dx$

**23** $\displaystyle\int_{\frac{\pi}{6}}^{\frac{\pi}{3}} \frac{\sin(2x - \pi)}{\cos(2x - \pi)}\,dx$

**24** $\displaystyle\int_2^4 \frac{1}{x(\ln x)}\,dx$

## Using partial fractions

When a rational function cannot be integrated by inspection or by substitution, it may be easy to integrate when expressed in partial fractions. Remember, though, that only proper fractions can be converted directly into partial fractions. An improper fraction must first be rearranged so that it is made up of non-fractional terms and a proper fraction.

## Example 14

Find $\int \dfrac{2x-3}{(x-1)(x-2)}\,dx$.

First decompose into partial fractions

$$\dfrac{2x-3}{(x-1)(x-2)} = \dfrac{A}{x-1} + \dfrac{B}{x-2} \quad\Rightarrow\quad 2x-3 = A(x-2) + B(x-1) \quad\Rightarrow\quad A=1 \text{ and } B=1$$

So $\quad \dfrac{2x-3}{(x-1)(x-2)} = \dfrac{1}{x-1} + \dfrac{1}{x-2}$

Therefore $\quad \displaystyle\int \dfrac{2x-3}{(x-1)(x-2)}\,dx = \int \dfrac{1}{x-1}\,dx + \int \dfrac{1}{x-2}\,dx$

$$= \ln|x-1| + \ln|x-2| + K$$

$$= \ln|(x-1)(x-2)| + K$$

## Example 15

Find $\int \dfrac{x^2+1}{x^2-1}\,dx$.

This fraction is improper, so first express it as a sum of non-fractional terms and a proper fraction. Then decompose the proper fraction into partial fractions.

$$\dfrac{x^2+1}{x^2-1} = \dfrac{(x^2-1)+2}{x^2-1} = 1 + \dfrac{2}{x^2-1} = 1 + \dfrac{2}{(x-1)(x+1)}$$

$$= 1 + \dfrac{1}{x-1} - \dfrac{1}{x+1}$$

Then $\quad \displaystyle\int \dfrac{x^2+1}{x^2-1}\,dx = \int 1\,dx + \int \dfrac{1}{x-1}\,dx - \int \dfrac{1}{x+1}\,dx$

$$= x + \ln|x-1| - \ln|x+1| + K$$

When you are asked to integrate a fraction, look at the fraction carefully, as fractions needing different integration techniques often *look* very similar. Example 16 shows this clearly.

## Example 16

Integrate with respect to $x$.

a  $\dfrac{x+1}{x^2+2x-8}$

b  $\dfrac{x+1}{(x^2+2x-8)^2}$

c  $\dfrac{x+2}{x^2+2x-8}$.

a  This fraction is basically of the form $\dfrac{f'(x)}{f(x)}$.

$$\int \dfrac{x+1}{x^2+2x-8}\,dx = \dfrac{1}{2}\int \dfrac{2x+2}{x^2+2x-8}\,dx = \dfrac{1}{2}\ln|x^2+2x-8| + K$$

b  This time the numerator is basically the derivative of the function *within* the denominator.

Let $u = x^2+2x-8 \quad\Rightarrow\quad \ldots du = \ldots(2x+2)\,dx = \ldots 2(x+1)\,dx$ *(continued)*

*(continued)*

$$\Rightarrow \int \frac{x+1}{(x^2+2x-8)^2}\,dx = \frac{1}{2}\int\frac{1}{u^2}\,du = -\frac{1}{2u}+K = -\frac{1}{2(x^2+2x-8)}+K$$

c   In this fraction the numerator is not related to the derivative of the
    denominator so, since the denominator factorises, use partial fractions.

$$\int\frac{x+2}{x^2+2x-8}\,dx = \int\frac{\frac{1}{3}}{x+4}\,dx + \int\frac{\frac{2}{3}}{x-2}\,dx$$

$$= \frac{1}{3}\ln|x+4| + \frac{2}{3}\ln|x-2| + K$$

## Exercise 8

In questions 1 to 12, integrate the expression with respect to $x$.

**1** $\dfrac{2}{x(x+1)}$      **2** $\dfrac{4}{(x-2)(x+2)}$      **3** $\dfrac{x}{(x-1)(x+1)}$

**4** $\dfrac{x-1}{x(x+2)}$      **5** $\dfrac{x-1}{(x-2)(x-3)}$      **6** $\dfrac{1}{x(x-1)(x+1)}$

**7** $\dfrac{x}{x+1}$      **8** $\dfrac{x+4}{x}$      **9** $\dfrac{x}{x+4}$

**10** $\dfrac{3x-4}{x(1-x)}$      **11** $\dfrac{x^2-2}{x^2-1}$      **12** $\dfrac{x^2}{(x+1)(x+2)}$

In questions 13 to 18, choose the best method to integrate the expression with respect to $x$.

**13** $\dfrac{x}{x^2-1}$      **14** $\dfrac{2x}{(x^2-1)^2}$      **15** $\dfrac{2}{x^2-1}$

**16** $\dfrac{2x-5}{x^2-5x+6}$      **17** $\dfrac{2x}{x^2-5x+6}$      **18** $\dfrac{2x-3}{x^2-5x+6}$

In questions 19 to 24, evaluate the definite integral.

**19** $\displaystyle\int_0^4 \frac{x+2}{x+1}\,dx$      **20** $\displaystyle\int_{-1}^1 \frac{5}{x^2+x-6}\,dx$

**21** $\displaystyle\int_1^2 \frac{x+2}{x(x+4)}\,dx$      **22** $\displaystyle\int_0^1 \frac{2}{3+2x}\,dx$

**23** $\displaystyle\int_{\frac{1}{2}}^3 \frac{2}{(3+2x)^2}\,dx$      **24** $\displaystyle\int_1^2 \frac{2x}{3+2x}\,dx$

## 6.5 Special techniques for integrating some trigonometric functions

### Integrating a function containing an odd power of sin $x$ or cos $x$

When the power of sin $x$ or cos $x$ is an *odd power* other than 1, the formula $\cos^2 x + \sin^2 x = 1$ can convert the given function to one that can be integrated.

For example, $\sin^3 x$ is converted to $(\sin^2 x)(\sin x) \;\Rightarrow\; (1-\cos^2 x)(\sin x)$

$$\Rightarrow\; \sin x - \cos^2 x \sin x$$

## Example 17

Integrate with respect to $x$.

a $\cos^5 x$

b $\sin^3 x \cos^2 x$.

a $\cos^5 x = (\cos^2 x)^2 \cos x = (1 - \sin^2 x)^2 \cos x = (1 - 2\sin^2 x + \sin^4 x)\cos x$

Therefore $\displaystyle\int \cos^5 x \, dx = \int \cos x \, dx - 2\int \sin^2 x \cos x \, dx + \int \sin^4 x \cos x \, dx$

Hence $\displaystyle\int \cos^5 x \, dx = \sin x - 2\left(\frac{1}{3}\right)\sin^3 x + \left(\frac{1}{5}\right)\sin^5 x + K$

$$= \sin x - \frac{2}{3}\sin^3 x + \frac{1}{5}\sin^5 x + K$$

b $\sin^3 x \cos^2 x = \sin x (1 - \cos^2 x)\cos^2 x$

so $\displaystyle\int \sin^3 x \cos^2 dx = \int \cos^2 x \sin x \, dx - \int \cos^4 x \sin x \, dx$

$$= -\frac{1}{3}\cos^3 x + \frac{1}{5}\cos^5 x + K$$

> **Note**
>
> For any value of $n$,
> $$\int \cos x \sin^n x \, dx$$
> $$= \frac{1}{n+1}\sin^{n+1} x + K.$$

## Integrating a function containing only even powers of sin x or cos x

When integrating even powers of $\sin x$ or $\cos x$, the double angle identities are useful. For example

$$\cos^4 x = (\cos^2 x)^2 = \left[\frac{1}{2}(1 + \cos 2x)\right]^2 = \frac{1}{4}[1 + 2\cos 2x + \cos^2 2x]$$

Then use a double angle identity again, which gives

$$\cos^4 x = \frac{1}{4}(1 + 2\cos 2x) + \frac{1}{4}\left[\frac{1}{2}(1 + \cos 4x)\right] = \frac{3}{8} + \frac{1}{2}\cos 2x + \frac{1}{8}\cos 4x$$

Now each of these terms can be integrated.

## Example 18

Integrate with respect to $x$.

a $\sin^2 x$

b $16\sin^2 x \cos^2 x$.

a $\displaystyle\int \sin^2 x \, dx = \int \frac{1}{2}(1 - \cos 2x)dx = \frac{1}{2}x - \frac{1}{4}\sin 2x + K$

b $\displaystyle 16\int \sin^2 x \cos^2 x \, dx = 16\int (\sin x \cos x)^2 \, dx$

$$= \frac{16}{4}\int \sin^2 2x \, dx$$

$$= 4\int \frac{1}{2}(1 - 4\cos 4x)dx$$

$$= 2\left(x - \frac{1}{4}\sin 4x\right) + k$$

## Integrating any power of tan *x*

The formula $\tan^2 x = \sec^2 x - 1$ is useful when integrating a power of tan *x*.

For example, $\tan^3 x$ becomes $\tan x(\sec^2 x - 1) = \sec^2 x \tan x - \tan x$

then $\int \tan^3 x \, dx$

$$= \int \sec^2 x \tan x - \int \tan x \, dx$$

$$= \frac{1}{2}\tan^2 x + \ln|\cos x| + c$$

## Example 19

**Question**

Integrate these expressions with respect to *x*.

a   $\tan^4 x$                    b   $\tan^5 x$

**Answer**

a   $\int \tan^4 x \, dx = \int \tan^2 x(\sec^2 x - 1)\, dx = \int \sec^2 x \tan^2 x \, dx - \int \tan^2 x \, dx$

$$= \frac{1}{3}\tan^3 x - \int (\sec^2 x - 1)\, dx$$

$$= \frac{1}{3}\tan^3 x - \tan x + x + K$$

b   $\int \tan^5 x \, dx = \int \tan^3 x(\sec^2 x - 1)\, dx$

$$= \int \sec^2 x \tan^3 x - \int \tan x(\sec^2 x - 1)\, dx$$

$$= \int \sec^2 x \tan^3 x \, dx - \int \sec^2 x \tan x \, dx + \int \tan x \, dx$$

$$= \frac{1}{4}\tan^4 x - \frac{1}{2}\tan^2 x + \ln|\sec x| + K$$

> **Note**
>
> To integrate *any* power of tan*x*, the formula $\tan^2 x = \sec^2 x - 1$ is used to convert $\tan^2 x$ *only, one step at a time*. Converting $\tan^4 x$ to $(1 - \sec^2 x)^2$ does not help.

## Exercise 9

Integrate each expression with respect to *x*.

1  $\cos^2 x$                         2  $\cos^3 x$

3  $\sin^5 x$                         4  $\tan^2 x$

5  $\sin^4 x$                         6  $\tan^3 x$

7  $\cos^4 x$                        8  $\sin^3 x$

## 6.6 Volume of revolution

When a region is rotated about a straight line, the three-dimensional object formed is called a **solid of revolution**, and its volume is a **volume of revolution**.

The line about which rotation takes place is always a line of symmetry for the solid of revolution. Also, any cross-section of the solid which is perpendicular to the axis of rotation is circular.

The diagram shows the solid of revolution formed when the shaded region is rotated about the $x$-axis.

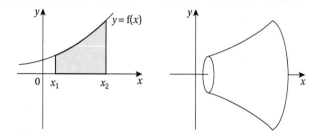

To calculate the volume of this solid, it can be divided it into 'slices' by making cuts perpendicular to the axis of rotation.

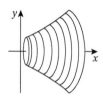

If the cuts are reasonably close together, each slice is approximately a cylinder and the approximate volume of the solid can be found by summing the volumes of these cylinders.

Look at an element formed by one cut through the point $P(x, y)$ and the other cut a distance $\delta x$ from the first.

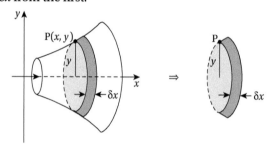

The volume, $\delta V$, of this element is approximately that of a cylinder of radius $y$ and 'height' $\delta x$,

so $\delta V \approx \pi y^2 \delta x$

Then the total volume of the solid is $V$, where $V \approx \sum_{x_1}^{x_2} \pi y^2 \, \delta x$

The smaller $\delta x$ is, the closer is this approximation to $V$,

Therefore $\quad V = \lim_{\delta x \to 0} \sum_{x_1}^{x_2} \pi y^2 \, \delta x = \pi \int_{x_1}^{x_2} y^2 \, \mathrm{d}x$

When the equation of the rotated curve is known, this integral can be evaluated and the volume of the solid of revolution found.

For example, to find the volume generated when the area between part of curve $y = \mathrm{e}^x$ and the $x$-axis is rotated about the $x$-axis, use $\pi \int (\mathrm{e}^x)^2 \, \mathrm{d}x = \pi \int \mathrm{e}^{2x} \, \mathrm{d}x$.

When a region rotates about the $y$-axis we can use a similar method based on slices perpendicular to the $y$-axis, giving

$$V = \pi \int_{y_1}^{y_2} x^2 \, \mathrm{d}y$$

## Example 20

Find the volume generated when the region bounded by the $x$- and $y$-axes, the line $x = 1$ and the curve $y = e^x$ is rotated through one revolution about the $x$-axis.

$$V = \pi \int_0^1 y^2 \, \mathrm{d}x = \pi \int_0^1 (e^x)^2 \, \mathrm{d}x$$

$$= \pi \int_0^1 e^{2x} \, \mathrm{d}x = \pi \left[ \frac{1}{2} e^{2x} \right]_0^1 = \frac{1}{2}\pi(e^2 - e^0)$$

Therefore the volume is $\frac{1}{2}\pi(e^2 - 1)$ cubic units.

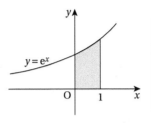

> **Note**
>
> $\int y \, \mathrm{d}x$ means integrate $y$ with respect to $x$, so $y$ must be expressed in terms of $x$.

## Example 21

The region enclosed by the curve $y = x^2 + 1$, the line $y = 2$ and the $y$-axis is rotated completely about the $y$-axis. Find the volume of the solid generated.

Rotating the shaded region about the $y$-axis gives the solid shown.

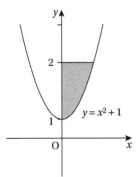
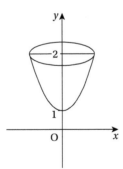

$$V = \pi \int_1^2 x^2 \, \mathrm{d}y$$

The integral is with respect to $y$, so $x^2$ must be expressed in terms of $y$.

Using the equation $y = x^2 + 1$ gives $x^2 = y - 1$

$$V = \pi \int_1^2 (y-1)\mathrm{d}y = \pi \left[ \frac{1}{2} y^2 - y \right]_1^2$$

$$= \pi \left[ (2-2) - \left( \frac{1}{2} - 1 \right) \right]$$

The volume of the solid is $\frac{1}{2}\pi$ cubic units.

## Exercise 10

In questions 1 to 5, find the volume generated when the region enclosed by the equations is rotated completely about the $x$-axis.

**1** The $x$-axis and $y = x(4-x)$

**2** $y = e^x$, the $y$-axis, the $x$-axis and $x = 3$

**3** $y = \dfrac{1}{x}$, $x = 1$, $x = 2$ and the $x$-axis

**4** $y = x^2$, the $y$-axis, $x = -2$ and $x = 2$

**5** $y = \cos x$, the $y$-axis, the $x$-axis, $x = 0$ and $x = \dfrac{\pi}{2}$

In questions 6 to 9, the region enclosed by the curve and line(s) given is rotated about the $y$-axis to form a solid. Find the volume generated.

**6** $y = x^2$, $y = 4$

**7** $y = 4 - x^2$, $y = 0$

**8** $y = x^3$, $y = 1$, $y = 2$, for $x \geq 0$

**9** $y = \ln x$, $x = 0$, $y = 0$, $y = 1$

## Summary

### Integrals that can be found by inspection

$$\int e^x \, dx = e^x + K, \text{ and } \int f'(x) e^{f(x)} \, dx = e^{f(x)} + K$$

$$\int \frac{1}{x} \, dx = \ln|x| + K, \text{ and } \int \frac{f'(x)}{f(x)} \, dx = \ln|f(x)| + K$$

$$\int (ax+b)^n \, dx = \frac{1}{(a)(n+1)} (ax+b)^{n+1} + K$$

$$\int \cos x \, dx = \sin x + K, \text{ and } \int f'(x) \cos f(x) \, dx = -\sin f(x) + K$$

$$\int \sin x \, dx = -\cos x + K, \text{ and } \int -f'(x) \sin f(x) \, dx = \cos f(x) + K$$

$$\int \sec^2 x \, dx = \tan x + K, \text{ and } \int f'(x) \sec^2 f(x) x \, dx = \tan f(x) + K$$

### Integrals that can be found by substitution

▶ Functions of the form $f'(x) g(f(x))$ or $\dfrac{f'(x)}{g(f(x))}$ can be integrated using the substitution $u = f(x)$ and

$$\ldots \frac{du}{dx} \, dx = \ldots du$$

### Integration by parts

▶ A product, $f(x)g(x)$, may be integrated using $\displaystyle\int v \frac{du}{dx} \, dx = uv - \int u \frac{dv}{dx} \, dx$ where

$$v = f(x) \quad \text{and} \quad \frac{du}{dx} = g(x).$$

## Integrating partial fractions

▶ A rational function that cannot be integrated by recognition or by substitution may be integrated by first expressing it in partial fractions.

## Integrals containing powers of sin $x$, cos $x$ or tan $x$

▶ When the power of sin $x$ or cos $x$ is an *odd power* other than 1, use the formula $\cos^2 x + \sin^2 x = 1$.

▶ When only even powers of sin $x$ or cos $x$ are involved, the double angle formulae can transform the function into one that can be integrated.

▶ When any power of tan $x$ is involved, the formula $\tan^2 x = \sec^2 x - 1$ can be used to transform the function into one that can be integrated.

## Volume of revolution

▶ The volume generated when the area enclosed by the $x$-axis, the ordinates at $a$ and $b$, and part of the curve $y = f(x)$ rotates completely about the $x$-axis, is given by $\pi \displaystyle\int_{x=a}^{x=b} y^2 \, dx$.

▶ The volume generated when the region enclosed by the $y$-axis and the lines $y = a$ and $y = b$ and the curve $y = f(x)$ is rotated completely about the $y$-axis, is given by $\pi \displaystyle\int_{y=a}^{y=b} x^2 \, dy$.

## Review

In questions 1 to 11, integrate the expression with respect to $x$.

**1** $3e^{-x}$

**2** $2xe^{x^2}$

**3** $\sec^2 x(3 \tan x - 4)$

**4** $\dfrac{4}{3x}$

**5** $\sec^2 x \tan^3 x$

**6** $(3x + 4)^5$

**7** $(\sin x)e^{\cos x}$

**8** $\cos\left(4x + \dfrac{\pi}{7}\right)$

**9** $(x-1)e^{(x^2 - 2x + 3)}$

**10** $x^2(1 - x^3)^9$

**11** $\sqrt{2x+1}$

**12** Find $\displaystyle\int x\sqrt{x^2 + 1}\, dx$ using the substitution $u = x^2 + 1$.

**13** Find $\displaystyle\int \cos x(2 - \sin x)^2 dx$ using the substitution $u = 2 - \sin x$.

In questions 14 to 16 use integration by parts to integrate the function.

**14** $x^2 e^{2x}$

**15** $(x + 1)\ln(x + 1)$

**16** $x^2 \cos x$

In questions 17 to 32 choose the best method and find the integral in each case.

**17** $\displaystyle\int e^{2x+3} dx$

**18** $\displaystyle\int x\sqrt{2x^2 - 5}\, dx$

**19** $\displaystyle\int xe^x \, dx$

**20** $\displaystyle\int \ln x \, dx$

**21** $\displaystyle\int \sin^2 3x \, dx$

**22** $\displaystyle\int xe^{-x^2} \, dx$

**23** $\displaystyle\int \cos x \sin^2 x\, dx$     **24** $\displaystyle\int u(u+7)^9\, du$     **25** $\displaystyle\int \frac{x^2}{(x^3+9)^5}\, dx$

**26** $\displaystyle\int \frac{\sin 2y}{1-\cos 2y}\, dy$     **27** $\displaystyle\int \frac{1}{2x+7}\, dx$     **28** $\displaystyle\int \sin 3x\sqrt{1+\cos 3x}\, dx$

**29** $\displaystyle\int \frac{x+2}{x^2+4x-5}\, dx$     **30** $\displaystyle\int \frac{x+1}{x^2+4x-5}\, dx$     **31** $\displaystyle\int x\cos 3x\, dx$

**32** $\displaystyle\int \ln 5x\, dx$

**33** The diagram shows part of the curve $y = 1 - \cos^2 x$ between $x = 0$ and $x = \pi$.

Find the volume generated when the region shown is rotated about the $x$-axis.

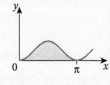

**34** The region shown in the diagram is enclosed by the curve $y = \ln x$, the $x$-axis, the $y$-axis and the line $y = 1$.

Show that the volume of the solid formed when the area is rotated completely about the $y$-axis is $\frac{1}{2}\pi(e^2 - 1)$.

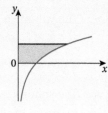

## Assessment

**1** Integrate

  **a** $(1-x)e^{(1-x)^2}$

  **b** $\dfrac{2x}{x^2+1}$

  **c** $\cos x \sin^4 x$

**2** Integrate

  **a** $\sin x(4+\cos x)^3$

  **b** $x(2x+3)^7$

  **c** $xe^{(2-3x)}$

**3** Find $\displaystyle\int \frac{3x}{\sqrt{x^2-3}}\, dx$ using the substitution $u = x^2 - 3$.

**4**  **a** Express $\dfrac{2}{(x^2-2x-3)}$ in partial fractions.

  **b** Find $\displaystyle\int \frac{2}{(x^2-2x-3)}\, dx$.

**5**  **a** Find $\displaystyle\int 3y\sqrt{9-y^2}\, dy$.

  **b** Find $\displaystyle\int_0^{\frac{\pi}{2}} x\sin 4x\, dx$.

**6** The diagram shows the region enclosed by the curve $y = x^2 - 2x + 1$, the $x$-axis and the lines $x = 0$ and $x = 3$.

Find the volume generated when these regions are rotated about the $x$-axis.

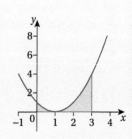

**7** The diagram shows the region enclosed by the curve $y = \ln(x+1)$, the $y$-axis and the line $y = 1$.

Find the volume generated when the region is rotated about the $y$-axis.

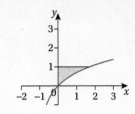

**8 a** Given that $\dfrac{4x^3 - 2x^2 + 16x - 3}{2x^2 - x + 2}$ can be expressed as $Ax + \dfrac{B(4x-1)}{2x^2 - x + 2}$, find the values of the constants $A$ and $B$.

   **b** The gradient of a curve is given by $\dfrac{dy}{dx} = \dfrac{4x^3 - 2x^2 + 16x - 3}{2x^2 - x + 2}$

   The point $(-1, 2)$ lies on the curve. Find the equation of the curve.

AQA MPC4 June 2014

**9 a** By using integration by parts twice, find $\displaystyle\int x^2 \sin 2x \, dx$

   **b** A curve has equation $y = x\sqrt{\sin 2x}$ for $0 \le x \le \dfrac{\pi}{2}$.

   The region bounded by the curve and the $x$-axis is rotated through $2\pi$ radians about the $x$-axis to generate a solid. Find the exact value of the volume of the solid generated.

AQA MPC3 June 2014

# 7 Differential Equations

## Introduction

Any equation that contains a derivative, for example $x\dfrac{dy}{dx}=2$, is called a differential equation. This chapter looks at how to find a direct relationship between $x$ and $y$. There are many different types of differential equation, with each type requiring a specific technique to find its solution. This chapter looks at first order linear differential equations where you can solve the equation by separating the variables.

## Objectives

By the end of this chapter, you should know how to...

▶ Solve first order linear differential equations by separating the variables.

▶ Form differential equations from given information.

## Recap

You will need to remember...

▶ When two quantities, $x$ and $y$, are such that $y$ is proportional to $x$, then $y \propto x \Rightarrow y = kx$ where $k$ is a constant.

▶ When two quantities $x$ and $y$, are such that $y$ is inversely proportional to $x$, then $y \propto \dfrac{1}{x} \Rightarrow y = \dfrac{k}{x}$ where $k$ is a constant.

## Applications

Differential equations occur whenever the variation in one quantity, $p$, depends upon the changing value of another quantity, $q$,

the rate of increase of $p$ with respect to $q$ can be expressed as $\dfrac{dp}{dq}$.

There are many everyday situations where such relationships exist, for example:

▶ Liquid expands when heated so, if $V$ is the volume of a quantity of liquid and $T$ is the temperature, then the rate at which the volume increases with temperature can be written $\dfrac{dV}{dT}$.

▶ When the profit, $P$, made by a company selling phones depends upon the number, $n$, of phones sold, then $\dfrac{dP}{dn}$ represents the rate at which profit increases with respect to sales.

## 7.1 First order differential equations with separable variables

An equation with at least one term containing $\dfrac{dy}{dx}, \dfrac{d^2y}{dx^2}, \ldots$ is called a **differential equation**. When the equation contains only $\dfrac{dy}{dx}$ it is a **first order linear differential equation**.

For example, $x + 2\dfrac{dy}{dx} = 3y$ is a first order linear differential equation.

A differential equation represents a relationship between two variables. The same relationship can often be expressed in a form that does not contain a derivative, for example, $\dfrac{dy}{dx} = 2x$ and $y = x^2 + K$ expresses the same relationship between $x$ and $y$.

Converting a differential equation into a direct one is called solving the differential equation and this process involves integration.

Look at the differential equation $3y\dfrac{dy}{dx} = 5x^2$. $\hfill$ [1]

Integrating both sides of the equation with respect to $x$ gives $\displaystyle\int 3y\dfrac{dy}{dx}\,dx = \int 5x^2\,dx$.

Using $\dots \dfrac{dy}{dx}dx = \dots dy$ gives

$$\int 3y\,dy = \int 5x^2\,dx \hfill [2]$$

Temporarily removing the integral signs from this equation gives

$$3y\,dy = 5x^2\,dx \hfill [3]$$

This relationship can be found directly from equation [1] by separating the variables – that is, by separating $dy$ from $dx$ and collecting on one side all the terms involving $y$ together with $dy$, while collecting all the $x$ terms along with $dx$ on the other side.

The relationship shown in [3] does not have any meaning, and it *should not be written down as a step in the solution.* It only provides a way of making a quick mental conversion from the differential equation [1] to the form [2] in which each side can be integrated separately.

Returning to equation [2] and integrating each side gives

$$\frac{3}{2}y^2 = \frac{5}{3}x^3 + A$$

A constant of integration does not need to be added on both sides. A constant on one side only is enough.

When solving differential equations, the constant of integration is usually denoted by $A$, $B$, etc. and is called the arbitrary constant.

## Example 1

Solve the differential equation $\dfrac{1}{x}\dfrac{dy}{dx} = \dfrac{2y}{x^2+1}$.

$$\frac{1}{x}\frac{dy}{dx} = \frac{2y}{x^2+1} \quad\Rightarrow\quad \frac{1}{y}\frac{dy}{dx} = \frac{2x}{x^2+1}$$

Separating the variables gives

$$\int \frac{1}{y}\,dy = \int \frac{2x}{x^2+1}\,dx$$

$$\Rightarrow\quad \ln|y| = \ln|x^2+1| + A$$

# Exercise 1

Solve each differential equation.

**1** $y\dfrac{dy}{dx} = \sin x$

**2** $x^2\dfrac{dy}{dx} = y^2$

**3** $\dfrac{1}{x}\dfrac{dy}{dx} = \dfrac{1}{y^2-2}$

**4** $\tan y\dfrac{dy}{dx} = \dfrac{1}{x}$

**5** $\dfrac{dy}{dx} = y^2$

**6** $\dfrac{1}{x}\dfrac{dy}{dx} = \dfrac{1}{1-x^2}$

**7** $(x-3)\dfrac{dy}{dx} = y$

**8** $\tan y\dfrac{dx}{dy} = 4$

**9** $u\dfrac{du}{dv} = v+2$

**10** $\dfrac{y^2}{x^3}\dfrac{dy}{dx} = \ln x$

**11** $e^x\dfrac{dy}{dx} = \dfrac{x}{y}$

**12** $\sec x\dfrac{dy}{dx} = e^y$

**13** $r\dfrac{dr}{d\theta} = \sin^2\theta$

**14** $\dfrac{dv}{du} = \dfrac{v+1}{u+2}$

**15** $xy\dfrac{dy}{dx} = \ln x$

**16** $y(x+1) = (x^2+2x)\dfrac{dy}{dx}$

**17** $v^2\dfrac{dv}{dt} = (2+t)^3$

**18** $x\dfrac{dy}{dx} = \dfrac{1}{y}+y$

**19** $r\dfrac{d\theta}{dr} = \cos^2\theta$

**20** $y\sin^3 x\dfrac{dy}{dx} = \cos x$

**21** $\dfrac{uv}{u-1} = \dfrac{du}{dv}$

**22** $e^x\dfrac{dy}{dx} = e^{y-1}$

**23** $\tan x\dfrac{dy}{dx} = 2y^2\sec^2 x$

**24** $\dfrac{dy}{dx} = \dfrac{x(y^2-1)}{(x^2+1)}$

## Calculation of the constant of integration

Look again at $3y\dfrac{dy}{dx} = 5x^2 \quad \Leftrightarrow \quad \dfrac{3}{2}y^2 = \dfrac{5}{3}x^3 + A$

The value of $A$ cannot be found from the differential equation alone, but further information is needed – for example a point on the curve.

To find the equation of the curve that satisfies the differential equation $2\dfrac{dy}{dx} = \dfrac{\cos x}{y}$ and which passes through the point $(0, 2)$, first solve the differential equation.

Separating the variables gives $\displaystyle\int 2y\,dy = \int \cos x\,dx \quad \Rightarrow \quad y^2 = \sin x + A$

When $x = 0$ and $y = 2$, $\quad 4 = A + 0 \quad \Rightarrow \quad A = 4$

Hence the equation of the specified curve is $y^2 = 4 + \sin x$.

## Example 2

Find the equation of the curve satisfied by the differential equation $y = x\dfrac{dy}{dx}$ and which passes through the point $(1, 2)$.

Separating the variables in the equation $y = x\dfrac{dy}{dx}$ gives $\displaystyle\int\dfrac{1}{y}\,dy = \int\dfrac{1}{x}\,dx$

$$\Rightarrow \quad \ln|y| = \ln|x| + A$$

When $x = 1$ and $y = 2$ then $\quad \ln 2 = A + \ln 1 = A + 0, \quad \Rightarrow \quad A = \ln 2$

So $\qquad\qquad\qquad\qquad \ln y = \ln x + \ln 2 \quad \Rightarrow \quad \ln y = \ln 2x$

Therefore $\qquad\qquad\qquad y = 2x$

## Exercise 2

In questions 1 to 6, find the equation of the curve satisfied by the differential equation and passing through the given point.

1. $y^2 \dfrac{dy}{dx} = x^2 + 1$; (2, 1)

2. $e^t \dfrac{ds}{dt} = \sqrt{s}$; $t = 0$ when $s = 4$

3. $\dfrac{y}{x} \dfrac{dy}{dx} = \dfrac{y^2 - 1}{x^2 - 1}$; (2, 3)

4. $e^{-x} \dfrac{dy}{dx} = 1$; (0, −1)

5. $\dfrac{2y}{3} \dfrac{dy}{dx} = e^{-3x}$; (1, 2)

6. $\dfrac{dy}{dx} = \dfrac{y + 1}{x^2 - 1}$; (−3, 1)

7. Solve the differential equation $(1 + x^2)\dfrac{dy}{dx} - y(y + 1)x = 0$, given that $y = 1$ when $x = 0$.

## 7.2 Natural occurrence of differential equations

Differential equations often arise when a physical situation is interpreted mathematically (that is, when a mathematical model is made to represent a physical situation). For example:

► A body falls from rest in a medium which causes the velocity to decrease at a rate proportional to the velocity.

As the velocity is *decreasing* with time, its rate of increase is *negative*.

Using $v$ for velocity and $t$ for time, the rate of change of velocity with respect to time is negative so $\dfrac{dv}{dt} \propto -v$.

The motion of the body satisfies the differential equation $-\dfrac{dv}{dt} = kv$.

► During the first stages of the growth of yeast cells in a culture, the number of cells present increases in proportion to the number already formed.

Therefore the rate of change of the number of cells, $n$, with respect to time is such that $\dfrac{dn}{dt} \propto n$, and $n$ can be found from the differential equation $\dfrac{dn}{dt} = kn$.

> **Note**
>
> To form (and then solve) differential equations from naturally occurring information, you do not need to understand the background to the situation or experiment.

## Exercise 3

In questions 1 to 6 form, but *do not solve,* the differential equations representing the given information.

1. A body moves with a velocity $v$ which is inversely proportional to its displacement $s$ from a fixed point.

2. The rate at which the height $h$ of a plant increases is proportional to the natural logarithm of the difference between its present height and its final height $H$.

3. When water enters a tank, the rate at which the depth of water, $h$ metres, in the tank is increasing is proportional to the volume, $V$ cubic metres, which is already in the tank.

4. The rate at which a radioactive material loses mass is proportional to the mass, $m$ kilograms, at time $t$ seconds.

**5** The depth of a sump in an engine is $H$ cm. At time $t$ hours the depth of oil in the sump is $h$ cm. Oil is leaking from the sump so that $h$ is changing at a rate proportional to $H - h$.

**6** In freezing weather, the ice on a lake is $d$ cm thick $t$ hours after freezing starts. The depth of ice is increasing at a rate that is inversely proportional to itself.

**7** The manufacturers of soap powder are concerned that the number, $n$, of people buying their product at any time $t$ months has remained constant for some months. An advertisement results in the number of customers increasing at a rate proportional to the square root of $n$.

Express as differential equations the progress of sales

**a** before advertising

**b** after advertising.

**8** In a community, the number, $n$, of people suffering from an infectious disease is $N_1$ at a particular time. The disease then becomes epidemic and spreads so that the number of sick people increases at a rate proportional to $n$, until the total number of sufferers is $N_2$. The rate of increase then becomes inversely proportional to $n$ until $N_3$ people have the disease. After this, the total number of sick people decreases at a constant rate. Write down the differential equation governing the incidence of the disease

**a** for $N_1 \le n \le N_2$

**b** for $N_2 \le n \le N_3$

**c** for $n \ge N_3$.

## Solving naturally occurring differential equations

We have seen that when one naturally occurring quantity varies with another, the relationship between them often involves a constant of proportion. Consequently, a differential equation that represents the relationship contains a constant of proportion whose value is not always known. So the initial solution of the differential equation contains both this constant and the arbitrary constant. Extra given information may allow either or both constants to be evaluated.

## Example 3

**a** Form a differential equation to represent this data.

**b** Given that the acceleration is 2 ms$^{-2}$ when the velocity is 5 ms$^{-1}$, solve the differential equation.

**a** Using $\dfrac{dv}{dt}$ for acceleration gives $\dfrac{dv}{dt} \propto \dfrac{1}{v} \quad \Rightarrow \quad \dfrac{dv}{dt} = \dfrac{k}{v}$

**b** $v = 5$ when $\dfrac{dv}{dt} = 2$, so $2 = \dfrac{k}{5} \quad \Rightarrow \quad k = 10$

Therefore $\dfrac{dv}{dt} = \dfrac{10}{v}$

Separating the variables gives $\displaystyle\int v \, dv = \int 10 \, dt \quad \Rightarrow \quad \dfrac{1}{2} v^2 = 10t + A$

## Natural growth and decay

A naturally occurring relationship arises when the rate of change of a quantity $Q$ is proportional to the value of $Q$. Often (but not always) the rate of change is with respect to time, so the relationship can be expressed as

$$\frac{dQ}{dt} \propto Q.$$

This relationship of proportionality can be represented by the differential equation

$$\frac{dQ}{dt} = kQ \quad \text{where } k \text{ is a constant of proportion.}$$

Solving this differential equation by separating the variables gives

$$\int \frac{1}{Q}\, dQ = \int k\, dt \quad \Rightarrow \quad \ln Q = kt + A$$

giving $Q = e^{kt+A} = e^A e^{kt}$ or $Q = Be^{kt}$ (letting $B = e^A$ which is a constant).

This equation shows that $Q$ varies exponentially with time.

The diagram shows a sketch of the corresponding graph.

Quantities that behave in this way undergo exponential growth. For example, the number of yeast cells undergoes exponential growth when the rate of increase of the number of cells is proportional to the number of cells present.

When it is the rate of *decrease* of $Q$ that is proportional to $Q$, then

$$\frac{dQ}{dt} = -kQ \quad \Rightarrow \quad Q = Be^{-kt}$$

This graph is typical of a quantity undergoing exponential decay.

If, when $t = 0$, the value of $Q$ is $Q_0$, the equations representing exponential growth and exponential decay respectively become

$$Q = Q_0 e^{kt} \quad \text{and} \quad Q = Q_0 e^{-kt}.$$

## Half-life

When a substance is decaying exponentially, the time taken for one-half of the original quantity to decay is called the **half-life** of the substance. So if the original amount is $Q_0$, the half-life is given by

$$\frac{1}{2}Q_0 = Q_0 e^{-kt} \quad \Rightarrow \quad e^{kt} = 2 \quad \Rightarrow \quad t = \frac{1}{k}\ln 2$$

therefore the value of the half-life is $\frac{1}{k}\ln 2$.

## Example 4

When a uniform rod is heated it expands so that the rate of increase of its length, $l$, with respect to the temperature, $\theta°\,C$, is proportional to the length.

a Form the differential equation that models this information.

When the temperature is $0°\,C$ the length of the rod is $L$.

b Solve the differential equation using the information given.

c Given that the length of the rod has increased by 1% when the temperature is $20°\,C$, find the value of $\theta$ at which the length of the rod has increased by 5%.

**a** From the given information, $\dfrac{dl}{d\theta} = kl$.

**b** $\dfrac{dl}{d\theta} = kl \quad \Rightarrow \quad \displaystyle\int \dfrac{1}{l}\, dl = \int k\, d\theta$

Therefore $\quad \ln l = k\theta + A \quad \Rightarrow \quad l = Be^{k\theta}$

When $\qquad \theta = 0, l = L$, so $B = L$

Therefore $\quad l = Le^{k\theta}$

**c** When $\qquad \theta = 20$, the length has increased by 1% so $l = L + 0.01L = 1.01L$

Then $\qquad 1.01L = Le^{20k}$

$\qquad\qquad e^{20k} = 1.01 \quad \Rightarrow \quad 20k = \ln 1.01$

$\Rightarrow \qquad\quad k = 0.0004975...$

Therefore $l = Le^{k\theta} \quad \Rightarrow \quad 0.0004975\, \theta = \ln(l/L)$

When $\qquad l = L + 0.05\,L = 1.05\,L, \quad 0.0004975\, \theta = \ln 1.05$

$\Rightarrow \qquad\quad \theta = 98°$ (to the nearest degree)

## Example 5

The rate at which the atoms in a mass of radioactive material are disintegrating is proportional to $N$, the number of atoms present at any time. Initially the number of atoms is $M$.

**a** Form and solve the differential equation that represents this data.

**b** Given that half of the original mass disintegrates in 152 days, evaluate the constant of proportion in the differential equation.

**a** The rate at which the atoms are disintegrating is $-\dfrac{dN}{dt}$.

Therefore $\quad -\dfrac{dN}{dt} = kN$

Separating the variables gives $\displaystyle\int \dfrac{1}{N}\, dN = -\int k\, dt$

Therefore $\quad \ln N = -kt + A \quad \Rightarrow \quad N = Be^{-kt}$

When $\qquad t = 0, N = M \quad$ so $\quad B = M$

Therefore $\quad N = Me^{-kt}$

**b** When $\qquad N = \dfrac{1}{2}M, t = 152$

So $\qquad \dfrac{1}{2}M = M\,e^{-152k} \quad \Rightarrow \quad \ln\left(\dfrac{1}{2}\right) = -152k$

Therefore $\quad 152k = \ln 2 \quad \Rightarrow \quad k = \dfrac{\ln 2}{152} = 0.00456$ (to 3 significant figures)

## Exercise 4

**1** Grain is pouring on to a barn floor where it forms a pile whose height $h$ is increasing at a rate that is inversely proportional to $h^3$.

   **a** Form a differential equation from the information given.

   **b** The initial height of the pile is 1m and the height doubles after 3 minutes. Use this information to solve the differential equation.

   **c** Find the time after which the height of the pile has grown to 3 m.

**2** The gradient of any point of a curve is proportional to the square root of the *x*-coordinate.

   **a** Form a differential equation from this information.

   **b** Given that the curve passes through the point (1, 2) and at that point the gradient is 0.6, solve the differential equation.

   **c** Show that the curve passes through the point (4, 4.8) and find the gradient at this point.

**3** The number of micro-organisms in a liquid is growing at a rate proportional to the number *n* of organisms present at any time *t*. Initially there are *N* organisms.

   **a** Form a differential equation that models the growth in the size of the colony.

   **b** Given that the colony increases by 50% in 10 hours, find the time that elapses from the start of the reaction before the size of the colony doubles.

**4** The half-life of a radioactive element that is decaying exponentially is 500 years.

   **a** Form a differential equation that models the mass remaining, *m* g, of a sample which initially contained 1000 g, at time *t* years after decay began.

   **b** Solve the differential equation.

   **c** How many years it will be before the original mass of the element is reduced by 75%?

**5** In a chemical reaction, a substance is transformed into a compound. The mass of the substance after any time *t* is *m*, and the substance is being transformed at a rate that is proportional to the mass of the substance at that time. Given that the original mass is 50 g and that 20 g is transformed after 200 seconds

   **a** Form and solve the differential equation relating *m* and *t*.

   **b** Find the mass of the substance transformed in 300 seconds.

## Summary

▶ An equation with at least one term containing $\dfrac{dy}{dx}$ or $\dfrac{d^2y}{dx^2}$, ... is called a differential equation.

▶ Solving a differential equation means converting a differential equation into a relationship that does not contain a derivative.

▶ You can solve a differential equation with separable variables by: separating d*y* from d*x*; and collecting on one side all the terms involving *y* together with d*y*, while collecting all the *x* terms, along with d*x*, on the other side.

## Review

1 The radius, $r$ metres, of a circular patch of oil pollution is increasing at a rate that is proportional to $r^2$.

Form, but do not solve, a differential equation for this situation.

2 Solve the differential equation $\dfrac{dy}{dx} = 3x^2y^2$ given that $y = 1$ when $x = 0$.

3 Solve the differential equation $\dfrac{dy}{dx} = x(y^2 + 1)$ given that $y = 0$ when $x = 2$.

4 **a** Find the equation of the curve which passes through the point $\left(\dfrac{1}{2}, 1\right)$ and is defined by the differential equation $ye^{y^2}\dfrac{dy}{dx} = e^{2x}$.

**b** Show that the curve also passes through the point $(2, 2)$.

5 A virus has infected the population of rabbits on an isolated island and the evidence suggests that the growth in the number of rabbits infected is proportional to the number already infected. Initially 20 rabbits were recorded as being infected.

**a** Form a differential equation that models the growth in the number infected.

**b** Thirty days after the initial evidence was collected, 60 rabbits were infected. After how many further days does the model predict that 200 rabbits will be infected?

**c** In the event, only 100 rabbits were infected by that time. Give one reason why the model turned out to be unreliable.

6 A student models the spending power of a given sum of money over several years. The student assumes that the rate of decrease in spending power is proportional to the spending power at any given time.

**a** Use the information that $100 in January 2000 buys only $90 worth of goods in January 2010. Find and solve the differential equation connecting the value of goods, $y$, that $100 in January 2000 will buy in January $x$ years later.

**b** Give one reason why this model might be unsuitable for predicting the spending power that $100 in January 2000 will have in January 2050.

## Assessment

1 Solve the differential equation $\dfrac{dy}{dx} = \dfrac{4x}{y}$ given that $y = 3$ when $x = 1$.

2 Find the equation of the curve defined by $\dfrac{dy}{dx} = \dfrac{\sin x}{\cos y}$ which passes through the point $\left(\dfrac{\pi}{2}, \dfrac{3\pi}{2}\right)$.

3 The rate of increase of $y$ with respect to $x$ is inversely proportional to $xy$.

**a** Form a differential equation relating $x$ and $y$.

**b** Solve the differential equation given that $y = 2$ when $x = 4$ and that $y = 10$ when $x = 1$.

**4** The rate of decrease of the temperature of a liquid is proportional to the amount by which this temperature exceeds the temperature of its surroundings. (This is Newton's Law of Cooling.)

Taking $\theta$ as the excess temperature at any time $t$, and $\theta_0$ as the initial excess

**a** show that $\theta = \theta_0 e^{-kt}$.

A pan of water at 65°C is standing in a kitchen whose temperature is a steady 15°C.

**b** Show that, after cooling for $t$ minutes, the water temperature, $\phi$, can be modelled by the equation $\phi = 15 + 50e^{-kt}$ where $k$ is a constant.

**c** Given that after 10 minutes the temperature of the water has fallen to 50°C, find the value of $k$.

**d** Find the temperature of the water after 15 minutes.

**5** Use the knowledge gained from the last question to undertake the following piece of detective work.

You are a forensic doctor called to a murder scene. When the victim was discovered, the body temperature was measured and found to be 20°C. You arrive one hour later and find the body temperature at that time to be 18°C. Assuming that the ambient temperature remained constant at 17°C in that intervening hour, give the police an estimate of the time of death. (Take 37°C as normal body temperature. You may assume also that the human body is largely made up of water.)

**6** Solve the differential equation $\dfrac{dy}{dx} = y^2 x \sin 3x$ given that $y = 1$ when $x = \dfrac{\pi}{6}$.

Give your answer in the form $y = \dfrac{9}{f(x)}$.

AQA MPC4 January 2012

**7** **a** A pond is initially empty and is then filled gradually with water. After $t$ minutes, the depth of the water, $x$ metres, satisfies the differential

equation $\dfrac{dx}{dt} = \dfrac{\sqrt{4+5x}}{5(1+t)^2}$

Solve this differential equation to find $x$ in terms of $t$.

**b** Another pond is gradually filling with water. After $t$ minutes, the surface of the water forms a circle of radius $r$ metres. The rate of change of the radius is inversely proportional to the area of the surface of the water.

**i** Write down a differential equation, in the variables $r$ and $t$ and a constant of proportionality, which represents how the radius of the surface of the water is changing with time.

**ii** When the radius of the pond is 1 metre, the radius is increasing at a rate of 4.5 metres per second. Find the radius of the pond when the radius is increasing at a rate of 0.5 metres per second.

AQA MPC4 June 2015

# 8 Numerical Methods

## Introduction

There are many situations when it's not possible to find an exact solution to a problem. For example, the roots of many equations cannot be found exactly. There are also many integrals that cannot be found exactly. This chapter gives some methods for finding approximate values for roots of equations and approximate values of definite integrals.

## Objectives

By the end of this chapter, you should know how to...

▶ Approximately locate the roots of an equation.

▶ Use an iteration formula to find a root of an equation to a given number of decimal places.

▶ Use the mid-ordinate rule and Simpson's rule to find an approximate value for the area under a curve.

## Recap

You will need to remember...

▶ The shapes of the graphs of quadratic and cubic functions.

▶ How to find the stationary values on a curve.

▶ The shapes of the graphs of trigonometric functions, exponential functions and logarithmic functions.

▶ The shape of the graph of $f(x) = \dfrac{1}{x}$ shown in the diagram.

▶ The value of the definite integral $\displaystyle\int_{x_1}^{x_2} f'(x)\,dx = \left[f(x)\right]_{x_1}^{x_2} = f(x_2) - f(x_1)$.

▶ The trapezium rule for finding the approximate value of the area under a curve.

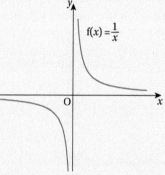

## 8.1 Approximately locating the roots of an equation

When the roots of an equation cannot be found exactly, you can locate the roots roughly. You can do this by sketching graphs.

For example, look at the equation $e^x = 4x$.

The roots of this equation are the values of $x$ where the curve $y = e^x$ and the line $y = 4x$ intersect.

The sketch shows that there is one root between $x = 0$ and $x = 1$ and another root somewhere near $x = 2$.

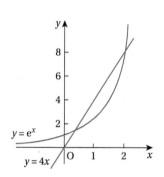

To see how to locate the roots more accurately look at the graph of a general curve $y = f(x)$.

The roots of the equation $f(x) = 0$ are the values of $x$ where this curve crosses the $x$-axis.

As you can see from the graph each time that the curve crosses the $x$-axis, the sign of $y$ changes.

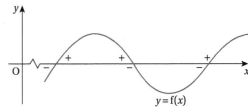

> So when one root only of the equation $f(x) = 0$ lies between $x_1$ and $x_2$ and when the curve $y = f(x)$ is unbroken between the points where $x = x_1$ and $x = x_2$ then $f(x_1)$ and $f(x_2)$ are opposite in sign.

The condition that the curve $y = f(x)$ must be unbroken in the interval between $x_1$ and $x_2$ is essential.

This curve in the diagram crosses the $x$-axis between $x_1$ and $x_2$ but $f(x_1)$ and $f(x_2)$ have the same sign, because the curve is broken between these values.

Returning to the equation $e^x = 4x$, the larger root can be located a little more precisely.

First write the equation in the form $f(x) = 0$, that is $e^x - 4x = 0$. This curve is unbroken for all real $x$. Then find where there is a change in the sign of $f(x)$.

There is a root near $x = 2$, so see if it lies between 1.8 and 2.2.

Using $f(x) = e^x - 4x$, gives $f(1.8) = e^{1.8} - 4(1.8) = -1.1...$

and $\qquad\qquad f(2.2) = e^{2.2} - 4(2.2) = 0.2...$

Therefore the larger root of the equation lies between 1.8 and 2.2 (and is likely to be nearer to 2.2 as $f(2.2)$ is nearer to zero than $f(1.8)$ is).

## Example 1

**Question**

Show that the equation $x^3 - 2x^2 + x + 1 = 0$ has a root between $x = -1$ and $x = 0$.

**Answer**

Using $f(x) = x^3 - 2x^2 + x + 1$ gives $f(-1) = -3$ and $f(0) = 1$.

$f(-1)$ and $f(0)$ are opposite in sign, so a root of $x^3 - 2x^2 + x + 1 = 0$ lies between $x = -1$ and $x = 0$.

## Exercise 1

1. Show that the equation $\cos x = x^2 - 1$ has a root between $x = 1$ and $x = 2$.
2. Show that the equation $(x^2 - 4) = \dfrac{1}{x}$ has a root between $x = 2$ and $x = 3$.
3. Show that the equation $x(2^x) = 1$ has a root between $x = 0$ and $x = 1$.
4. Show that the equation $x \ln x = 1$ has a root between $x = 1$ and $x = 2$.
5. Show that the equation $\sin x = x^2$ has a root between $x = 0$ and $x = 1$.

**6** Show that the equation $\ln x + 2^x = 0$ has a root between $x = 0.1$ and $x = 0.5$.

**7** Show that the equation $4 + 5x^2 - x^3 = 0$ has a root between 5 and 5.2.

**8** Show that the equation $x^4 - 4x^3 - x^2 + 4x - 10 = 0$ has a root between 4 and 4.2.

**9** Sketch the curves $y = x^3$ and $y = 3x^2 - 1$ and use your sketch to find the number of real roots of the equation $x^3 - 3x^2 + 1 = 0$.

**10** **a** Sketch graphs to show that the equation $3^{-x} = x^2 + 2$ has just one root.

    **b** Show that this root is exactly $-1$.

**11** Show that one root of the equation $2^x = \dfrac{1}{2}(x + 3)$ lies between $x = -3$ and $x = -2$.

**12** Sketch the graphs of $y = x^3$ and $y = 5 - x^2$. Hence find two consecutive integer values of $x$ between which there is a root of the equation $x^3 + x^2 - 5 = 0$.

**13** Sketch graphs to show that there is only one positive root of the equation $3 \sin x - x = 0$, where the angle is measured in radians.
Show that, if this root is $\alpha$, then $2 < \alpha < 3$.

**14** On the same axes, sketch the graphs of $y = 4x$ and $y = 2^x$ for values of $x$ from $-1$ to 4. Hence show that one root of the equation $2^x - 4x = 0$ is 4, and that there is another root between 0 and 1.

# 8.2 Using the iteration $x_{n+1} = g(x_n)$

This method of iteration, can often be used to find successive approximations to a root of an equation $f(x) = 0$. It can be written in the form $x = g(x)$. The roots of the equation $x = g(x)$ are the values of $x$ at the points of intersection of the line $y = x$ and the curve $y = g(x)$.

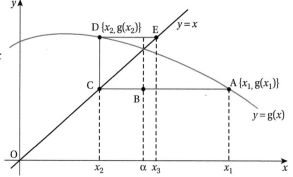

Taking $x_1$ as a first approximation to a root $\alpha$ then, as you can see in the diagram:

▶ A is the point on the *curve* where $x = x_1$ and $y = g(x_1)$

▶ B is the point where $x = \alpha$ and $y = g(x_1)$

▶ C is the point on the *line* where $y = g(x_1)$ and $x = x_2$.

When, in the region of $\alpha$, the slope of $y = g(x)$ is less steep than that of the line $y = x$ (that is provided that $|g'(x)| < 1$) then CB < BA.

Therefore, $x_2$ is closer to $\alpha$ than $x_1$ is, so $x_2$ is a better approximation to $\alpha$.

The point C is on the line $y = x$ and therefore $x_2 = g(x_1)$.

Next, taking the point D on the curve where $x = x_2$, $y = g(x_2)$ and repeating the argument above, shows that $x_3$ is a better approximation to $\alpha$ than $x_2$ is, where $x_3 = g(x_2)$.

This process can be repeated as often as necessary to achieve the required degree of accuracy and is called **iteration**.

The rate at which these approximations converge to $\alpha$ depends on the value of $|g'(x)|$ near $\alpha$. The smaller $|g'(x)|$ is, the more rapid the convergence. However this method fails if $|g'(x)| > 1$ near $\alpha$, as $x_1$, $x_2$, ... diverge from $\alpha$.

These diagrams illustrate some of the factors which determine the success, or failure, of this iterative method to find a solution to the equation $x = g(x)$.

> **Note**
>
> The diagrams on the left are called cobweb diagrams. The diagrams on the right are called staircase diagrams.

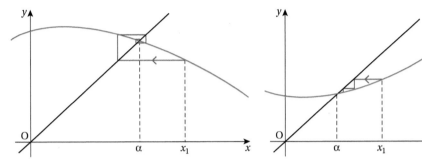

Rapid rate of convergence ($|g'(x)|$ small).

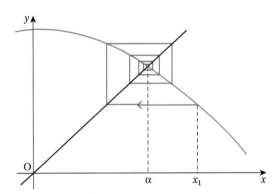

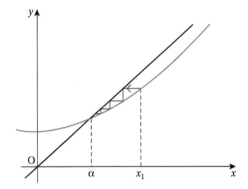

Slow rate of convergence ($|g'(x)| < 1$ but close to 1).

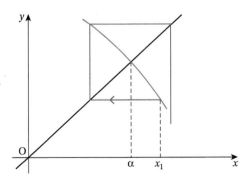

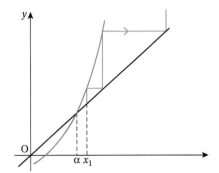

Divergence, that is, the method fails to converge ($|g'(x)| > 1$).

## Example 2

**Question**

The curve $y = x^3$ intersects the curve $y = 1 - 2x^2 - 5x$ at a single point where $x = \alpha$.

**a** Show that $\alpha$ lies between 0 and 0.2.

**b** Show that the equation $x^3 + 2x^2 + 5x - 1 = 0$ can be written in the form
$$x = -\frac{1}{5}(x^3 + 2x^2 - 1).$$

**c** Show that iteration $x_{n+1} = -\frac{1}{5}(x_n^3 + 2x_n^2 - 1)$ converges.

**d** Use the iteration $x_{n+1} = -\frac{1}{5}(x_n^3 + 2x_n^2 - 1)$ with $x_1 = 0$ to find the value of $\alpha$ correct to 3 decimal places.

**Answer**

**a** At the point of intersection $x^3 = 1 - 2x^2 - 5x$.

Therefore $\alpha$ is a root of the equation $x^3 = 1 - 2x^2 - 5x \Rightarrow x^3 + 2x^2 + 5x - 1 = 0$

Using $f(x) = x^3 + 2x^2 + 5x - 1 \Rightarrow f(0) = -1$ and $f(0.2) = 0.088$.

As $f(0) < 0$ and $f(0.2) > 0$, $\alpha$ lies between 0 and 0.2.

**b** $x^3 + 2x^2 + 5x - 1 = 0 \Rightarrow 5x = 1 - 2x^2 - x^3$

Therefore the equation can be written in the form
$$x = -\frac{1}{5}(x^3 + 2x^2 - 1).$$

**c** Using $g(x) = -\frac{1}{5}(x^3 + 2x^2 - 1)$ gives $g'(x) = -\frac{1}{5}(3x^2 + 4x)$ and

$g'(0.2) = -\frac{1}{5}(0.12 + 0.8) = -0.184$ so $|g'(x)| < 1$ near the root.

Therefore the iteration converges.

**d** Using $x_{n+1} = -\frac{1}{5}(x_n^3 + 2x_n^2 - 1)$ with $x_1 = 0$

then $x_2 = g(x_1) = -\frac{1}{5}[0^3 + 2(0)^2 - 1] = 0.2$

and $x_3 = g(x_2) = -\frac{1}{5}[(0.2)^3 + 2(0.2)^2 - 1] = 0.1824$

$x_4 = g(x_3) = \frac{1}{5}[(0.1824)^3 + 2(0.1824)^2 - 1] = 0.1854...$

$x_5 = g(x_4) = -\frac{1}{5}[(0.1854...)^3 + 2(0.1854...)^2 - 1] = 0.1849...$

It looks as if $\alpha = 0.185$ is correct to 3 decimal places, that is $0.1846 < \alpha < 0.1854$. The degree of accuracy can be checked by finding the signs of $f(0.1846)$ and $f(0.1854)$.

$f(0.1846)$ is negative and $f(0.1854)$ is positive, so $x = 0.185$ correct to 3 decimal places.

> **Note**
>
> This does *not* show that $x = 0.1850$ to 4 d.p.

## Exercise 2

In questions 1 to 4

**a** show that each equation has a root between $x = 0$ and $x = 1$

**b** express each equation in the form $x = g(x)$

**c** determine, without doing the iteration, whether your form for $x = g(x)$ gives an iteration formula that converges.

**1** $x^3 - x^2 + 10x - 2 = 0$      **2** $3x^3 - 2x^2 - 9x + 2 = 0$

**3** $2x^3 + x^2 + 6x - 1 = 0$      **4** $x^2 + 8x - 8 = 0$

5 The equation $x(2^x) = 1$ has a root between $x = 0$ and $x = 1$.

  a Show that the equation can be written in the form $x = 2^{-x}$.

  b Use the iteration $x_{n+1} = 2^{-x_n}$ with $x_1 = 0.5$ to find $x_2$ and $x_3$, giving your answers to 2 decimal places.

6 The equation $x \ln(x + 2) = 1$ has a root between $x = 0$ and $x = 1$.

  a Show that the equation can be written in the form $x = \dfrac{1}{\ln(x+2)}$.

  b Use the iteration $x_{n+1} = \dfrac{1}{\ln(x_n + 2)}$ with $x_1 = 0.8$ to find $x_2$ and $x_3$, giving your answers to 2 decimal places.

7 a Show that the equation $x^3 - x^2 - 5x + 1 = 0$ can be written in the form
$x = \dfrac{1}{5}(x^3 - x^2 + 1)$.

  b The equation $x^3 - x^2 - 5x + 1 = 0$ has a root between 0.1 and 0.2.

  Use the iteration $x_{n+1} = \dfrac{1}{5}(x_n^3 - x_n^2 + 1)$ with $x_1 = 0.1$ to find $x_2$ and $x_3$ giving your answers to 3 decimal places.

  c The equation $x^3 - x^2 - 5x + 1 = 0$ also has a root between 2.7 and 2.8.

  Use the iteration $x_{n+1} = \dfrac{1}{5}(x_n^3 - x_n^2 + 1)$ with $x_1 = 2.7$ to find four successive values of $x_n$ and hence show that this iteration does not converge to this root.

8 a Show that the equation $e^{x+5} = \sqrt{x^2 - 1}$ has a root $\alpha$ between $-4$ and $-3.5$.

  b Show that the equation $e^{x+5} = \sqrt{x^2 - 1}$ can be written as $x = -5 + \dfrac{1}{2}\ln(x^2 - 1)$.

  c Use the iteration $x_{n+1} = -5 + \dfrac{1}{2}\ln(x_n^2 - 1)$ to show that $\alpha = -3.72$ correct to 2 decimal places.

# 8.3 Rules to find the approximate value of an area under a curve

## The mid-ordinate rule

The **mid-ordinate rule** is similar to the trapezium rule but it uses rectangular strips instead of trapeziums. The area is divided into equal width strips where the top boundary of each strip is a horizontal line through the point on the curve at the centre of the strip.

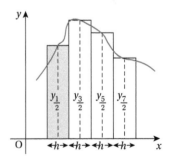

The sum of the areas of these rectangles gives an approximate value for the area under the curve, so

$$A \approx h\left[ y_{\frac{1}{2}} + y_{\frac{3}{2}} + \cdots + y_{\frac{(2n-1)}{2}} \right]$$

The mid-ordinate value is usually more accurate than the trapezium rule. This is because the top of the trapezium is often entirely on one side of the curve, so that part of the area under the curve is not allowed for. The top line of the rectangular strip, however, usually cuts across the curve this means that the part of the area included in the rectangle, but which is not under the curve, tends to balance the area under the curve that is not included in the rectangle.

The diagrams showing the tops of two enlarged strips show this.

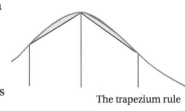

The trapezium rule

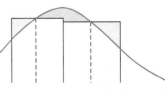

The mid-ordinate rule

## Simpson's rule

**Simpson's rule** is a formula that gives a better approximation to the area under a curve than that obtained from the trapezium and the mid-ordinate rules.

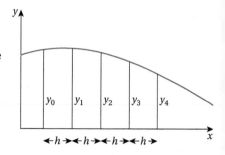

Using the same notation as before, Simpson's rule states that

$$A \approx \frac{1}{3}h\left[y_0 + 4y_1 + 2y_2 + 4y_3 + 2y_4 + \ldots + 2y_{n-2} + 4y_{n-1} + y_n\right], \text{ so}$$

$$A \approx \frac{1}{3}h\left[(y_0 + y_n) + 4(y_1 + y_3 + \cdots) + 2(y_2 + y_4 + \cdots)\right]$$

This formula is based on dividing the required area into equal width strips and, for each *pair* of strips, finding a parabola which passes through the top of the three ordinates bounding the two strips.

Because this formula is based on pairs of strips it follows that it can be used only when the number of strips is even, that is, when the number of ordinates is odd.

Simpson's rule gives an even more accurate approximation than the other rules because, as it is based on a parabola, the tops of the strips are even nearer to the shape of the curve.

The degree of accuracy of an area under a curve, given by either of these rules, depends upon the number of strips into which the required area is divided. This is because the narrower the strip, the nearer its shape at the top becomes to the shape of the curve.

## Example 3

**Question**

Use four strips to find an approximate value for the definite integral $\int_1^5 x^3 \mathrm{d}x$ using

a the mid-ordinate rule

b Simpson's rule.

**Answer**

The integral represents the area of the region bounded by the $x$-axis, the lines $x = 1$ and $x = 5$, and the curve $y = x^3$.

Five ordinates are used when there are four strips whose widths must all be the same.

From $x = 1$ to $x = 5$ there are four units, so the width of each strip must be 1 unit. Hence the five ordinates are where $x = 1$, $x = 2$, $x = 3$, $x = 4$ and $x = 5$.

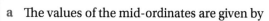

a The values of the mid-ordinates are given by

$$y_1 = 1.5^3, \, y_2 = 2.5^3, \, y_3 = 3.5^3, \, y_4 = 4.5^3$$

The width of each strip is 1 unit so using the mid-ordinate rule gives

$$A = (1)[3.375 + 15.63 + 42.88 + 91.13] = 153.0$$

The area is approximately 153 square units.

b There is an odd number of ordinates, so Simpson's rule can be used.

$$A \approx \frac{1}{3}(1)[(1 + 125) + 4(8 + 64) + 2(27)] = 156$$

The area is approximately 156 square units.

The exact value of the area is given by $\int_1^5 x^3 \, \mathrm{d}x = \left[\frac{1}{4}x^4\right]_1^5 = 156.25 - 0.25 = 156$

so Simpson's rule gave an accurate answer for this area.

# Exercise 3

**1** Estimate the value of $\int_0^4 x^2 \, dx$ using 5 ordinates and

    **a** the mid-ordinate rule

    **b** Simpson's rule.

**2** Estimate the value of $\int_1^3 \frac{1}{x^2} \, dx$ using 5 ordinates and

    **a** the mid-ordinate rule

    **b** Simpson's rule.

**3** Estimate the value of $\int_0^{\frac{2\pi}{3}} \sqrt{\sin x} \, dx$ using 3 ordinates and

    **a** the mid-ordinate rule

    **b** Simpson's rule.

**4** Estimate the value of $\int_1^3 \ln x \, dx$ using 3 ordinates and

    **a** the mid-ordinate rule

    **b** Simpson's rule.

**5** Estimate the value of $\int_0^\pi 2 + (\cos x) \ln(x+1) \, dx$ using 3 ordinates and

    **a** the mid-ordinate rule

    **b** Simpson's rule.

# Summary

## Locating the root of an equation between two values

▶ When a root of the equation $f(x) = 0$ lies between $x_1$ and $x_2$ and when the curve $y = f(x)$ is unbroken between the points where $x = x_1$ and $x = x_2$ then $f(x_1)$ and $f(x_2)$ are opposite in sign.

## Using the iteration $x_{n+1} = g(x_n)$

▶ The iteration formula $x_{n+1} = g(x_n)$ can be used to find a root of the equation $f(x) = 0$ provided that the equation can be rearranged as $x = g(x)$ for some function $g(x)$, and that $|g'(x)| < 1$ near the root.

## The mid-ordinate rule

▶ $A = \int_a^b f(x) \, dx \approx h \left[ y_{\frac{1}{2}} + y_{\frac{3}{2}} + \cdots + y_{\frac{(2n-1)}{2}} \right]$

where the values of $y$ are heights of the rectangles, that is the $y$-coordinates of the points on the curve at the centre of each strip.

## Simpson's Rule

▶ To use this rule there must be an odd number of ordinates.

$A = \int_a^b f(x) \, dx \approx \frac{1}{3} h[(y_0 + y_n) + 4(y_1 + y_3 + \cdots) + 2(y_2 + y_4 + \cdots)]$

## Review

**1** The function f is given by $f(x) = 12 \ln x - x^{\frac{3}{2}}$ for positive values of $x$. The curve $y = f(x)$ crosses the $x$-axis at the point A. Show that the value of $x$ at the point A lies between 1.1 and 1.2.

**2** **a** Show that the equation $x^3 - 2x^2 - 1 = 0$ has a root between 2 and 3.

**b** Show that $x^3 - 2x^2 - 1 = 0$ can be written as $x = 2 + \dfrac{1}{x^2}$.

**c** Taking $x_1 = 2$ as the first approximation to this root, use the iteration $x_{n+1} = 2 + \dfrac{1}{x_n^2}$ twice to obtain a better approximation.

**3** By sketching graphs in the interval $0 < x < 4$, find two consecutive integers between which there is a root of the equation $\ln(x - 2) = \dfrac{1}{x}$. Use the change of sign of $f(x) = \ln(x - 2) - \dfrac{1}{x}$ to verify that your answer is correct.

**4** Use the mid-ordinate rule with 5 ordinates to find an approximate value of $\displaystyle\int_1^5 (e^{2x} - 1)\, dx$.

**5** Use Simpson's rule with 3 ordinates to find an approximate value of $\displaystyle\int_1^5 (\ln x)(\cos x)\, dx$.

**6** Use the mid-ordinate rule with 3 ordinates to find an approximate value of $\displaystyle\int_0^{\frac{2\pi}{3}} \dfrac{1}{\sin x + \cos x}\, dx$.

## Assessment

**1** **a** Sketch graphs of $y = e^x$ and $y = x^2 + 2$ to show that the curves intersect at a single point where $x = \alpha$.

**b** Show that $\alpha$ lies between $x = 1.2$ and $x = 1.4$.

**c** Show that the equation $e^x = x^2 + 2$ can be written in the form $x = \ln(x^2 + 2)$.

**d** Use the iteration $x_{n+1} = \ln(x_n^2 + 2)$ with $x_1 = 1.2$ to find $x_2$ and $x_3$.

**2** **a** Show that the equation $x^4 + x^2 - x = 0$ has one root exactly equal to 0 and another root between 0.6 and 0.7.

**b** By expressing the equation in part **a** in the form $x = \dfrac{1}{x^2 + 1}$ show that the iteration formula $x_{n+1} = \dfrac{1}{(x_n^2 + 1)}$ converges to the larger root and find this root correct to 3 significant figures.

**3** **a** Use the mid-ordinate rule with 4 ordinates to find an approximate value for $\displaystyle\int_1^4 \left(x + \dfrac{4}{\sqrt{x}}\right)^2 dx$.

**b** Find the value of $\displaystyle\int_1^4 \left(x + \dfrac{4}{\sqrt{x}}\right)^2 dx$.

**4** **a** The equation of a curve is $y = x \ln(x^2 + 1)$.

Use Simpson's rule with 3 ordinates to find an approximate value of

$$\int_0^2 x \ln(x^2 + 1) \, dx.$$

**b** Use the substitution $u = x^2 + 1$ to find $\int x \ln(x^2 + 1) \, dx$.

**c** Hence find the exact value of $\int_0^2 x \ln(x^2 + 1) \, dx$.

**5** Use Simpson's rule, with five ordinates (four strips), to calculate an estimate

for $\int_0^\pi x^{\frac{1}{2}} \sin x \, dx$.

Give your answer to four significant figures.

AQA MPC3 June 2014

**6** Use the mid-ordinate rule with four strips to find an estimate for

$\int_{1.5}^{5.5} e^{2-x} \ln(3x - 2) \, dx$, giving your answer to three decimal places.

AQA MPC3 June 2015 (part question)

**7** A curve is defined by the equation $y = (x^2 - 4)\ln(x + 2)$ for $x > 3$.

The curve intersects the line $y = 15$ at a single point, where $x = \alpha$.

**a** Show that $\alpha$ lies between 3.5 and 4.

**b** Show that the equation $(x^2 - 4)\ln(x + 2) = 15$ can be arranged into the

form $x = \pm\sqrt{4 + \dfrac{15}{\ln(x + 2)}}$.

**c** Use the iteration $x_{n+1} = \sqrt{4 + \dfrac{15}{\ln(x_n + 2)}}$ with $x_1 = 3.5$ to find the values of

$x_2$ and $x_3$, giving your answers to three decimal places.

AQA MPC3 January 2011

# 9 Vectors

## Introduction

Many quantities need a size and a direction to define them. For example, to give directions to get from a point A to a point B, giving the distance between the points does not help. The direction of B from A is also needed. An aircraft flying in a wind needs to know how to set its course to take into account not only the speed and but also the direction of the wind. This chapter shows how to work with quantities that need both size and direction to define them.

## Recap

You will need to remember...

▶ The modulus of a quantity means its magnitude irrespective of its sign, for example when $x = 2$ or $-2$, $|x| = 2$.
▶ Pythagoras' theorem.
▶ A parameter is a variable that can take any value.
▶ $\cos^{-1}x$ means the angle whose cosine is $x$.

## Objectives

By the end of this chapter, you should know how to...

▶ Find the position vector of the midpoint of a line.
▶ Locate a vector in space using $x$, $y$ and $z$ coordinates.
▶ Add and subtract vectors in three dimensions.
▶ Find the magnitude of a vector.
▶ Understand and find a vector equation of a line.
▶ Determine whether two lines are parallel, intersect or are skew.
▶ Find the scalar product of two vectors.
▶ Find the coordinates of the foot of the perpendicular from a point to a line.
▶ Find the perpendicular distance of a point from a line.

## 9.1 Properties of vectors

A vector is a quantity which has both magnitude and a specific direction in space.

A **scalar** is a quantity that is fully defined by magnitude alone.

### Vector representation

A vector can be represented by a section of a straight line, whose length represents the magnitude of the vector and whose direction, indicated by an arrow, represents the direction of the vector.

### The modulus of a vector

The **modulus** of a vector **a** is its magnitude and is written $|\mathbf{a}|$ or $a$. In other words, $|\mathbf{a}|$ is the length of the line representing **a**.

# Equal vectors

Two vectors with the same magnitude and the same direction are equal.

$$\mathbf{a} = \mathbf{b} \iff \begin{cases} |\mathbf{a}| = |\mathbf{b}| \text{ and} \\ \text{the direction of } \mathbf{a} \text{ and } \mathbf{b} \text{ are the same} \end{cases}$$

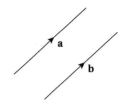

Therefore a vector can be represented by *any* line of the right length and direction, regardless of the line's position, so each of the lines in the diagram represents the vector **c**.

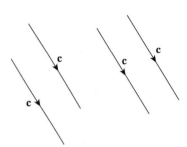

# Equal magnitude and opposite direction

When two vectors, **a** and **b**, have the same magnitude but opposite directions then $\mathbf{b} = -\mathbf{a}$.

$-\mathbf{a}$ is a vector of magnitude $|\mathbf{a}|$ in the opposite direction to **a**.

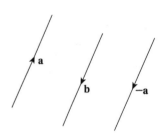

# Multiplication of a vector by a scalar

When $\lambda$ is a positive real number, then $\lambda\mathbf{a}$ is a vector in the same direction as **a** with magnitude $\lambda|\mathbf{a}|$. It follows that $-\lambda\mathbf{a}$ is a vector in the opposite direction to **a**, with magnitude $\lambda|\mathbf{a}|$.

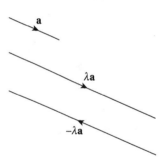

# Addition of vectors

When the sides $\overrightarrow{AB}$ and $\overrightarrow{BC}$ of a triangle ABC represent the vectors **p** and **q** then the third side $\overrightarrow{AC}$ represents the vector sum, or **resultant**, of **p** and **q**, which is denoted by $\mathbf{p} + \mathbf{q}$.

Note that **p** and **q** follow each other round the triangle (in this case in the clockwise sense), whereas the resultant, $\mathbf{p} + \mathbf{q}$, goes the opposite way round (anticlockwise in the diagram).

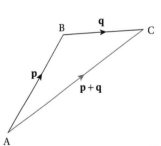

This is called the **triangle law** for addition of vectors. It can be extended to cover the addition of more than two vectors.

Let $\overrightarrow{AB}, \overrightarrow{BC}, \overrightarrow{CD}$ and $\overrightarrow{DE}$ represent the vectors **a**, **b**, **c** and **d** respectively.

The triangle law gives   $\overrightarrow{AB} + \overrightarrow{BC} = \mathbf{a} + \mathbf{b} = \overrightarrow{AC}$

then                      $\overrightarrow{AC} + \overrightarrow{CD} = (\mathbf{a} + \mathbf{b}) + \mathbf{c} = \overrightarrow{AD}$

and                       $\overrightarrow{AD} + \overrightarrow{DE} = (\mathbf{a} + \mathbf{b} + \mathbf{c}) + \mathbf{d} = \overrightarrow{AE}$

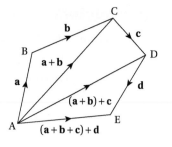

Note that the vectors **a**, **b**, **c** and **d** may not be coplanar so the polygon may not be two-dimensional.

AE completes the polygon of which AB, BC, CD and DE are four sides taken in order (that is, they follow each other round the polygon in the *same sense)*. Again, the side representing the resultant closes the polygon in the *opposite* sense. The vectors **a**, **b**, **c** and **d** may not be coplanar so the polygon may not be two-dimensional.

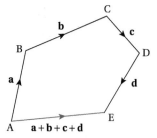

## Order of vector addition

The order in which the addition is performed does not matter as is shown by looking at the parallelogram ABCD.

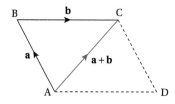
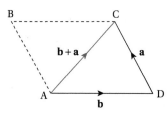

Because the opposite sides of a parallelogram are equal and parallel, $\overrightarrow{AB}$ and $\overrightarrow{DC}$ both represent **a** and $\overrightarrow{BC}$ and $\overrightarrow{AD}$ both represent **b**.

In $\triangle ABC$, $\overrightarrow{AC} = \mathbf{a} + \mathbf{b}$ and in $\triangle ADC$, $\overrightarrow{AC} = \mathbf{b} + \mathbf{a}$.

Therefore $\mathbf{a} + \mathbf{b} = \mathbf{b} + \mathbf{a}$.

## The angle between two vectors

There are two angles between two lines, $\alpha$ and $180° - \alpha$.

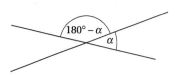

The angle between two vectors, however, is defined uniquely.

It is the angle between their directions, when the lines representing them *both converge* or *both diverge* (see diagrams below).

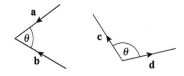

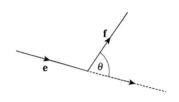

In some cases one of the lines may have to be produced in order to mark the correct angle (see diagram below).

## 9.2 Position vectors

Usually a vector has no specific location in space and is called a **free vector**. Some vectors, however, are constrained to a specific position, such as the vector $\overrightarrow{OA}$ where O is a fixed origin.

$\overrightarrow{OA}$ is called the position of A relative to O.

This displacement is unique and *cannot* be represented by any other line of equal length and direction.

Vectors such as $\overrightarrow{OA}$, representing quantities that have a specific location, are called **position vectors**.

### The position vector of the midpoint of a line

Look at the line AB where the position vectors of A and B relative to O are **a** and **b** respectively, and C is the midpoint of AB.

In the diagram, $\overrightarrow{AB} = \mathbf{b} - \mathbf{a}$

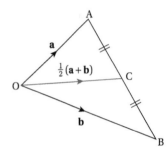

Therefore $\qquad \overrightarrow{AC} = \dfrac{1}{2}(\mathbf{b} - \mathbf{a})$

Hence $\qquad \overrightarrow{OC} = \overrightarrow{OA} + \overrightarrow{AC} = \mathbf{a} + \dfrac{1}{2}(\mathbf{b} - \mathbf{a}) = \dfrac{1}{2}\mathbf{a} + \dfrac{1}{2}\mathbf{b}$

Therefore the position vector of C is $\dfrac{1}{2}(\mathbf{a} + \mathbf{b})$.

### Exercise 1

The position vectors, relative to O, of A, B, C and D are **a**, **b**, **c** and **d** respectively. P, Q and R are the midpoints of AB, BC and CD respectively.

**1** **a** Find the position vector of the midpoint of AC.

   **b** Find the position vector of the midpoint of BD.

**2** **a** Find the position vector of the midpoint of PQ.

   **b** Find the position vector of the midpoint of QR.

**3** Show that PQ is parallel to AC.

## 9.3 The location of a point in space

Any point P in a plane can be located by giving its distances from a fixed point O, in each of two perpendicular directions. These distances are the Cartesian coordinates of the point.

In three-dimensional space, when O is a fixed point, any other point P can be located by giving its distances from O in each of *three* mutually perpendicular directions, so three coordinates are needed to locate a point in space. The familiar *x*- and *y*-axes are used, together with a third *z*-axis. Then any point has coordinates $(x, y, z)$ relative to the origin O.

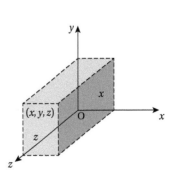

## Cartesian unit vectors

A **unit vector** is a vector whose magnitude is one unit.

When **i** is a unit vector in the direction of the $x$-axis

**j** is a unit vector in the direction of the $y$-axis

**k** is a unit vector in the direction of the $z$-axis

then the position vector, relative to O, of any point P can be given in terms of **i**, **j** and **k**.

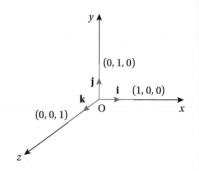

The point P distant

3 units from O in the $x$ direction

4 units from O in the $y$ direction

5 units from O in the $z$ direction

has coordinates $(3, 4, 5)$ and $\overrightarrow{OP} = 3\mathbf{i} + 4\mathbf{j} + 5\mathbf{k}$.

This can also be written as $\overrightarrow{OP} = \begin{bmatrix} 3 \\ 4 \\ 5 \end{bmatrix}$ and this is the form used in this book.

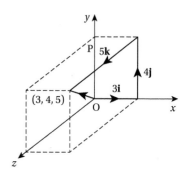

In general, when P is a point, $(x, y, z)$ and $\overrightarrow{OP} = \mathbf{r}$, then $\mathbf{r} = \begin{bmatrix} x \\ y \\ z \end{bmatrix}$ is the position vector of P.

Free vectors can be given in the same form. For example, the vector $\begin{bmatrix} 3 \\ 4 \\ 5 \end{bmatrix}$ can represent the position vector of the point $P(3, 4, 5)$ but it can equally well represent *any* vector of length and direction equal to those of $\overrightarrow{OP}$.

> **Note**
>
> Note that, unless a vector is *specified* as a position vector, it is taken to be free.

## 9.4 Operations on cartesian vectors

### Addition and subtraction

To add or subtract vectors given in the form $\begin{bmatrix} x \\ y \\ z \end{bmatrix}$ the coordinates are added separately. For example

when $\mathbf{v}_1 = \begin{bmatrix} 3 \\ 2 \\ 2 \end{bmatrix}$ and $\mathbf{v}_2 = \begin{bmatrix} 1 \\ 2 \\ -3 \end{bmatrix}$

then $\mathbf{v}_1 + \mathbf{v}_2 = \begin{bmatrix} 3 \\ 2 \\ 2 \end{bmatrix} + \begin{bmatrix} 1 \\ 2 \\ -3 \end{bmatrix} = \begin{bmatrix} 3+1 \\ 2+2 \\ 2-3 \end{bmatrix} = \begin{bmatrix} 4 \\ 4 \\ -1 \end{bmatrix}$

And $\mathbf{v}_1 - \mathbf{v}_2 = \begin{bmatrix} 3-1 \\ 2-2 \\ 2-(-3) \end{bmatrix} = \begin{bmatrix} 2 \\ 0 \\ 5 \end{bmatrix}$

# Modulus

The magnitude or the modulus of **v**, where $\mathbf{v} = \begin{bmatrix} 12 \\ -3 \\ 4 \end{bmatrix}$ is the length of OP where P is the point $(12, -3, 4)$.

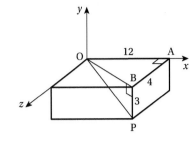

Using Pythagoras twice we have

$$OB^2 = OA^2 + AB^2 = 12^2 + 4^2$$
$$OP^2 = OB^2 + BP^2 = (12^2 + 4^2) + (-3)^2$$
$$\therefore \quad OP = \sqrt{12^2 + 4^2 + 3^2} = 13$$

In general, when $\mathbf{v} = \begin{bmatrix} a \\ b \\ c \end{bmatrix}$ then $|\mathbf{v}| = \sqrt{a^2 + b^2 + c^2}$.

# Parallel vectors

Two vectors $\mathbf{v}_1$ and $\mathbf{v}_2$ are parallel when $\mathbf{v}_1 = \lambda \mathbf{v}_2$ where $\lambda$ is a constant.

For example, $\begin{bmatrix} 2 \\ -3 \\ -1 \end{bmatrix}$ is parallel to $\begin{bmatrix} 4 \\ -6 \\ -2 \end{bmatrix}$ (by taking $\lambda = 2$), and

$\begin{bmatrix} 1 \\ 1 \\ 1 \end{bmatrix}$ is parallel to $\begin{bmatrix} -3 \\ -3 \\ -3 \end{bmatrix}$ (by taking $\lambda = -3$).

# Equal vectors

When two vectors $\mathbf{v}_1 = \begin{bmatrix} a_1 \\ b_1 \\ c_1 \end{bmatrix}$ and $\mathbf{v}_2 = \begin{bmatrix} a_2 \\ b_2 \\ c_2 \end{bmatrix}$ are equal then $a_1 = a_2$ and $b_1 = b_2$ and $c_1 = c_2$.

# Example 1

Given the vector $\mathbf{v} = \begin{bmatrix} 5 \\ -2 \\ 4 \end{bmatrix}$ state whether

each of these vectors is parallel to **v**, equal to **v** or neither.

a $\begin{bmatrix} 10 \\ -4 \\ 8 \end{bmatrix}$     b $-\dfrac{1}{2}\begin{bmatrix} -10 \\ 4 \\ -8 \end{bmatrix}$

c $\begin{bmatrix} -5 \\ 2 \\ -4 \end{bmatrix}$     d $\begin{bmatrix} 4 \\ -2 \\ 5 \end{bmatrix}$

a $\begin{bmatrix} 10 \\ -4 \\ 8 \end{bmatrix} = 2\begin{bmatrix} 5 \\ -2 \\ 4 \end{bmatrix}$ (taking $\lambda = 2$), therefore $\begin{bmatrix} 10 \\ -4 \\ 8 \end{bmatrix}$ is parallel to **v**.

*(continued)*

*(continued)*

**Answer**

**b** $-\dfrac{1}{2}\begin{bmatrix} -10 \\ 4 \\ -8 \end{bmatrix} = \begin{bmatrix} 5 \\ -2 \\ 4 \end{bmatrix}$ therefore $-\dfrac{1}{2}\begin{bmatrix} -10 \\ 4 \\ -8 \end{bmatrix}$ is equal to **v**.

**c** $\begin{bmatrix} -5 \\ 2 \\ -4 \end{bmatrix} = -\begin{bmatrix} 5 \\ -2 \\ 4 \end{bmatrix}$ (taking $\lambda = -1$), therefore $\begin{bmatrix} 5 \\ -2 \\ 4 \end{bmatrix}$ is parallel to **v**.

**d** $\begin{bmatrix} 4 \\ -2 \\ 5 \end{bmatrix}$ is not a multiple of $\begin{bmatrix} 5 \\ -2 \\ 4 \end{bmatrix}$ therefore $\begin{bmatrix} 4 \\ -2 \\ 5 \end{bmatrix}$ is not equal or parallel to **v**.

## Example 2

**Question**

A triangle ABC has its vertices at the points A(2, −1, 4), B(3, −2, 5) and

C(−1, 6, 2). Find, in the form $\begin{bmatrix} a \\ b \\ c \end{bmatrix}$ the vectors $\overrightarrow{AB}, \overrightarrow{BC}$ and $\overrightarrow{CA}$ and hence find

the lengths of the sides of the triangle.

**Answer**

The coordinate axes are not drawn in the diagram, as they can cause confusion when two or more points are illustrated. The origin should always be included as it provides a reference point.

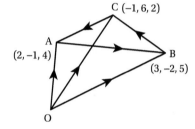

$\overrightarrow{AB} = \overrightarrow{OB} - \overrightarrow{OA} = \begin{bmatrix} 3 \\ -2 \\ 5 \end{bmatrix} - \begin{bmatrix} 2 \\ -1 \\ 4 \end{bmatrix} = \begin{bmatrix} 1 \\ -1 \\ 1 \end{bmatrix}$

$\overrightarrow{BC} = \overrightarrow{OC} - \overrightarrow{OB} = \begin{bmatrix} -1 \\ 6 \\ 2 \end{bmatrix} - \begin{bmatrix} 3 \\ -2 \\ 5 \end{bmatrix} = \begin{bmatrix} -4 \\ 8 \\ -3 \end{bmatrix}$

$\overrightarrow{CA} = \overrightarrow{OA} - \overrightarrow{OC} = \begin{bmatrix} 2 \\ -1 \\ 4 \end{bmatrix} - \begin{bmatrix} -1 \\ 6 \\ 2 \end{bmatrix} = \begin{bmatrix} 3 \\ -7 \\ 2 \end{bmatrix}$

Therefore $AB = \left| \overrightarrow{AB} \right| = \sqrt{(1)^2 + (-1)^2 + (1)^2} = \sqrt{3}$

$BC = \left| \overrightarrow{BC} \right| = \sqrt{(-4)^2 + (8)^2 + (-3)^2} = \sqrt{89}$

$CA = \left| \overrightarrow{CA} \right| = \sqrt{(3)^2 + (-7)^2 + (2)^2} = \sqrt{62}$

## Exercise 2

**1** Write down, in the form $\begin{bmatrix} a \\ b \\ c \end{bmatrix}$ the vector represented by $\overrightarrow{OP}$ if P is a point

with coordinates.

**a** (3, 6, 4)          **b** (1, −2, −7)          **c** (1, 0, −3).

**2** $\overrightarrow{OP}$ represents a vector **r**. Write down the coordinates of P when

**a** $\mathbf{r} = \begin{bmatrix} 5 \\ -7 \\ 2 \end{bmatrix}$  **b** $\mathbf{r} = \begin{bmatrix} 1 \\ 4 \\ 0 \end{bmatrix}$  **c** $\mathbf{r} = \begin{bmatrix} 0 \\ 1 \\ -1 \end{bmatrix}$.

**3** Find the length of the line OP when P is the point

**a** $(2, -1, 4)$  **b** $(3, 0, 4)$  **c** $(-2, -2, 1)$.

**4** Find the modulus of the vector **v** when

**a** $\mathbf{v} = \begin{bmatrix} 2 \\ -4 \\ 4 \end{bmatrix}$  **b** $\mathbf{v} = \begin{bmatrix} 6 \\ 2 \\ -3 \end{bmatrix}$  **c** $\mathbf{v} = \begin{bmatrix} 11 \\ -7 \\ -6 \end{bmatrix}$.

**5** When $\mathbf{p} = \begin{bmatrix} 1 \\ 1 \\ 1 \end{bmatrix}$, $\mathbf{q} = \begin{bmatrix} 2 \\ -1 \\ 3 \end{bmatrix}$ and $\mathbf{r} = \begin{bmatrix} -1 \\ 3 \\ -1 \end{bmatrix}$ find

**a** $\mathbf{p} + \mathbf{q}$  **b** $\mathbf{p} - \mathbf{r}$  **c** $\mathbf{p} + \mathbf{q} + \mathbf{r}$  **d** $\mathbf{p} - 2\mathbf{q} + 3\mathbf{r}$.

**6** Given $\overrightarrow{OA} = \mathbf{a} = \begin{bmatrix} 4 \\ -12 \\ 0 \end{bmatrix}$, $\overrightarrow{OB} = \mathbf{b} = \begin{bmatrix} 1 \\ 6 \\ 0 \end{bmatrix}$ and $\overrightarrow{OD} = \lambda\,\overrightarrow{OA}$,

find the value of $\lambda$ for which $\overrightarrow{OD} + \overrightarrow{OB}$ is parallel to the $x$-axis.

**7** State which of the following vectors are parallel to $\begin{bmatrix} 3 \\ -1 \\ -2 \end{bmatrix}$.

**a** $\begin{bmatrix} 6 \\ -3 \\ -4 \end{bmatrix}$  **b** $\begin{bmatrix} -9 \\ 3 \\ 6 \end{bmatrix}$  **c** $\begin{bmatrix} -3 \\ -1 \\ -2 \end{bmatrix}$

**d** $-2\begin{bmatrix} 3 \\ 1 \\ 2 \end{bmatrix}$  **e** $\begin{bmatrix} \frac{3}{2} \\ -\frac{1}{2} \\ -1 \end{bmatrix}$  **f** $\begin{bmatrix} -1 \\ \frac{1}{3} \\ \frac{2}{3} \end{bmatrix}$

**8** Given that $\mathbf{a} = \begin{bmatrix} 4 \\ 1 \\ -6 \end{bmatrix}$, state whether each of these vectors

is parallel to **a**, equal to **a** or neither parallel or equal to **a**.

**a** $\begin{bmatrix} 8 \\ 2 \\ -10 \end{bmatrix}$  **b** $\begin{bmatrix} -4 \\ -1 \\ 6 \end{bmatrix}$  **c** $2\begin{bmatrix} 2 \\ \frac{1}{2} \\ -3 \end{bmatrix}$

**9** The triangle ABC has its vertices at the points A($-1$, 3, 0), B($-3$, 0, 7) and C($-1$, 2, 3).

Find, in the form $\begin{bmatrix} a \\ b \\ c \end{bmatrix}$, the vectors representing

**a** $\overrightarrow{AB}$  **b** $\overrightarrow{AC}$  **c** $\overrightarrow{CB}$.

**10** Find the lengths of the sides of the triangle described in question 9.

11 Find $|\mathbf{a} - \mathbf{b}|$ where $\mathbf{a} = \begin{bmatrix} 1 \\ -1 \\ 2 \end{bmatrix}$ and $\mathbf{b} = \begin{bmatrix} 2 \\ -1 \\ 0 \end{bmatrix}$.

12 A, B, C and D are the points $(0, 0, 2)$, $(-1, 3, 2)$, $(1, 0, 4)$ and $(-1, 2, -2)$ respectively. Find the vectors representing $\overrightarrow{AB}, \overrightarrow{BD}, \overrightarrow{CD}$ and $\overrightarrow{AD}$.

## 9.5 Properties of a line joining two points

The points A and B have coordinates $(x_1, y_1, z_1)$ and $(x_2, y_2, z_2)$ respectively.

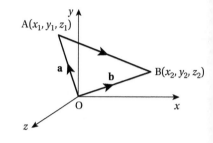

$$\overrightarrow{OA} = \begin{bmatrix} x_1 \\ y_1 \\ z_1 \end{bmatrix} \quad \text{and} \quad \overrightarrow{OB} = \begin{bmatrix} x_2 \\ y_2 \\ z_2 \end{bmatrix} \quad \text{so} \quad \overrightarrow{AB} = \overrightarrow{AO} + \overrightarrow{OB} = \overrightarrow{OB} - \overrightarrow{OA}$$

Therefore $\overrightarrow{AB} = \begin{bmatrix} x_2 - x_1 \\ y_2 - y_1 \\ z_2 - z_1 \end{bmatrix}$

### The length of AB

Since $AB = \left\| \begin{bmatrix} x_2 - x_1 \\ y_2 - y_1 \\ z_2 - z_1 \end{bmatrix} \right\|$ then

the length of the line joining $(x_1, y_1, z_1)$ and $(x_2, y_2, z_2)$ is

$$\sqrt{(x_2 - x_1)^2 + (y_2 - y_1)^2 + (z_2 - z_1)^2}$$

### The position vector of the midpoint of AB

When C is the midpoint of AB then the position vector of C is $\dfrac{1}{2}(\overrightarrow{OA} + \overrightarrow{OB})$ (you saw this in section 9.2).

So when A is the point $(x_1, y_1, z_1)$ and B is the point $(x_2, y_2, z_2)$

the coordinates of C are $\left( \dfrac{1}{2}(x_1 + x_2), \dfrac{1}{2}(y_1 + y_2), \dfrac{1}{2}(z_1 + z_2) \right)$

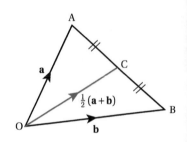

Therefore the coordinates of the midpoint are the arithmetic means of the respective coordinates of the end points.

### Example 3

**Question**

The coordinates of the midpoint R of a line PQ are $(3, -2, 6)$. P is the point $(4, 1, -3)$ and Q is the point $(a, b, c)$. Find the values of $a$, $b$ and $c$.

R is the point $(3, -2, 6)$.

But R is the midpoint of PQ, so R is the point $\left(\frac{1}{2}(4+a), \frac{1}{2}(1+b), \frac{1}{2}(-3+c)\right)$

Therefore $\frac{1}{2}(4+a)=3$, $\frac{1}{2}(1+b)=-2$, $\frac{1}{2}(-3+c)=6$

So $\quad a=2, b=-5, c=15$.

## Example 4

Find the length of the median through O of the triangle OAB, where A is the point $(2, 7, -1)$ and B is the point $(4, 1, 2)$.

The median of a triangle is the line joining a vertex to the midpoint of the opposite side.

The coordinates of M, the midpoint of AB, are

$$\left(\frac{1}{2}(2+4), \frac{1}{2}(7+1), \frac{1}{2}(-1+2)\right) = \left(3, 4, \frac{1}{2}\right)$$

So the length of OM is $\sqrt{3^2 + 4^2 + \frac{1}{2}^2} = \frac{1}{2}\sqrt{101}$.

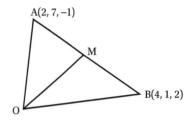

## Example 5

The points A, B and C have coordinates $(3, -1, 5)$, $(7, 1, 3)$ and $(-5, 9, -1)$ respectively. L is the midpoint of AB and M is the midpoint of BC. Find the length of LM.

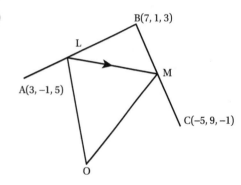

L is the point $\left(\frac{1}{2}(3+7), \frac{1}{2}(-1+1), \frac{1}{2}(5+3)\right) = (5, 0, 4)$

M is the point $\left(\frac{1}{2}(7-5), \frac{1}{2}(1+9), \frac{1}{2}(3-1)\right) = (1, 5, 1)$

Therefore $LM = \sqrt{(1-5)^2 + (5-0)^2 + (1-4)^2} = \sqrt{50} = 5\sqrt{2}$.

## Example 6

A, B and C are the points with position vectors $\begin{bmatrix} 2 \\ -1 \\ 5 \end{bmatrix}$, $\begin{bmatrix} 1 \\ -2 \\ 1 \end{bmatrix}$ and $\begin{bmatrix} 3 \\ 1 \\ -2 \end{bmatrix}$ respectively.

D and E are the respective midpoints of BC and AC. Show that DE is parallel to AB.

Using $\mathbf{a} = \begin{bmatrix} 2 \\ -1 \\ 5 \end{bmatrix}$, $\mathbf{b} = \begin{bmatrix} 1 \\ -2 \\ 1 \end{bmatrix}$ and $\mathbf{c} = \begin{bmatrix} 3 \\ 1 \\ -2 \end{bmatrix}$ then

In $\triangle$ OBC, $\overrightarrow{OD} = \frac{1}{2}(\mathbf{b}+\mathbf{c}) = \frac{1}{2}\begin{bmatrix} 4 \\ -1 \\ -1 \end{bmatrix}$

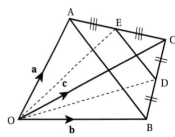

*(continued)*

*(continued)*

and in $\triangle OAC$, $\overrightarrow{OE} = \dfrac{1}{2}(\mathbf{a} + \mathbf{c}) = \dfrac{1}{2}\begin{bmatrix} 5 \\ 0 \\ 3 \end{bmatrix}$

Therefore $\overrightarrow{DE} = \overrightarrow{OE} - \overrightarrow{OD} = \dfrac{1}{2}\begin{bmatrix} 1 \\ 1 \\ 4 \end{bmatrix}$

Also $\overrightarrow{AB} = \mathbf{b} - \mathbf{a} = \begin{bmatrix} -1 \\ -1 \\ -4 \end{bmatrix}$

So $\overrightarrow{AB} = -2\overrightarrow{DE}$ therefore AB and DE are parallel.

## Exercise 3

In this exercise A, B, C and D are the points with position vectors

$\begin{bmatrix} 1 \\ 1 \\ -1 \end{bmatrix}$, $\begin{bmatrix} 1 \\ -1 \\ 2 \end{bmatrix}$, $\begin{bmatrix} 0 \\ 1 \\ 1 \end{bmatrix}$ and $\begin{bmatrix} 2 \\ 1 \\ 0 \end{bmatrix}$ respectively.

**1** Find $|\overrightarrow{AB}|$ and $|\overrightarrow{BD}|$.

**2** Determine whether any of these pairs of lines are parallel.

    **a**  AB and CD

    **b**  AC and BD

    **c**  AD and BC

**3** When L and M are the position vectors of the midpoints of AD and BD respectively, show that $\overrightarrow{LM}$ is parallel to $\overrightarrow{AB}$.

**4** When H and K are the midpoints of AC and CD respectively, show that $\overrightarrow{HK} = \dfrac{1}{2}\overrightarrow{AD}$.

**5** When L, M, N and P are the midpoints of AD, BD, BC and AC respectively, show that $\overrightarrow{LM}$ is parallel to $\overrightarrow{NP}$.

## 9.6 The equation of a straight line

A particular line is uniquely located in space if

▶   it has a known direction and passes through a known fixed point, or

▶   it passes through two known fixed points.

### A line with known direction passing through a fixed point

Look at a line that is parallel to a vector $\mathbf{m}$ and which passes through a fixed point A with position vector $\mathbf{a}$.

When $\mathbf{r}$ is the position vector, $\overrightarrow{OP}$, of a point P then

      P is a point on this line  $\Leftrightarrow$  $\overrightarrow{AP} = \lambda\mathbf{m}$

where $\lambda$ is a parameter (a variable scalar).

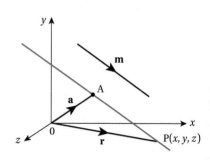

Now $\overrightarrow{OP} = \overrightarrow{OA} + \overrightarrow{AP}$

So $\mathbf{r} = \mathbf{a} + \lambda \mathbf{m}$

Therefore P is on the line $\Leftrightarrow$ $\mathbf{r} = \mathbf{a} + \lambda \mathbf{m}$.

In general, if a line passes through A$(x_1, y_1, z_1)$ and is parallel to $\begin{bmatrix} a \\ b \\ c \end{bmatrix}$ its

equation is $\mathbf{r} = \begin{bmatrix} x_1 \\ y_1 \\ z_1 \end{bmatrix} + \lambda \begin{bmatrix} a \\ b \\ c \end{bmatrix}$.

The point $(x_1, y_1, z_1)$ is only one of an infinite number of points on the line.
Therefore the equation representing a given line is *not* unique.

## Example 7

A line passes through the point with position vector $\begin{bmatrix} 2 \\ -1 \\ 4 \end{bmatrix}$

and is parallel to the vector $\begin{bmatrix} 1 \\ 1 \\ -2 \end{bmatrix}$.

Find a vector equation of the line.

The vector equation of a line is $\mathbf{r} = \mathbf{a} + \lambda \mathbf{m}$ where $\mathbf{a}$ is the position vector
of a point on the line and $\mathbf{m}$ is parallel to the line.

For this line, $\mathbf{a} = \begin{bmatrix} 2 \\ -1 \\ 4 \end{bmatrix}$ and $\mathbf{m} = \begin{bmatrix} 1 \\ 1 \\ -2 \end{bmatrix}$.

A vector equation of the line is $\mathbf{r} = \begin{bmatrix} 2 \\ -1 \\ 4 \end{bmatrix} + \lambda \begin{bmatrix} 1 \\ 1 \\ -2 \end{bmatrix}$.

## Example 8

Find a vector equation for the line through the points A(3, 4, −7) and B(1, −1, 6).

To find a vector equation of a line, you need to know a point on the line (you
could use either A or B) and a vector parallel to the line.
As A and B are on the line, $\overrightarrow{AB}$ is parallel to the line and $\overrightarrow{AB} = \overrightarrow{OB} - \overrightarrow{OA}$.

$$\overrightarrow{OA} = \begin{bmatrix} 3 \\ 4 \\ -7 \end{bmatrix} \text{ and } \overrightarrow{OB} = \begin{bmatrix} 1 \\ -1 \\ 6 \end{bmatrix}$$

Hence $\overrightarrow{AB} = \overrightarrow{OB} - \overrightarrow{OA} = \begin{bmatrix} 1 \\ -1 \\ 6 \end{bmatrix} - \begin{bmatrix} 3 \\ 4 \\ -7 \end{bmatrix} = \begin{bmatrix} -2 \\ -5 \\ 13 \end{bmatrix}$

*(continued)*

A vector equation of the line is $\mathbf{r} = \begin{bmatrix} 3 \\ 4 \\ -7 \end{bmatrix} + \lambda \begin{bmatrix} -2 \\ -5 \\ 13 \end{bmatrix}$.

This equation is not unique; we could have used $\overrightarrow{OB}$ instead of $\overrightarrow{OA}$ as the position vector. Furthermore, you could use any multiple of $\overrightarrow{OB} - \overrightarrow{OA}$ as the direction vector, since all are parallel to the line.

$\mathbf{r} = \begin{bmatrix} 1 \\ -1 \\ 6 \end{bmatrix} + \lambda \begin{bmatrix} -2 \\ -5 \\ 13 \end{bmatrix}$ is an equally valid vector equation for this line.

## Example 9

Find the coordinates of the point where the line $\mathbf{r} = \begin{bmatrix} 2 \\ -3 \\ 2 \end{bmatrix} + s \begin{bmatrix} 1 \\ -1 \\ 4 \end{bmatrix}$ cuts the $xy$-plane.

The $z$-coordinate of any point P on the $xy$-plane is zero.

Rearranging the equation of the line as $\mathbf{r} = \begin{bmatrix} 2+s \\ -3-s \\ 2+4s \end{bmatrix}$ shows that

it cuts the $xy$-plane where $2 + 4s = 0$, that is where $s = -\dfrac{1}{2}$.

When $s = -\dfrac{1}{2}$, then $\mathbf{r} = \begin{bmatrix} \frac{3}{2} \\ -\frac{7}{2} \\ 0 \end{bmatrix}$.

Therefore the line cuts the $xy$-plane at the point $\left( \dfrac{3}{2}, -\dfrac{7}{2}, 0 \right)$.

## Exercise 4

**1** Write down a vector which is parallel to each of these lines.

**a** $\mathbf{r} = \begin{bmatrix} 1 \\ -2 \\ 4 \end{bmatrix} + t \begin{bmatrix} 2 \\ -1 \\ -5 \end{bmatrix}$ 　　　　**b** $\mathbf{r} = \begin{bmatrix} 2 \\ 0 \\ -1 \end{bmatrix} + s \begin{bmatrix} 0 \\ 3 \\ -5 \end{bmatrix}$

**c** $\mathbf{r} = \begin{bmatrix} 1-2s \\ 4s-3 \\ 1-s \end{bmatrix}$

**2** Write down an equation in vector form for the line through a point A with position vector **a** and in the direction of vector **b** where

**a** $\mathbf{a} = \begin{bmatrix} 1 \\ -3 \\ 2 \end{bmatrix}, \mathbf{b} = \begin{bmatrix} 5 \\ 4 \\ -1 \end{bmatrix}$ 　　　　**b** $\mathbf{a} = \begin{bmatrix} 2 \\ 1 \\ 0 \end{bmatrix}, \mathbf{b} = \begin{bmatrix} 0 \\ 3 \\ -1 \end{bmatrix}$

**c** A is the origin, $\mathbf{b} = \begin{bmatrix} 1 \\ -1 \\ -1 \end{bmatrix}$

**3** The points A(4, 5, 10), B(2, 3, 4) and C(1, 2, −1) are three vertices of a parallelogram ABCD. Find vector equations for the sides AB, BC and AD.

**4** **a** Write down a vector equation for the line through A and B when

**i** $\overrightarrow{AB} = \begin{bmatrix} 3 \\ 1 \\ -4 \end{bmatrix}$ and $\overrightarrow{OB} = \begin{bmatrix} 1 \\ 7 \\ 8 \end{bmatrix}$

**ii** A and B have coordinates (1, 1, 7) and (3, 4, 1).

**b** Find, for each line in part **a**, the coordinates of the points where the line crosses the *xy*-plane, the *yz*-plane and the *zx*-plane.

# 9.7 Pairs of lines

The location of two lines in space can be such that

**a** the lines are parallel

**b** the lines are not parallel and intersect

**c** the lines are not parallel and do not intersect (such lines are called **skew**).

## Parallel lines

If two lines are parallel, the vector giving the direction of one line will be a multiple of the vector giving the direction of the other line.

## Non-parallel lines

When two lines whose vector equations are $\mathbf{r}_1 = \mathbf{a}_1 + \lambda\mathbf{b}_1$ and $\mathbf{r}_2 = \mathbf{a}_2 + \mu\mathbf{b}_2$ intersect, there must be unique values of $\lambda$ and $\mu$ such that $\mathbf{a}_1 + \lambda\mathbf{b}_1 = \mathbf{a}_2 + \mu\mathbf{b}_2$.

If no such values can be found, the lines do not intersect.

## Example 10

Find out whether these pairs of lines are parallel, intersecting or skew.

**a** $\mathbf{r}_1 = \begin{bmatrix} 1 \\ 1 \\ 2 \end{bmatrix} + \lambda \begin{bmatrix} 3 \\ -2 \\ 4 \end{bmatrix}$ and $\mathbf{r}_2 = \begin{bmatrix} 2 \\ -1 \\ 3 \end{bmatrix} + \mu \begin{bmatrix} -6 \\ 4 \\ -8 \end{bmatrix}$

**b** $\mathbf{r}_1 = \begin{bmatrix} 1 \\ -1 \\ 3 \end{bmatrix} + s \begin{bmatrix} 1 \\ -1 \\ 1 \end{bmatrix}$ and $\mathbf{r}_2 = \begin{bmatrix} 2 \\ 4 \\ 6 \end{bmatrix} + t \begin{bmatrix} 2 \\ 1 \\ 3 \end{bmatrix}$

**c** $\mathbf{r}_1 = \begin{bmatrix} 1 \\ 0 \\ 1 \end{bmatrix} + \lambda \begin{bmatrix} 1 \\ 3 \\ 4 \end{bmatrix}$ and $\mathbf{r}_2 = \begin{bmatrix} 2 \\ 3 \\ 0 \end{bmatrix} + \mu \begin{bmatrix} 4 \\ -1 \\ 1 \end{bmatrix}$

**a** Check first whether the lines are parallel by comparing their directions.

The first line is parallel to $\begin{bmatrix} 3 \\ -2 \\ 4 \end{bmatrix}$.

The second line is parallel to $\begin{bmatrix} -6 \\ 4 \\ -8 \end{bmatrix} = -2 \begin{bmatrix} 3 \\ -2 \\ 4 \end{bmatrix}$.

Therefore these two lines are parallel.

**b** The directions of the lines are $\begin{bmatrix} 1 \\ -1 \\ 1 \end{bmatrix}$ and $\begin{bmatrix} 2 \\ 1 \\ 3 \end{bmatrix}$.

These are not multiples of one another, so the two lines are not parallel.

If the lines intersect it will be at a point where $\mathbf{r}_1 = \mathbf{r}_2$ that is, where

$$\begin{bmatrix} 1+s \\ -1-s \\ 3+s \end{bmatrix} = \begin{bmatrix} 2+2t \\ 4+t \\ 6+3t \end{bmatrix}.$$

Equating the coordinates of $x$ and $y$ gives

$$1 + s = 2(1 + t) \text{ and } -(1 + s) = 4 + t$$

so $t = -2$, $s = -3$.

With these values for $s$ and $t$, the $z$-coordinates are

$$\left. \begin{array}{ll} \text{first line} & 3 + s = 0 \\ \text{second line} & 6 + 3t = 0 \end{array} \right\} \text{ equal values.}$$

So $\mathbf{r}_1 = \mathbf{r}_2$ when $s = -3$ and $t = -2$.

Therefore the lines *do* intersect at the point with position vector

$$\begin{bmatrix} 1-3 \\ -1+3 \\ 3-3 \end{bmatrix} = \begin{bmatrix} -2 \\ 2 \\ 0 \end{bmatrix} \text{ (using } s = -3 \text{ in } \mathbf{r}_1).$$

**c** The directions of these two lines are not multiples of one another, so the lines are not parallel.

If the lines intersect it will be where $\mathbf{r}_1 = \mathbf{r}_2$ that is, where

$$\begin{bmatrix} 1+\lambda \\ 3\lambda \\ 1+4\lambda \end{bmatrix} = \begin{bmatrix} 2+4\mu \\ 3-\mu \\ \mu \end{bmatrix}.$$

Equating the coordinates of $x$ and $y$ gives

$$\left. \begin{array}{l} 1 + \lambda = 2 + 4\mu \\ 3\lambda = 3 - \mu \end{array} \right\} \Rightarrow \mu = 0, \lambda = 1.$$

With these values of $\lambda$ and $\mu$, the $z$-coordinates are

$$\left. \begin{array}{ll} \text{first line} & 1 + 4\lambda = 5 \\ \text{second line} & \mu = 0 \end{array} \right\} \text{ not equal values.}$$

So there are no values of $\lambda$ and $\mu$ for which $\mathbf{r}_1 = \mathbf{r}_2$. Hence these lines do not intersect and are therefore skew.

## Exercise 5

**1** Find whether the following pairs of lines are parallel, intersecting or skew. In the case of intersection state the position vector of the common point.

**a** $\mathbf{r}_1 = \begin{bmatrix} 1 \\ -1 \\ 1 \end{bmatrix} + \lambda \begin{bmatrix} 3 \\ -4 \\ 1 \end{bmatrix}$ and $\mathbf{r}_2 = \mu \begin{bmatrix} -9 \\ 12 \\ -3 \end{bmatrix}$

**b** $\mathbf{r}_1 = \begin{bmatrix} 4 - t \\ 8 - 2t \\ 3 - t \end{bmatrix}$ and $\mathbf{r}_2 = \begin{bmatrix} 7 + 6s \\ 6 + 4s \\ 5 + 5s \end{bmatrix}$

**c** $\mathbf{r}_1 = \begin{bmatrix} 1 \\ 0 \\ 3 \end{bmatrix} + \lambda \begin{bmatrix} 2 \\ 1 \\ 1 \end{bmatrix}$ and $\mathbf{r}_2 = \begin{bmatrix} 2 \\ -1 \\ 1 \end{bmatrix} + \mu \begin{bmatrix} 1 \\ -2 \\ 0 \end{bmatrix}$

**2** Two lines which intersect have equations

$$\mathbf{r} = \begin{bmatrix} 2 \\ 9 \\ 13 \end{bmatrix} + \lambda \begin{bmatrix} 1 \\ 2 \\ 3 \end{bmatrix} \text{ and } \mathbf{r} = \begin{bmatrix} a \\ 7 \\ -2 \end{bmatrix} + \mu \begin{bmatrix} -1 \\ 2 \\ -3 \end{bmatrix}.$$

Find the value of $a$ and the position vector of the point of intersection.

**3** Show that the lines $\mathbf{r} = \begin{bmatrix} 2 \\ -1 \\ 1 \end{bmatrix} + \lambda \begin{bmatrix} 1 \\ -2 \\ 2 \end{bmatrix}$ and $\mathbf{r} = \begin{bmatrix} 1 \\ -3 \\ 4 \end{bmatrix} + \mu \begin{bmatrix} 2 \\ 3 \\ -6 \end{bmatrix}$ are skew.

## 9.8 The scalar product

This section involves an operation on two vectors and the angle between them. This operation is called a product but, because it involves vectors, it is not related to the product of real numbers.

### The definition of the scalar product

The scalar product of two vectors **a** and **b** is denoted by $\mathbf{a} \cdot \mathbf{b}$ and defined as

$$\mathbf{a} \cdot \mathbf{b} = ab \cos \theta$$

where $\theta$ is the angle between **a** and **b**.

Since $ab \cos \theta = ba \cos \theta$, $\qquad \mathbf{a} \cdot \mathbf{b} = \mathbf{b} \cdot \mathbf{a}$.

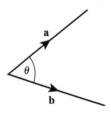

### Parallel vectors

When **a** and **b** are parallel then either

$$\mathbf{a} \cdot \mathbf{b} = ab \cos 0 \qquad \text{or} \qquad \mathbf{a} \cdot \mathbf{b} = ab \cos \pi$$

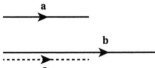 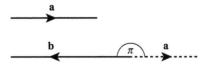

For parallel vectors in the same direction $\mathbf{a} \cdot \mathbf{b} = ab$ and for parallel vectors in the opposite direction $\mathbf{a} \cdot \mathbf{b} = -ab$.

In particular, for the unit vectors $\mathbf{i}$, $\mathbf{j}$, $\mathbf{k}$,

$$\mathbf{i} \cdot \mathbf{i} = \mathbf{j} \cdot \mathbf{j} = \mathbf{k} \cdot \mathbf{k} = 1$$

In the special case when $\mathbf{a} = \mathbf{b}$

$$\mathbf{a} \cdot \mathbf{b} = \mathbf{a} \cdot \mathbf{a} = a^2$$

## Perpendicular vectors

If $\mathbf{a}$ and $\mathbf{b}$ are perpendicular then $\theta = \dfrac{1}{2}\pi \Rightarrow \mathbf{a} \cdot \mathbf{b} = ab \cos \dfrac{1}{2}\pi = 0$.

**For perpendicular vectors, $\mathbf{a} \cdot \mathbf{b} = 0$.**

In particular, for the unit vectors $\mathbf{i}$, $\mathbf{j}$, $\mathbf{k}$,

$$\mathbf{i} \cdot \mathbf{j} = \mathbf{i} \cdot \mathbf{k} = \mathbf{j} \cdot \mathbf{k} = 0$$

## Calculating $\mathbf{a} \cdot \mathbf{b}$ in Cartesian form

When $\mathbf{a} = \begin{bmatrix} x_1 \\ y_1 \\ z_1 \end{bmatrix}$ and $\mathbf{b} = \begin{bmatrix} x_2 \\ y_2 \\ z_2 \end{bmatrix}$ then $\mathbf{a} \cdot \mathbf{b} = (x_1 x_2 + y_1 y_2 + z_1 z_2)$

You can see why this is true by writing the dot product in $\mathbf{i}$, $\mathbf{j}$, $\mathbf{k}$ notation:

$$(x_1 \mathbf{i} + y_1 \mathbf{j} + z_1 \mathbf{k}) \cdot (x_2 \mathbf{i} + y_2 \mathbf{j} + z_2 \mathbf{k})$$

$$= x_1 x_2 \mathbf{i} \cdot \mathbf{i} + x_1 y_2 \, \mathbf{i} \cdot \mathbf{j} + x_1 z_2 \mathbf{i} \cdot \mathbf{k} + y_1 x_2 \mathbf{j} \cdot \mathbf{i} + y_1 y_2 \mathbf{j} \cdot \mathbf{j} + y_1 z_2 \mathbf{j} \cdot \mathbf{k} + z_1 x_2 \mathbf{k} \cdot \mathbf{i} + z_1 y_2 \mathbf{k} \cdot \mathbf{j} + z_1 z_2 \mathbf{k} \cdot \mathbf{k}$$

$$= x_1 x_2 + y_1 y_2 + z_1 z_2$$

For example, $\begin{bmatrix} 2 \\ -3 \\ 4 \end{bmatrix} \cdot \begin{bmatrix} 1 \\ 3 \\ -2 \end{bmatrix} = (2)(1) + (-3)(3) + (4)(-2) = -15$

## Example 11

**Question**

Find the scalar product of $\mathbf{a} = \begin{bmatrix} 2 \\ -3 \\ 5 \end{bmatrix}$ and $\mathbf{b} = \begin{bmatrix} 1 \\ -3 \\ 1 \end{bmatrix}$.

Hence find the cosine of the angle between $\mathbf{a}$ and $\mathbf{b}$.

**Answer**

$$\mathbf{a} \cdot \mathbf{b} = (2)(1) + (-3)(-3) + (5)(1) = 16$$

But $\mathbf{a} \cdot \mathbf{b} = |\mathbf{a}||\mathbf{b}| \cos \theta$

$$|\mathbf{a}| = \sqrt{4+9+25} = \sqrt{38} \text{ and } |\mathbf{b}| = \sqrt{1+9+1} = \sqrt{11}$$

Hence $\cos \theta = \dfrac{\mathbf{a} \cdot \mathbf{b}}{|\mathbf{a}||\mathbf{b}|} = \dfrac{16}{\sqrt{11}\sqrt{38}} = \dfrac{16}{\sqrt{418}}$.

## Example 12

**Question**

Given $\mathbf{a} = \begin{bmatrix} 10 \\ -3 \\ 5 \end{bmatrix}$, $\mathbf{b} = \begin{bmatrix} 2 \\ 6 \\ -3 \end{bmatrix}$ and $\mathbf{c} = \begin{bmatrix} 1 \\ 10 \\ -2 \end{bmatrix}$, show that $\mathbf{a} \cdot \mathbf{b} + \mathbf{a} \cdot \mathbf{c} = \mathbf{a} \cdot (\mathbf{b} + \mathbf{c})$.

$$\mathbf{a} \cdot \mathbf{b} = (10)(2) + (-3)(6) + (5)(-3) = -13$$

$$\mathbf{a} \cdot \mathbf{c} = (10)(1) + (-3)(10) + (5)(-2) = -30$$

$$\mathbf{b} + \mathbf{c} = \begin{bmatrix} 3 \\ 16 \\ -5 \end{bmatrix}$$

Therefore $\mathbf{a} \cdot (\mathbf{b} + \mathbf{c}) = (10)(3) + (-3)(16) + (5)(-5) = -43$

and $\quad\quad \mathbf{a} \cdot \mathbf{b} + \mathbf{a} \cdot \mathbf{c} = -13 - 30 = -43$

Therefore $\mathbf{a} \cdot \mathbf{b} + \mathbf{a} \cdot \mathbf{c} = \mathbf{a} \cdot (\mathbf{b} + \mathbf{c})$.

## Example 13

Find the acute angle between the lines $\mathbf{r} = \begin{bmatrix} 2 \\ -1 \\ 4 \end{bmatrix} + \lambda \begin{bmatrix} 1 \\ 1 \\ -2 \end{bmatrix}$ and $\mathbf{r} = \begin{bmatrix} 3 \\ 4 \\ -7 \end{bmatrix} + \mu \begin{bmatrix} -2 \\ -5 \\ 13 \end{bmatrix}$.

The angle between the lines is the angle between their directions,

that is the angle between the vectors $\begin{bmatrix} 1 \\ 1 \\ -2 \end{bmatrix}$ and $\begin{bmatrix} -2 \\ -5 \\ 13 \end{bmatrix}$ which can be found

using the scalar product.

If $\theta$ is the angle between the lines, then

$$\begin{bmatrix} 1 \\ 1 \\ -2 \end{bmatrix} \cdot \begin{bmatrix} -2 \\ -5 \\ 13 \end{bmatrix} = \left( \sqrt{1^2 + 1^2 + (-2)^2} \times \sqrt{(-2)^2 + (-5)^2 + 13^2} \right) \cos \theta$$

$$\Rightarrow \quad -33 = \left( \sqrt{6} \right) \left( \sqrt{198} \right) \cos \theta$$

$$\Rightarrow \quad \cos \theta = -\frac{33}{6\sqrt{33}}, \text{ so } \theta = \cos^{-1} -\frac{\sqrt{33}}{6} = 162.2°$$

so the acute angle between the lines is $16.8°$ to 1 decimal place.

# 9.9 The coordinates of the foot of the perpendicular from a point to a line

To find the coordinates of D, the foot of the perpendicular from a point Q to a line $\mathbf{r} = \mathbf{a} + t\mathbf{b}$, we can use the fact that the angle between the vector from Q to D is 90°. The scalar product of a vector from Q to a general point P on the line is $\overrightarrow{QP} \cdot \mathbf{b}$, so D is the point where $\overrightarrow{QP} \cdot \mathbf{b} = 0$.

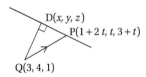

For example, an equation of a line is $\mathbf{r} = \begin{bmatrix} 1 \\ 0 \\ 3 \end{bmatrix} + t \begin{bmatrix} 2 \\ 1 \\ 1 \end{bmatrix} = \begin{bmatrix} 1+2t \\ t \\ 3+t \end{bmatrix}$

The coordinates of a general point P on the line are $(1 + 2t, t, 3 + t)$.

The point Q(3, 4, 1) is not on the line and D(x, y, z) is the foot of the perpendicular from Q to the line.

$$\overrightarrow{QP} = \begin{bmatrix} 3-(2t+1) \\ 4-t \\ 1-(3+t) \end{bmatrix} = \begin{bmatrix} 2-2t \\ 4-t \\ -2-t \end{bmatrix} \text{ and } \begin{bmatrix} 2 \\ 1 \\ 1 \end{bmatrix} \text{ is parallel to the line.}$$

Therefore $\overrightarrow{QP} \cdot \begin{bmatrix} 2 \\ 1 \\ 1 \end{bmatrix} = \begin{bmatrix} 2-2t \\ 4-t \\ -2-t \end{bmatrix} \cdot \begin{bmatrix} 2 \\ 1 \\ 1 \end{bmatrix} = 2(2-2t)+(4-t)+(-2-t)=6-6t.$

When the scalar product is zero, $6-6t=0$, so $t=1$.

Therefore D is the point on the line where $t=1$

So the coordinates of D are $(3, 1, 4)$.

## The distance of a point from a line

To find the distance between the foot $D(x, y, z)$ of the perpendicular from a point $Q(a, b, c)$ to a line, first find the coordinates of D.

Then the distance between Q and D is $\sqrt{(a-x)^2+(b-y)^2+(b-z)^2}$

## Exercise 6

**1** Calculate $\mathbf{m} \cdot \mathbf{n}$ when

**a** $\mathbf{m} = \begin{bmatrix} 2 \\ -4 \\ 5 \end{bmatrix}$ and $\mathbf{n} = \begin{bmatrix} 1 \\ 3 \\ 8 \end{bmatrix}$      **b** $\mathbf{m} = \begin{bmatrix} 3 \\ -7 \\ 2 \end{bmatrix}$ and $\mathbf{n} = \begin{bmatrix} 5 \\ 1 \\ -4 \end{bmatrix}$

**c** $\mathbf{m} = \begin{bmatrix} 2 \\ -3 \\ 6 \end{bmatrix}$ and $\mathbf{n} = \begin{bmatrix} 1 \\ 1 \\ 0 \end{bmatrix}$.

**2** Find $\mathbf{p} \cdot \mathbf{q}$ and the cosine of the angle between $\mathbf{p}$ and $\mathbf{q}$ when

**a** $\mathbf{p} = \begin{bmatrix} 2 \\ 4 \\ 1 \end{bmatrix}$ and $\mathbf{q} = \begin{bmatrix} 1 \\ 1 \\ 1 \end{bmatrix}$      **b** $\mathbf{p} = \begin{bmatrix} -1 \\ 3 \\ -2 \end{bmatrix}$ and $\mathbf{q} = \begin{bmatrix} 1 \\ 1 \\ -6 \end{bmatrix}$

**c** $\mathbf{p} = \begin{bmatrix} -2 \\ 5 \\ 0 \end{bmatrix}$ and $\mathbf{q} = \begin{bmatrix} 1 \\ 1 \\ 0 \end{bmatrix}$      **d** $\mathbf{p} = \begin{bmatrix} 2 \\ 1 \\ 0 \end{bmatrix}$ and $\mathbf{q} = \begin{bmatrix} 0 \\ 1 \\ -2 \end{bmatrix}$.

**3** The cosine of the angle between two vectors $\mathbf{v}_1$ and $\mathbf{v}_2$ is $\dfrac{4}{21}$

where $\mathbf{v}_1 = \begin{bmatrix} 6 \\ 3 \\ -2 \end{bmatrix}$ and $\mathbf{v}_2 = \begin{bmatrix} -2 \\ \lambda \\ -4 \end{bmatrix}$.

Find the positive value of $\lambda$.

**4** In a triangle ABC, $\overrightarrow{AB} = \begin{bmatrix} 1 \\ 2 \\ 3 \end{bmatrix}$ and $\overrightarrow{BC} = \begin{bmatrix} -1 \\ 4 \\ 0 \end{bmatrix}$.

**a** Find the cosine of angle ABC.

**b** Find the vector $\overrightarrow{AC}$ and use it to calculate the angle BAC.

**5** Show that $\begin{bmatrix} 1 \\ 7 \\ 3 \end{bmatrix}$ is perpendicular to both $\begin{bmatrix} 1 \\ -1 \\ 2 \end{bmatrix}$ and $\begin{bmatrix} 2 \\ 1 \\ -3 \end{bmatrix}$.

**6** Show that $\begin{bmatrix} 13 \\ 23 \\ 7 \end{bmatrix}$ is perpendicular to both $\begin{bmatrix} 2 \\ 1 \\ -7 \end{bmatrix}$ and $\begin{bmatrix} 3 \\ -2 \\ 1 \end{bmatrix}$.

**7** The magnitudes of two vectors **p** and **q** are 5 and 4 units respectively.
The angle between **p** and **q** is 30°.

  **a**  Find $\mathbf{p} \cdot \mathbf{q}$

  **b**  Find the magnitude of the vector $\mathbf{p} - \mathbf{q}$.

**8** Calculate the angle between the vectors $\begin{bmatrix} 2 \\ -1 \\ 3 \end{bmatrix}$ and $\begin{bmatrix} 0 \\ -1 \\ -1 \end{bmatrix}$.

**9** The diagram shows a cube where the length of each edge is 4 cm.

  **a**  Express in the form $\begin{bmatrix} a \\ b \\ c \end{bmatrix}$ the vectors

    **i**  $\overrightarrow{AE}$         **ii**  $\overrightarrow{AG}$.

  **b**  H is the midpoint of AB.

    Find the angles of the triangle AEH, giving your answers to the
nearest degree.

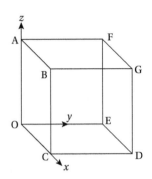

**10** Is it true to say that if $\mathbf{a} \cdot \mathbf{b} = 0$ then either $\mathbf{a} = \mathbf{0}$ or $\mathbf{b} = \mathbf{0}$?

Explain your answer.

**11** In triangle OAB, O is the origin, $\overrightarrow{OA} = \begin{bmatrix} 4 \\ -3 \\ 4 \end{bmatrix}$ and $\overrightarrow{OB} = \begin{bmatrix} 1 \\ 6 \\ -2 \end{bmatrix}$.

  **a**  Show that triangle OAB is isosceles.

  **b**  Find angle AOB correct to the nearest degree.

  **c**  Hence or otherwise find the area of triangle OAB.

**12** Find the acute angle between the lines

  **a**  $\mathbf{r} = \begin{bmatrix} 1 \\ -1 \\ 1 \end{bmatrix} + \lambda \begin{bmatrix} 3 \\ -4 \\ 1 \end{bmatrix}$ and $\mathbf{r} = \mu \begin{bmatrix} -9 \\ 12 \\ -3 \end{bmatrix}$

  **b**  $\mathbf{r} = \begin{bmatrix} 4-t \\ 8-2t \\ 3-t \end{bmatrix}$ and $\mathbf{r} = \begin{bmatrix} 7+6s \\ 6+4s \\ 5+5s \end{bmatrix}$

  **c**  $\mathbf{r} = \begin{bmatrix} 1 \\ 0 \\ 3 \end{bmatrix} + \lambda \begin{bmatrix} 2 \\ 1 \\ 1 \end{bmatrix}$ and $\mathbf{r} = \begin{bmatrix} 2 \\ -1 \\ 1 \end{bmatrix} + \mu \begin{bmatrix} 1 \\ -2 \\ 0 \end{bmatrix}$.

**13** The line $l$ has equation $\mathbf{r} = \begin{bmatrix} 1 \\ 0 \\ 3 \end{bmatrix} + \lambda \begin{bmatrix} 2 \\ 1 \\ 1 \end{bmatrix}$. The coordinates of a point A are (2, 1, 5).

    **a** Show that A does not lie on the line $l$.

    **b** The point B on the line $l$ is the foot of the perpendicular from A to $l$. Find the coordinates of B.

    **c** Hence find the perpendicular distance of the point A from the line $l$.

**14** The line $l$ has equation $\mathbf{r} = \begin{bmatrix} 1 \\ -1 \\ 1 \end{bmatrix} + \lambda \begin{bmatrix} 3 \\ -4 \\ 1 \end{bmatrix}$. The coordinates of a point A are (−2, 1, 0).

    **a** Show that A does not lie on the line $l$.

    **b** The point B on the line $l$ is the foot of the perpendicular from A to $l$. Find the coordinates of B.

    **c** Hence find the perpendicular distance of the point A from the line $l$.

## Summary

### Vectors

A vector is a quantity with both magnitude and direction. It can be represented by a straight line segment whose length represents the magnitude and whose direction represents the direction of the vector.

When lines representing several vectors are drawn 'head to tail' in order; then the line (in the opposite sense) which completes a closed polygon, represents the sum of the vectors and is called the resultant vector.

A position vector has a fixed location in space.

### Cartesian vectors

▶ The coordinates of a point P in three dimensions are $(x, y, z)$.

▶ The position vector of P is given by $\overline{OP} = \begin{bmatrix} x \\ y \\ z \end{bmatrix}$.

▶ For vectors $\mathbf{v}_1 = \begin{bmatrix} a \\ b \\ c \end{bmatrix}$ and $\mathbf{v}_2 = \begin{bmatrix} p \\ q \\ r \end{bmatrix}$

$$\mathbf{v}_1 + \mathbf{v}_2 = \begin{bmatrix} a + p \\ b + q \\ c + r \end{bmatrix}$$

the magnitude of is $\mathbf{v}_1$ written as $|\mathbf{v}_1|$ and is equal to $\sqrt{a^2 + b^2 + c^2}$

$\mathbf{v}_1$ and $\mathbf{v}_2$ are parallel when $\mathbf{v}_1 = \lambda \mathbf{v}_2$

$\mathbf{v}_1$ and $\mathbf{v}_2$ are equal when $a = p$, $b = q$ and $c = r$.

# Equation of a line

▶ A line passing through a point with position vector **a** and in the direction of the vector **b** has vector equation $\mathbf{r} = \mathbf{a} + t\mathbf{b}$, where **r** is the position vector of a point on the line and $t$ is a parameter.

▶ Two lines with equations $\mathbf{r}_1 = \mathbf{a}_1 + \lambda\mathbf{b}_1$ and $\mathbf{r}_2 = \mathbf{a}_2 + \mu\mathbf{b}_2$

 are parallel if $\mathbf{b}_1$ is a multiple of $\mathbf{b}_2$

 intersect if there are values of $\lambda$ and $\mu$ for which $\mathbf{r}_1 = \mathbf{r}_2$

 are skew (do not intersect) in all other cases.

# The scalar product of two vectors

▶ If $\theta$ is the angle between two vectors $\mathbf{a} = \begin{bmatrix} x_1 \\ y_1 \\ z_1 \end{bmatrix}$ and $\mathbf{b} = \begin{bmatrix} x_2 \\ y_2 \\ z_2 \end{bmatrix}$ then

 $\mathbf{a} \cdot \mathbf{b} = |\mathbf{a}|\,|\mathbf{b}| \cos\theta$

 If **a** and **b** are perpendicular then $\mathbf{a} \cdot \mathbf{b} = 0$.

 $\mathbf{a} \cdot \mathbf{b} = x_1 x_2 + y_1 y_2 + z_1 z_2$

# Review

**1** ABCD is a square. Write down the single vector equivalent to

 **a**  $\overrightarrow{AB} + \overrightarrow{BC}$

 **b**  $\overrightarrow{AB} + \overrightarrow{BD}$

**2** OABC are four points such that $\overrightarrow{OA} = \mathbf{a}$, $\overrightarrow{OB} = \mathbf{b}$ and $\overrightarrow{OC} = \mathbf{c}$.

 Express in terms of **a**, **b** and **c**

 **a**  $\overrightarrow{AB}$          **b**  $\overrightarrow{BC}$          **c**  $\overrightarrow{AC}$.

**3** The vector **r** is the position vector of the point P where $\mathbf{r} = \begin{bmatrix} 2 \\ 1 \\ 0 \end{bmatrix}$.

 Write down the coordinates of P.

**4** Given $\overrightarrow{OA} = \begin{bmatrix} 3 \\ 1 \\ -1 \end{bmatrix}$ and $\overrightarrow{OB} = \begin{bmatrix} 5 \\ -1 \\ 0 \end{bmatrix}$ find

 **a**  $\left|\overrightarrow{AB}\right|$

 **b**  a vector that is parallel to $\overrightarrow{OA} - \overrightarrow{OB}$

 **c**  the position vector of the midpoint of AB.

**5** A line passes through the point with position vector $\begin{bmatrix} 2 \\ 1 \\ 0 \end{bmatrix}$ and is parallel to the vector $\begin{bmatrix} -1 \\ 1 \\ -2 \end{bmatrix}$.

 Find a vector equation of the line.

**6** A line $l$ passes through the points A and B whose coordinates are $(3, 2, -1)$ and $(0, -2, 4)$ respectively.

   **a** Find a vector equation of the line $l$.

   **b** Find the coordinates of the point where $l$ crosses the $xz$-plane.

**7** Two points A and B are on the line $l_1$ where $\overrightarrow{OA} = \begin{bmatrix} 3 \\ -1 \\ 2 \end{bmatrix}$ and $\overrightarrow{OB} = \begin{bmatrix} -1 \\ 1 \\ 9 \end{bmatrix}$.

   The equation of a line $l_2$ is $\mathbf{r} = \begin{bmatrix} 8 \\ 1 \\ -6 \end{bmatrix} + t \begin{bmatrix} 1 \\ -2 \\ -2 \end{bmatrix}$.

   **a** Show that $l_1$ and $l_2$ intersect.

   **b** Find the coordinates of the point of intersection.

**8** Calculate $\mathbf{a} \cdot \mathbf{b}$ where $\mathbf{a} = \begin{bmatrix} -2 \\ 4 \\ -1 \end{bmatrix}$ and $\mathbf{b} = \begin{bmatrix} 3 \\ 1 \\ -2 \end{bmatrix}$.

**9** Calculate the angle between the vectors $\mathbf{a}$ and $\mathbf{b}$ where

   $\mathbf{a} = \begin{bmatrix} 5 \\ 2 \\ -1 \end{bmatrix}$     $\mathbf{b} = \begin{bmatrix} -1 \\ 2 \\ -2 \end{bmatrix}$.

**10** Calculate the acute angle between the lines whose vector equations are

   $\mathbf{r} = \begin{bmatrix} 6 \\ -3 \\ 2 \end{bmatrix} + t \begin{bmatrix} 1 \\ 1 \\ -5 \end{bmatrix}$     $\mathbf{r} = \begin{bmatrix} -2 \\ 7 \\ 7 \end{bmatrix} + s \begin{bmatrix} 4 \\ 2 \\ -5 \end{bmatrix}$.

**11** The line $l$ has equation $\mathbf{r} = \begin{bmatrix} 1 \\ 3 \\ -1 \end{bmatrix} + t \begin{bmatrix} 1 \\ 3 \\ -4 \end{bmatrix}$. The coordinates of a point A are $(1, 1, 4)$.

   **a** The point B on the line $l$ is the foot of the perpendicular from A to $l$. Find the coordinates of B.

   **b** Hence find the perpendicular distance of the point A from the line $l$.

## Assessment

**1** The diagram shows a cube where the length of each edge is 2 cm.

   **a** Express in the form $\begin{bmatrix} a \\ b \\ c \end{bmatrix}$ the vectors

   **i** $\overrightarrow{AD}$

   **ii** $\overrightarrow{CE}$.

   **b** Find the angle between $\overrightarrow{AD}$ and $\overrightarrow{CE}$.

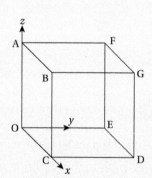

**2** The points A, B and C have coordinates (2, 1, 0), (4, 0, 3) and (1, 1, 5) respectively.

   **a** Find a vector equation of the line through A and B.

   **b** Find a vector equation of the line through A and C.

   **c** Find the acute angle between the lines AB and AC.

**3** The position vectors of the points A and B are $\begin{bmatrix} 13 \\ -4 \\ 2 \end{bmatrix}$ and $\begin{bmatrix} 18 \\ -4 \\ 3 \end{bmatrix}$ respectively.

   **a** Find a vector equation of the line through A and B.

   **b** Show that the vector $\begin{bmatrix} 1 \\ 6 \\ -5 \end{bmatrix}$ is perpendicular to the line through A and B.

**4** The line $l$ has equation $\mathbf{r} = \begin{bmatrix} 1+3s \\ 1+2s \\ -s \end{bmatrix}$. The coordinates of a point A are (−2, 2, 1).

Find the perpendicular distance of the point A from the line $l$.

**5** The triangle ABC is such that the coordinates A and B are (1, 0, 2) and (2, 1, −1) respectively.

The coordinates of C are ($a$, $b$, $c$) and angle CBA is a right angle. Find a relationship between $a$, $b$ and $c$.

**6** The points A and B have coordinates (3, 2, 10) and (5, −2, 4) respectively.

The line $l$ passes through A and has equation $\mathbf{r} = \begin{bmatrix} 3 \\ 2 \\ 10 \end{bmatrix} + \lambda \begin{bmatrix} 3 \\ 1 \\ -2 \end{bmatrix}$.

   **a** Find the acute angle between $l$ and the line AB.

   **b** The point C lies on $l$ such that angle ABC is 90°.

      Find the coordinates of C.

   **c** The point D is such that BD is parallel to AC and angle BCD is 90°. The point E lies on the line through B and D and is such that the length of DE is half that of AC.

      Find the coordinates of the two possible positions of E.

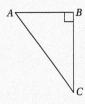

AQA MPC4 June 2015

**7** The points A and B have coordinates (5, 1, −2) and (4, −1, 3) respectively.

The line $l$ has equation $\mathbf{r} = \begin{bmatrix} -8 \\ 5 \\ -6 \end{bmatrix} + \mu \begin{bmatrix} 5 \\ 0 \\ -2 \end{bmatrix}$.

   **a** Find a vector equation of the line that passes through A and B.

   **b** **i** Show that the line that passes through A and B intersects the line $l$, and find the coordinates of the point of intersection, P.

      **ii** The point C lies on $l$ such that triangle PBC has a right angle at B. Find the coordinates of C.

AQA MPC4 June 2011

# 10 Mathematical Modelling and Kinematics

## Introduction

Mechanical problems involving people, cars, tow ropes, and so on need to be simplified to give accurate results. This chapter looks at those simplifications and extends the work on **kinematics** started in the AS book.

## Recap

You will need to remember...

▶ Newton's laws of motion, $F = ma$, and 'action and reaction are equal and opposite'.

▶ The tension in a string is the same at each end.

▶ The acceleration of each particle is the same in a system of connected particles.

▶ The equations for motion in a straight line with constant acceleration. M1 Chapter 17

▶ Vectors can be added or subtracted using the triangle law and a vector can be multiplied by a scalar. P2 Chapter 9

▶ When $s$, $v$ and $a$ are functions of time, $v = \dfrac{ds}{dt} \Rightarrow s = \int v \, dt$ and $a = \dfrac{dv}{dt} \Rightarrow v = \int a \, dt$.

▶ The differentials of polynomials, rational, exponential and trigonometric functions. P2

## Objectives

By the end of this chapter, you should know how to...

▶ Use assumptions to simplify real-life situations so that a mathematical model can be used to describe them.

▶ Use the vector notation $a\mathbf{i} + b\mathbf{j} + c\mathbf{k}$.

▶ Use vectors to solve problems of motion with variable acceleration in two and three dimensions.

## Applications

Rockets that launch satellites move in three dimensions with acceleration that varies with respect to time. When the acceleration can be expressed as a function of time, it is possible to work out the speed and distance of the rocket from the launch pad at a given instant after launch.

### Note

Throughout this module, unless specified, the value of $g$ is taken as 9.8 and answers are given correct to 3 significant figures or for angles correct to 1 decimal place.

## 10.1 Mathematical modelling

**Mathematical modelling** uses mathematics to solve real-life problems. In M1 problems involved 'real-life' situations such as cars pulling trailers and balls being dropped and so on. Some simplifications were made because a solution is more difficult when all the complications of size, irregular shape, air resistance, and so on, are included. So that mathematical techniques based on known physical laws can be used easily and directly, the effect of certain quantities is ignored or simplified. For example, a ball is treated as a particle, air resistance is ignored, strings are weightless and inextensible, pegs and pulleys offer no friction to strings moving over them and so on. When we do this we are making assumptions.

When a problem is solved using a mathematical model, all the assumptions should be stated clearly at the start. It is also sensible to judge whether these assumptions are reasonable in the context of the problem. The results found from such a model can only be approximate but they are accurate enough for most purposes.

When a practical problem has been modelled it becomes a simplified mathematical exercise. The diagram used in the solution does not need to be realistic; instead of large objects such as trees, vehicles, planks, crates, and so on, it can be made up of points (for example, particles, balls), blocks (for example, cars) and lines (for example, rods or ropes). Then forces, velocities, dimensions, and so on, can be marked clearly.

## Example 1

A tractor of mass 1000 kg is pulling a trailer of mass 750 kg. The tractor exerts a steady driving force of 5000 N.

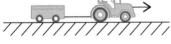

a   Construct a mathematical model stating all assumptions made and comment on the reasonableness of the assumptions.

b   Find the acceleration of the trailer.

c   Find the tension in the tow bar.


a   Model the tractor and the trailer each as a small block. Assume no resistance to motion. Assume that the tow bar is horizontal, light and does not stretch.

The first two assumptions depend on the conditions such as the friction in wheel bearings, the shape of the vehicles and the strength and direction of the wind.

The last assumption is reasonable as the mass of the tow bar is small compared to the mass of the tractor and of the car.

The assumption that the tow bar is light and does not stretch means that the acceleration of the tractor and trailer are equal and that the tension acting on the tractor and trailer are equal in magnitude and opposite in direction. These assumptions are shown in the diagram.

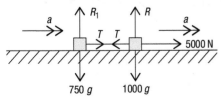

The tractor and trailer have the same acceleration.

For the trailer    $F = ma$   gives                $T = 750a$    [1]

For the tractor    $F = ma$   gives    $5000 - T = 1000a$   [2]

*(continued)*

**b** [1] + [2] gives $\qquad 5000 = 1750a$

The acceleration of the trailer is 2.86 ms$^{-2}$ (3 significant figures).

**c** From [1] $\qquad T = 750\left(\dfrac{5000}{1750}\right) = 2140$ (3 significant figures)

The tension in the tow rope is 2140 N.

## Exercise 1

In each question state any assumptions you make in order to form a mathematical model that can be used to solve the problem. Also state whether these assumptions are reasonable. Give answers in terms of $g$ where appropriate.

1. A lift of mass 500 kg carrying a load of 80 kg is drawn vertically up by a cable. The lift first accelerates at 1.5 ms$^{-2}$ from rest to its maximum speed which is maintained for a time, after which the lift decelerates to rest at 0.1 ms$^{-2}$. For each of these three stages of motion find

   **a** the tension in the cable

   **b** the force exerted by the load on the floor of the lift.

2. A car of mass 1 tonne is pulling a caravan of mass 800 kg along a level straight road. The resistance to motion on the car is 250 N and the resistance to motion on the caravan is 200 N. The combination accelerates uniformly from rest to 20 ms$^{-1}$ in 12.5 seconds.

   **a** Find the tension in the tow bar.

   **b** Find the driving force exerted by the car's engine.

3. In question 2 the tow bar snaps at the instant when the speed reaches 20 ms$^{-1}$ and the car continues with the same driving force.

   **a** Find the subsequent acceleration of the car.

   **b** Find the deceleration of the caravan.

   **c** Find how long it takes for the caravan to stop.

4. A car of mass 800 kg exerts a driving force of 2200 N and is pulling a trailer of mass 300 kg along a level road. There is no resistance to the motion of either the car or the trailer. Find the acceleration of the car and the tension in the towbar.

5. A lift of mass 800 kg is operated by a cable as shown in the diagram. A passenger of mass 70 kg is standing in the lift and the lift is accelerating upwards at 2ms$^{-2}$. Find, stating what object you can use to represent the passenger,

   **a** the force exerted by the passenger on the floor of the lift

   **b** the tension in the cable.

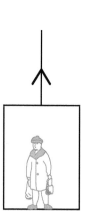

6. A brick of mass 500 kg and a crate of mass 750 kg are attached to each end of a rope passing over a pulley.

   The system is released from rest with the brick and the crate hanging freely. Find the tension in the rope.

7. A car of mass 300 kg has a trailer of mass 300 kg attached by a horizontal tow bar. When the car and trailer are traveling at 10 ms$^{-1}$ the brakes are applied bringing them to rest in 5 seconds.

   **a** Find the deceleration.

   **b** Find the thrust in the tow bar.

# 10.2 Vectors and kinematics

## Cartesian unit vectors

Vectors were introduced in Chapter 9. There are two ways of representing vectors in a plane or in three dimensions using a Cartesian frame of reference. Column vectors are used in the P2 module.

In this module, we use the **i**, **j**, **k** notation and a repeat of the introduction is included here.

A **unit vector** is a vector whose magnitude is one unit.

When   **i** is a unit vector in the direction of O$x$

        **j** is a unit vector in the direction of O$y$

        **k** is a unit vector in the direction of O$z$

then the position vector, relative to O, of any point P can be given in terms of **i**, **j** and **k**, so the point distant

        3 units from O in the direction O$x$
        4 units from O in the direction O$y$
        5 units from O in the direction O$z$

has coordinates (3, 4, 5) and position vector $\overrightarrow{OP} = 3\mathbf{i} + 4\mathbf{j} + 5\mathbf{k}$.

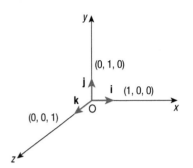

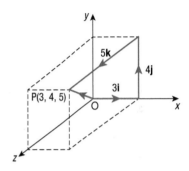

For work in two dimensions only the vectors **i** and **j** are used.

To add or subtract vectors given in **i**, **j**, **k** form, the coefficients of **i**, **j** and **k** are added separately.

For example when   $\mathbf{v}_1 = 3\mathbf{i} + 2\mathbf{j} + 2\mathbf{k}$   and   $\mathbf{v}_2 = \mathbf{i} + 2\mathbf{j} - 3\mathbf{k}$

then
$$\mathbf{v}_1 + \mathbf{v}_2 = (3\mathbf{i} + 2\mathbf{j} + 2\mathbf{k}) + (\mathbf{i} + 2\mathbf{j} - 3\mathbf{k})$$
$$= (3+1)\mathbf{i} + (2+2)\mathbf{j} + (2-3)\mathbf{k} = 4\mathbf{i} + 4\mathbf{j} - \mathbf{k}$$

And   $\mathbf{v}_1 - \mathbf{v}_2 = (3-1)\mathbf{i} + (2-2)\mathbf{j} + (2-\{-3\})\mathbf{k} = 2\mathbf{i} + 5\mathbf{k}$

The modulus of **v**, where $\mathbf{v} = 12\mathbf{i} - 3\mathbf{j} + 4\mathbf{k}$, is the length of OP where P is the point (12, –3, 4) so

$$OP = \sqrt{12^2 + 3^2 + 4^2} = 13$$

In general, when $\mathbf{v} = a\mathbf{i} + b\mathbf{j} + c\mathbf{k}$, $|\mathbf{v}| = \sqrt{a^2 + b^2 + c^2}$.

Two vectors $\mathbf{v}_1$ and $\mathbf{v}_2$ are parallel when $\mathbf{v}_1 = \lambda\mathbf{v}_2$.

For example $2\mathbf{i} - 3\mathbf{j} - \mathbf{k}$ is parallel to $4\mathbf{i} - 6\mathbf{j} - 2\mathbf{k}$ ($\lambda = 2$)
   and $\mathbf{i} + \mathbf{j} + \mathbf{k}$ is parallel to $-3\mathbf{i} - 3\mathbf{j} - 3\mathbf{k}$ ($\lambda = -3$).

## Exercise 2

1 Write down, in the form $a\mathbf{i} + b\mathbf{j} + c\mathbf{k}$, the vector represented by $\overrightarrow{OP}$ where P is the point with coordinates (3, 6, 4).

2 $\overrightarrow{OP}$ represents a vector **r**. Write down the coordinates of P when $\mathbf{r} = 5\mathbf{i} - 7\mathbf{j} + 2\mathbf{k}$.

3 Find the length of the line OP where $\overrightarrow{OP} = 2\mathbf{i} - \mathbf{j} + 4\mathbf{k}$.

4 Given $\mathbf{a} = \mathbf{i} + \mathbf{j} + \mathbf{k}$, $\mathbf{b} = 2\mathbf{i} - \mathbf{j} + 3\mathbf{k}$, $\mathbf{c} = -\mathbf{i} + 3\mathbf{j} - \mathbf{k}$ find

  **a**  $\mathbf{a} + \mathbf{b}$          **b**  $\mathbf{a} - \mathbf{c}$

5. The triangle ABC has its vertices at the points A(−1, 3, 0), B(−3, 0, 7), C(−1, 2, 3). Find in the form $a\mathbf{i} + b\mathbf{j} + c\mathbf{k}$ the vectors representing

a  $\overrightarrow{AB}$     b  $\overrightarrow{AC}$     c  $\overrightarrow{CB}$

## Motion in two or three dimensions with variable acceleration

When a particle is moving in a plane its motion in two perpendicular directions can be dealt with separately. When displacement, velocity and acceleration are functions of time, the calculus methods used in M1 can be applied to the components in each direction.

For example, look at the motion of the particle shown in the diagram.

In the direction O$x$, at any time $t$, $x = t^2$

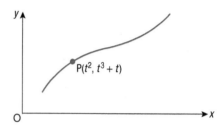

so        the displacement from O is $t^2$,

the velocity is $2t$

and       the acceleration is 2.

In the direction O$y$, $y = t^3 + t$

so        the displacement from O is $t^3 + t$,

the velocity is $3t^2 + 1$

and       the acceleration is $6t$.

These components can be expressed in terms of unit vectors $\mathbf{i}$ and $\mathbf{j}$ in the chosen directions and, by adding them, form a resultant vector.

Therefore the position vector of P is denoted by $\mathbf{r}$, and is given by $\mathbf{r} = t^2\,\mathbf{i} + (t^3 + t)\mathbf{j}$.

Each component of $\mathbf{v}$ is obtained by differentiating the corresponding component of $\mathbf{r}$ with respect to $t$, so

the velocity vector, $\mathbf{v}$, is $2t\mathbf{i} + (3t^2 + 1)\mathbf{j}$

and similarly the acceleration vector, $\mathbf{a}$, is $2\mathbf{i} + 6t\mathbf{j}$.

In general    $\mathbf{v} = \dfrac{d\mathbf{r}}{dt}$    and    $\mathbf{r} = \displaystyle\int \mathbf{v}\,dt$

Similarly    $\mathbf{a} = \dfrac{d\mathbf{v}}{dt}$    and    $\mathbf{v} = \displaystyle\int \mathbf{a}\,dt$

> **Note**
>
> The speed is equal to $|\mathbf{v}|$

For motion in three dimensions, when the position vector of a point is given by

$$\mathbf{r} = f(t)\mathbf{i} + g(t)\mathbf{j} + h(t)\mathbf{k}$$

then        $\mathbf{v} = f'(t)\mathbf{i} + g'(t)\mathbf{j} + h'(t)\mathbf{k}$

and        $\mathbf{a} = f''(t)\mathbf{i} + g''(t)\mathbf{j} + h''(t)\mathbf{k}$

## Example 2

A particle is moving in a plane in such a way that its velocity at any time $t$ is given by $2t\mathbf{i} + 3t^2\mathbf{j}$. Initially the position vector of the particle, relative to a fixed point O in the plane, is $5\mathbf{i} − 8\mathbf{j}$.

When $t = 3$, find

a  the speed of the particle

b  the acceleration of P

c  the position vector of P.

a When $t = 3$, $\mathbf{v} = 6\mathbf{i} + 27\mathbf{j}$.

$$|\mathbf{v}| = \sqrt{36 + 729} = \sqrt{765} = 27.65\ldots$$

Therefore the speed is 27.7 ms$^{-1}$.

b $\quad \mathbf{v} = 2t\mathbf{i} + 3t^2\mathbf{j}$

$$\Rightarrow \mathbf{a} = \frac{d\mathbf{v}}{dt} = 2\mathbf{i} + 6t\mathbf{j}$$

When $t = 3$, $\quad \mathbf{a} = 2\mathbf{i} + 18\mathbf{j}$

c $\quad \mathbf{r} = \int \mathbf{v}\, dt = \int (2t\mathbf{i} + 3t^2\mathbf{j})\, dt$

When a function is integrated, a constant of integration must be added. When integrating a vector function, the constant of integration must also be a vector. We will denote it by $\mathbf{A}$.

$$\mathbf{r} = t^2\mathbf{i} + t^3\mathbf{j} + \mathbf{A}$$

Initially (when $t = 0$) $\quad 5\mathbf{i} - 8\mathbf{j} = 0\mathbf{i} + 0\mathbf{j} + \mathbf{A}$

$\Rightarrow \qquad\qquad \mathbf{A} = 5\mathbf{i} - 8\mathbf{j}$

Therefore $\quad \mathbf{r} = t^2\mathbf{i} + t^3\mathbf{j} + 5\mathbf{i} - 8\mathbf{j} = (t^2 + 5)\mathbf{i} + (t^3 - 8)\mathbf{j}$

When $t = 3$ $\quad \mathbf{r} = (9 + 5)\mathbf{i} + (27 - 8)\mathbf{j} = 14\mathbf{i} + 19\mathbf{j}$

## Example 3

At any time $t$, the position vector of a particle moving in a plane, relative to a fixed point O in the plane, is $10t\mathbf{i} + (t^4 - 4t)\mathbf{j}$.

a Show that the particle has zero component of acceleration in the direction of $\mathbf{i}$.

b Find the time when the velocity is perpendicular to the acceleration.

c Find the Cartesian equation of the path of the particle.

a $\qquad \mathbf{r} = 10t\mathbf{i} + (t^4 - 4t)\mathbf{j}$

$$\Rightarrow \quad \mathbf{v} = \frac{d\mathbf{r}}{dt} = 10\mathbf{i} + (4t^3 - 4)\mathbf{j}$$

and $\quad \mathbf{a} = \frac{d\mathbf{v}}{dt} = 12t^2\mathbf{j}$

$\mathbf{a}$ has no term in $\mathbf{i}$, therefore the acceleration has no component in the direction of $\mathbf{i}$.

b The acceleration is always in the direction of $\mathbf{j}$. Therefore, in order to be perpendicular to the acceleration, the velocity must be parallel to $\mathbf{i}$.

$\mathbf{v} = 10\mathbf{i} + (4t^3 - 4)\mathbf{j}$

$\mathbf{v}$ is perpendicular to $\mathbf{a}$ when the coefficient of $\mathbf{j}$ is zero,

so when $\quad 4(t^3 - 1) = 0 \Rightarrow t = 1$

The velocity is perpendicular to the acceleration after 1 second.

c To find the Cartesian equation of the path we need a direct relationship between the coordinates of P at time $t$.

From $\mathbf{r} = 10t\mathbf{i} + (t^4 - 4t)\mathbf{j}$, the coordinates of the position of P at time $t$ are

$x = 10t$ $\qquad\qquad$ [1]

$y = (t^4 - 4t)$ $\qquad\qquad$ [2]

From [1] $\quad t = \dfrac{x}{10} \Rightarrow y = \dfrac{x^4}{10\,000} - \dfrac{2x}{5} = \dfrac{x^4 - 4000x}{10\,000}$

## Example 4

Question

A particle starts from a fixed point O and moves so that its velocity $t$ seconds after leaving O is given by

$$\mathbf{v} = e^{-t}\mathbf{i} + \mathbf{j} - (3t + t^2)\mathbf{k}$$

a  Find the speed of the particle as it leaves O.

b  Find the acceleration vector of the particle two seconds after it leaves O.

c  Find the distance of the particle from O two seconds after it leaves O.

Answer

a  When $t = 0$, $\mathbf{v} = \mathbf{i} + \mathbf{j}$  $\Rightarrow$  $|\mathbf{v}| = \sqrt{2}$

Therefore the speed of the particle as it leaves O is 1.41 ms$^{-1}$

b   $\mathbf{v} = e^{-t}\mathbf{i} + \mathbf{j} - (3t + t^2)\mathbf{k}$

$\Rightarrow$ $\mathbf{a} = -e^{-t}\mathbf{i} - (3 + 2t)\mathbf{k}$

When $t = 2$, $\mathbf{a} = -e^{-2}\mathbf{i} - 7\mathbf{k}$

c  At time $t$ the position vector of the particle is $\mathbf{r} = \int \mathbf{v}\,dt = \int (e^{-t}\mathbf{i} + \mathbf{j} - (3t + t^2)\mathbf{k})dt$

so   $\mathbf{r} = -e^{-t}\mathbf{i} + t\mathbf{j} - \left(\frac{3}{2}t^2 + \frac{1}{3}t^3\right)\mathbf{k} + \mathbf{A}$

When $t = 0$, $\mathbf{r} = 0$ so $\mathbf{A} = \mathbf{i}$, therefore $\mathbf{r} = (1 - e^{-t})\mathbf{i} + t\mathbf{j} - \left(\frac{3}{2}t^2 + \frac{1}{3}t^3\right)\mathbf{k}$

The position vector of the particle when $t = 2$ is $\mathbf{r} = (1 - e^{-2})\mathbf{i} + 2\mathbf{j} + \frac{26}{3}\mathbf{k}$

The distance of the particle from O when $t = 2$ is $|\mathbf{r}|$ when $t = 2$,

that is $|(1 - e^{-2})\mathbf{i} + 2\mathbf{j} - \frac{26}{3}\mathbf{k}| = \sqrt{\left(1 - e^{-2}\right)^2 + 4 + \frac{676}{9}} = \sqrt{79.85} = 8.94$ correct to 3 significant figures.

The particle is 8.94 m from O.

## Exercise 3

1  A particle P is moving in the $xy$ plane and at time $t$ the position vector of P is $\mathbf{r} = 2t^3\mathbf{i} + 3t^2\mathbf{j}$.

Find expressions for $\mathbf{v}$ and $\mathbf{a}$

  a  at any time $t$       b  when $t = 2$.

2  A particle P is moving in the $xy$ plane. Find the Cartesian equation of the path traced out by P when

  a  $\mathbf{r} = 2t\mathbf{i} + 3t^2\mathbf{j}$       b  $\mathbf{r} = (t + 1)^2\mathbf{i} + 4t\mathbf{j}$       c  $\mathbf{r} = 2t\mathbf{i} + \frac{3}{t}\mathbf{j}$

3  A particle P is moving in the $xy$ plane. Initially P is at rest at a point with position vector $3\mathbf{i} + \mathbf{j}$. Given that the acceleration of P after $t$ seconds is $\mathbf{i} - 2\mathbf{j}$, find expressions for $\mathbf{v}$ and $\mathbf{r}$ at any time $t$.

4  A particle P is moving in the $xy$ plane. At any time $t$, $\mathbf{v} = 3t^2\mathbf{i} + (t - 1)\mathbf{j}$. Given that P is initially at O, find

  a  the initial velocity       b  $\mathbf{a}$ when $t = 3$       c  $\mathbf{r}$ when $t = 2$.

5  A particle P is moving in three dimensions. At any time $t$, the position vector of P is given by $\mathbf{r} = (\cos t)\mathbf{i} + t^3\mathbf{j} - (\sin t)\mathbf{k}$.

  a  Find an expression for the velocity of P at any time $t$.

  b  Find the initial acceleration of P.

6. A particle P is moving in the $xy$ plane. When $t = 1$, $\mathbf{v} = \mathbf{i} + 3\mathbf{j}$ and $\mathbf{r} = 4\mathbf{i} - \mathbf{j}$.
The acceleration of P at any time $t$ is given by $\mathbf{a} = t\mathbf{i} + (2 - t)\mathbf{j}$.

   a. Find $\mathbf{r}$ when $t = 4$.

   b. Find the distance of P from O when $t = 4$.

7. A particle P is moving in the $xy$ plane with constant acceleration given by
$\mathbf{a} = p\mathbf{i} + q\mathbf{j}$. When $t = 0$ the velocity is zero and when $t = 1$, $\mathbf{v} = 3\mathbf{i} - 2\mathbf{j}$.

   a. Find $\mathbf{v}$ at any time $t$.

   b. Find the *speed* of P when $t = 3$.

8. The coordinates of a particle P at any time $t$ are $(t + t^2, 3t^2 - 2, 5t)$.

   Show that P has a constant acceleration and give its magnitude.

9. A particle P is moving in the $xy$ plane. The acceleration of P at time $t$ is given
by $\mathbf{a} = 2t\mathbf{i} + 3\mathbf{j}$. Initially P is at O with velocity $5\mathbf{j}$.

   Find $\mathbf{v}$ and $\mathbf{r}$ when $t = 3$.

10. A particle P is moving in space such that at any time $t$, $\mathbf{a} = \dfrac{32}{t^3}\mathbf{j}$.
When $t = 2$, $\mathbf{v} = 9\mathbf{i} - 4\mathbf{j} + \mathbf{k}$ and $\mathbf{r} = 18\mathbf{i} + 8\mathbf{j} + 4\mathbf{k}$.

   a. Find $\mathbf{v}$ in terms of $t$.

   b. Find $\mathbf{r}$ in terms of $t$.

11. A particle P is moving in the $xy$ plane. Initially P is at O with velocity
$(4 \cos \alpha)\mathbf{i} + (4 \sin \alpha)\mathbf{j}$.

   Given that $\mathbf{a} = -g\mathbf{j}$, where $g$ is the acceleration due to gravity,

   a. find, at any time $t$, expressions for   i  $\mathbf{v}$   ii  $\mathbf{r}$.

   b. Hence derive the equation of the path of the particle.

## Summary

A mathematical model uses assumptions to simplify a problem so that
mathematical techniques can be used to solve it.

$$\mathbf{v} = \frac{d\mathbf{r}}{dt} \quad \text{and} \quad \mathbf{r} = \int \mathbf{v}\, dt$$

Similarly
$$\mathbf{a} = \frac{d\mathbf{v}}{dt} \quad \text{and} \quad \mathbf{v} = \int \mathbf{a}\, dt$$

So when $\mathbf{r} = f(t)\mathbf{i} + g(t)\mathbf{j} + h(t)\mathbf{k}$
then $\mathbf{v} = f'(t)\mathbf{i} + g'(t)\mathbf{j} + h'(t)\mathbf{k}$
and $\mathbf{a} = f''(t)\mathbf{i} + g''(t)\mathbf{j} + h''(t)\mathbf{k}$.

## Review

1. A ball of mass 1 kg and a ball of mass 3 kg are connected by a string passing
over a peg. The balls are released from rest with the string taut and the 3 kg
ball one metre above a horizontal table.

   State the assumptions that you would make and how reasonable they are
in order to find the speed with which the 3 kg ball hits the table. You do not
have to solve the problem.

**2** Starting from O, a point P traces out consecutive displacement vectors of $2\mathbf{i} + 3\mathbf{j}$, $-\mathbf{i} + 4\mathbf{j}$, $7\mathbf{i} - 5\mathbf{j}$ and $\mathbf{i} + 3\mathbf{j}$.

Find the final displacement of P from O.

**3** Velocities of magnitudes $5\text{ ms}^{-1}$, $7\text{ ms}^{-1}$, $4\text{ ms}^{-1}$ and $6\text{ ms}^{-1}$ act in the directions north-east, north, south-east and west respectively. Take $\mathbf{i}$ and $\mathbf{j}$ as unit vectors east and north respectively.

  **a** Draw a sketch showing the separate velocities.

  **b** Find, in the form $a\mathbf{i} + b\mathbf{j}$, the resultant velocity.

  **c** Find the angle the resultant velocity makes with the north direction.

  **d** Find the resultant speed.

**4** A particle P is moving in the $xy$ plane and at time $t$ the position vector of P is $\mathbf{r} = 2t^3\mathbf{i} + 3t^2\mathbf{j}$. Find expressions for $\mathbf{v}$ and $\mathbf{a}$

  **a** at any time $t$

  **b** when $t = 2$.

**5** A particle P is moving in the $xy$ plane and at time $t$ the position vector of P is $\mathbf{r} = t(t+1)\mathbf{i} + (4 - t^2)\mathbf{j}$. Find expressions for $\mathbf{v}$ and $\mathbf{a}$

  **a** at any time $t$

  **b** when $t = 1$.

**6** A particle P is moving in space and at time $t$ the position vector of P is $\mathbf{r} = e^t\mathbf{i} - t^2\mathbf{j} + t^3\mathbf{k}$. Find expressions for $\mathbf{v}$ and $\mathbf{a}$

  **a** at any time $t$

  **b** when $t = 6$.

**7** A particle P is moving in the $xy$ plane. Initially P is at rest at a point with position vector $3\mathbf{i} + \mathbf{j}$. Given that the acceleration of P after $t$ seconds is $\mathbf{i} - 2\mathbf{j}$, find expressions for $\mathbf{v}$ and $\mathbf{r}$ at any time $t$.

**8** A particle P is moving in three dimensions. At any time $t$, the velocity of P is given by

$\mathbf{v} = (\sin t)\mathbf{i} + (\cos^2 t)\mathbf{j} - (\sin t)\mathbf{k}$. Find the initial acceleration of P.

**9** A particle moving in three dimensions starts from rest and $t$ seconds later its velocity is given by $\mathbf{v} = 2t^2\mathbf{i} + 4t\mathbf{j} - \mathbf{k}$.

Find the distance of the particle from O three seconds after leaving O.

## Assessment

**1** A particle P is moving in space such that at any time $t$, $\mathbf{a} = \dfrac{54}{t^3}\mathbf{j}$.

When $t = 1$, $\mathbf{v} = 8\mathbf{i} - 4\mathbf{j} + 2\mathbf{k}$ and $\mathbf{r} = 8\mathbf{i} + 40\mathbf{j} - 4\mathbf{k}$.

  **a** Find $\mathbf{v}$ in terms of $t$.

  **b** Find $\mathbf{r}$ in terms of $t$.

  **c** Find the distance of P from O when $t = 2$.

2 A boat travels with a constant acceleration of $0.2\mathbf{i} + 0.1\mathbf{j}$ where $\mathbf{i}$ and $\mathbf{j}$ are unit vectors in the direction north and east respectively.

At time $t = 0$ the boat starts from rest moving away from a tanker anchored at the point with position vector $5\mathbf{j}$.

  a  Find the velocity of the boat 60 seconds after leaving the tanker. State any assumptions that you make.

  b  Find the distance of the boat from the tanker 120 seconds after leaving the tanker.

3 A particle moves so that at time $t$ its position from a fixed point O is

$$\mathbf{r} = t\mathbf{i} - \cos\left(\frac{\pi t}{2}\right)\mathbf{j}.$$

  a  Find the velocity of the particle when $t = 3$.

  b  Find the acceleration of the particle when $t = 3$.

  c  Find the equation of the path of the particle as it moves.

4 The position vector of a particle P at time $t$ seconds is $\mathbf{r} = (\sin 3t)\mathbf{i} + (\cos 3t)\mathbf{j}$.

  a  Find the velocity of P at time $t$ seconds.

  b  Show that the speed of P is constant and state its value.

5 A particle moves with constant acceleration between the points A and B. At A, it has velocity $(4\mathbf{i} + 2\mathbf{j})$ ms$^{-1}$. At B, it has velocity $(7\mathbf{i} + 6\mathbf{j})$ ms$^{-1}$. It takes 10 seconds to move from A to B.

  a  Find the acceleration of the particle.

  b  Find the distance between A and B.

  c  Find the average velocity as the particle moves from A to B.

*AQA MM1 June 2015*

6 A helicopter travels at a constant height above the sea. It passes directly over a lighthouse with position vector $(500\mathbf{i} + 200\mathbf{j})$ metres relative to the origin, with a velocity of $(-17.5\mathbf{i} - 27\mathbf{j})$ ms$^{-1}$. The helicopter moves with a constant acceleration of $(0.5\mathbf{i} + 0.6\mathbf{j})$ ms$^{-2}$. The unit vectors $\mathbf{i}$ and $\mathbf{j}$ are directed east and north respectively.

  a  Find the position vector of the helicopter $t$ seconds after it has passed over the lighthouse.

  b  The position vector of a rock is $(200\mathbf{i} - 400\mathbf{j})$ metres relative to the origin. Show that the helicopter passes directly over the rock, and state the time that it takes for the helicopter to move from the lighthouse to the rock.

  c  Find the average velocity of the helicopter as it moves from the lighthouse to the rock.

  d  Is the magnitude of the average velocity equal to the average speed of the helicopter? Give a reason for your answer.

*AQA MM1B June 2013*

## Introduction

The study of the forces acting on objects that are at rest is called **statics**. This chapter looks at different methods for working with forces that are in equilibrium and all in one plane.

## Recap

You will need to remember...

▶ Types of force. M1 Chapter 18
▶ Pythagoras' theorem and trigonometry in right-angled triangles.
▶ Friction. M1 Chapter 18
▶ The resultant of a set of vectors. P2 Chapter 9

## Applications

When a ladder is placed against a wall it must be safe for a person to climb. This will depend on the weight of the ladder, the friction at each end of the ladder and the angle that the ladder makes with the wall.

## Objectives

By the end of this chapter, you should know how to...

▶ Resolve forces in two perpendicular directions.
▶ Find the resultant of a set of coplanar forces.
▶ Find the resultant moment of a set of coplanar forces.
▶ Solve problems on the equilibrium of an object that can be modelled as a point and larger objects under the action of coplanar forces.

## 11.1 Types of force and drawing diagrams

Here is a brief reminder of the different forces that may act on an object:

▶ The weight of the object always acts vertically downwards.
▶ When an object is in contact with another surface, a normal reaction always acts on the body perpendicular to the surface of contact. There may also be a frictional force unless the contact is smooth.
▶ When the object is attached to another by a string or hinge, a force acts on the body at the point of attachment.

Each of these examples describes a situation and shows how a working diagram can be drawn.

**Note**

A clear diagram is needed to solve any problem involving forces acting on an object. All the forces acting on the object need to be shown and the diagram should be large enough so that the force lines can be seen clearly.

① A small block is sliding down a smooth inclined plane.

The normal reaction $R$ is perpendicular to the plane, which is the surface of contact.

The contact is smooth so there is no friction.

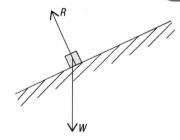

**2** A block is being pulled along a rough horizontal floor by a rope inclined at 50° to the floor.

The frictional force acts along the plane in the direction opposite to the motion of the block.

**3** A uniform ladder rests with its foot on rough ground and the top against a smooth wall.

At the foot, the normal reaction is perpendicular to the ground.

The frictional force acts along the ground and towards the wall because *if* the ladder moved, its foot would slip *away* from the wall.

At the top of the ladder the normal reaction is perpendicular to the wall and there is no friction.

The weight of the ladder acts through its centre.

> **Note**
>
> The mass of a uniform body is evenly distributed, so the weight of a uniform ladder acts through the midpoint. When an object is described as uniform it means that its weight acts through its geometric centre.

**4** A uniform beam is hinged at one end to a wall to which it is inclined at 60°. It is held in this position by a horizontal chain attached to the other end.

The direction of the force that the hinge exerts on the beam is not known at this stage so we cannot mark it at any specific angle.

**5** A particle is fastened to one end of a light string. The other end of the string is held and the particle is whirled round in a circle in a vertical plane.

The only forces acting on the particle are its weight and the tension in the string. There is *no force in the direction of motion of the particle.*

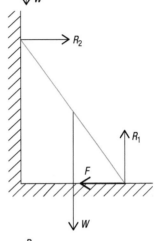
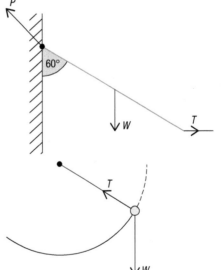

## Exercise 1

In questions 1 to 3, copy the diagram and draw the forces acting on the given object in the specified situation. Take any rod, ladder and so on. as being uniform, that is with its centre of gravity at its midpoint.

**1** A rod hinged to a wall and held in a horizontal position by a string.

**2** A light string fixed at one end has a particle tied to the other end being pulled aside by a horizontal force.

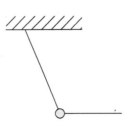
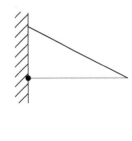

**3** A small block at rest on a rough inclined plane.

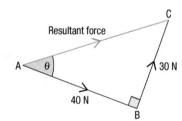

In questions 4 to 6, draw a diagram of the specified object and mark on the diagram all the forces acting on the object.

**4** A ladder with its foot on rough ground is leaning against a rough wall.

**5** A particle is attached to one end of a light string whose other end is fixed to a point A. The particle is

   **a** hanging at rest

   **b** rotating in a horizontal circle below A

   **c** held at 30° to the vertical by a horizontal force.

**6** A rod of length 1 m is hinged at one end to a wall at a point A. It is held in a horizontal position by a string joining the point of the rod that is 0.8 m from the wall, to a point on the wall 1 m vertically above A.

**7** A beam is hinged at one end, A, to a wall and is held horizontal by a rope attached to the other end, B, and to a point on the wall above A. The rope is at 45° to the wall. A crate hangs from B. Draw separate diagrams to show the forces acting on

   **a** the beam       **b** the crate.

**8** The diagram shows a rough plank resting on a cylinder and with one end of the plank on rough ground.

Draw diagrams to show

   **a** the forces acting on the plank

   **b** the forces acting on the cylinder.

# 11.2 The resultant of coplanar forces

Force has magnitude and direction so forces are vectors.

When two (or more) vectors are added, the single equivalent vector is called the **resultant vector**. The vectors that are combined are called **components**.

Consider the example of a heavy crate being pulled along by two ropes.

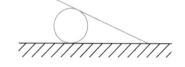

Although the ropes are pulling in different directions, the crate moves in only one direction. This is the direction of the resultant of the tensions in the ropes.

By drawing a triangle of vectors we can find both the magnitude and the direction of the resultant force.

The magnitude and direction of the resultant force can be found by calculation, which in this case is easy because the components are at right angles.

In triangle ABC: $AC^2 = AB^2 + BC^2$ so AC represents 50 N

and    $\tan A = \dfrac{3}{4}$  $\left( \tan = \dfrac{\text{opp}}{\text{adj}} \right)$

$\Rightarrow$    angle A = 37° (nearest degree)

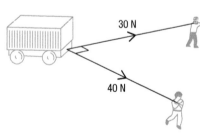

Hence, the resultant force is of magnitude 50 N acting at an angle of 37° to AB.

An object is often under the action of several **coplanar** forces (forces all in one plane) acting in different directions. To investigate the effect of these forces we need to be able to find their resultant.

To do this we **resolve** each force into the same two perpendicular components. Components in a chosen direction are positive while those in the opposite direction are negative.

By collecting each set of components, the original set of forces can be replaced by an equivalent pair of forces in perpendicular directions.

For example, a particle is resting on a rough plane inclined to the horizontal at 30°. The forces acting on the particle are shown in the diagram.

As the normal reaction and the frictional force are perpendicular to each other, it is sensible to resolve each force in these two directions, that is along ($\swarrow$) and perpendicular ($\nwarrow$) to the plane.

Using trigonometry in the triangle of forces shown gives

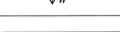

|  | Friction | Reaction | Weight |
|---|---|---|---|
| **Component $\swarrow$** | $-F$ | $0$ | $W \sin 30°$ |
| **Component $\nwarrow$** | $0$ | $R$ | $-W \cos 30°$ |

The components of force down and perpendicular to the plane can be added:

| Resolving $\swarrow$ gives | $W \sin 30° - F$ | [1] |
| Resolving $\nwarrow$ gives | $R - W \cos 30°$ | [2] |

## Calculating the resultant

Representing [1] by $X$ and [2] by $Y$, gives

$$X = W \sin 30° - F$$

$$Y = R - W \cos 30°$$

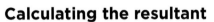

As $X$ and $Y$ are perpendicular, the magnitude of the resultant, $R$, of $X$ and $Y$ is $\sqrt{X^2 + Y^2}$

and $R$ makes an angle $\alpha$ with the plane where $\tan \alpha = \dfrac{Y}{X}$.

When each force is given in the form $a\mathbf{i} + b\mathbf{j}$ the forces are already expressed as components in the directions of $\mathbf{i}$ and $\mathbf{j}$ so it is easy to find $X$, $Y$ and $R$.

## Example 1

A uniform ladder of weight $W$ rests with its top against a rough wall and its foot on rough ground which slopes down from the base of the wall at $10°$ to the horizontal. Resolve to find horizontal and vertical components of the forces acting on the ladder.

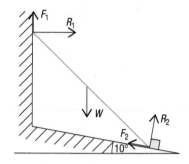

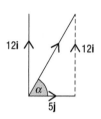

Resolving $\rightarrow$ gives $\qquad R_1 - F_2 \cos 10° + R_2 \sin 10°$

Resolving $\uparrow$ gives $\qquad F_1 - W + F_2 \sin 10° + R_2 \cos 10°$

### Note

Drawing the components of $R_2$ and $F_2$ on separate small diagrams can help.

## Example 2

Find the magnitude of the resultant of the set of forces $3\mathbf{i} + 5\mathbf{j}$, $-7\mathbf{j}$, $-4\mathbf{i} + 11\mathbf{j}$, $5\mathbf{i}$ and $\mathbf{i} + 3\mathbf{j}$. Each force is measured in newtons. Find the angle between the resultant and the unit vector $\mathbf{i}$.

The resultant is $R$ where

$$R = (3\mathbf{i} + 5\mathbf{j}) + (-7\mathbf{j}) + (-4\mathbf{i} + 11\mathbf{j}) + 5\mathbf{i} + (\mathbf{i} + 3\mathbf{j})$$

$$= (3 - 4 + 5 + 1)\mathbf{i} + (5 - 7 + 11 + 3)\mathbf{j}$$

$$= 5\mathbf{i} + 12\mathbf{j}$$

$$|5\mathbf{i} + 12\mathbf{j}| = \sqrt{(25 + 144)} = 13$$

$$\tan \alpha = \frac{12}{5} = 2.4$$

$$\alpha = 67° \text{ (nearest degree)}$$

The resultant is 13 N at $67°$ to $\mathbf{i}$.

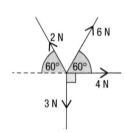

## Example 3

Find the resultant of the forces of 4 N, 6 N, 2 N and 3 N shown in the diagram.

Resolving the forces parallel and perpendicular to the 4 N force gives

$$\text{resolving} \rightarrow \quad X = 4 + 6 \cos 60° - 2 \cos 60° = 4 + 3 - 1 = 6$$

$$\text{resolving} \uparrow \quad Y = 6 \sin 60° - 3 + 2 \sin 60° = 8 \sin 60° - 3 = 3.928...$$

The resultant force, $R$ N is given by

$$R = \sqrt{(X^2 + Y^2)} = \sqrt{(6^2 + 3.928^2)}$$

$$= 7.17 \text{ (3 sf)}$$

and $\quad \tan \alpha = \dfrac{Y}{X} = \dfrac{3.928}{6} = 0.6546...$

$\Rightarrow \qquad \alpha = 33° \text{ (nearest degree)}$

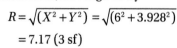

Therefore the resultant force is 7.17 N at $33°$ to the force of 4 N.

## Example 4

ABCDEF is a regular hexagon. Four forces act on a particle. The forces are of magnitudes 3 N, 4 N, 2 N and 6 N and they act in the directions of the sides AB, AC, EA and AF respectively.

Find the magnitude of the resultant force and the angle it makes with AB.

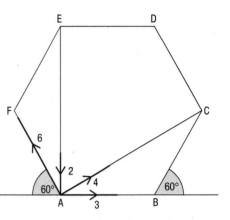

Using $X$ newtons and $Y$ newtons for the components of the resultant $R$ parallel and perpendicular to AB gives

resolving $\rightarrow$ $\quad X = 3 + 4\cos 30° - 6\cos 60°$
$$= 3 + 2\sqrt{3} - 3 = 2\sqrt{3}$$

resolving $\uparrow$ $\quad Y = 4\sin 30° - 2 + 6\sin 60°$
$$= 2 - 2 + 3\sqrt{3} = 3\sqrt{3}$$

Therefore $\quad R^2 = X^2 + Y^2 = 12 + 27 = 39 \Rightarrow R = \sqrt{39}$

and $\quad \tan\alpha = \dfrac{Y}{X} = \dfrac{3\sqrt{3}}{2\sqrt{3}} = 1.5$

The resultant force is $\sqrt{39}$ N at 56° to AB (nearest degree).

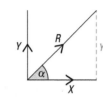

**Note**

The forces are *not* represented by the *lengths* of the lines in the hexagon.

## Exercise 2

In questions 1 to 6, find the magnitude of the resultant of the given vectors and give the angle between the resultant and the direction of the positive $x$-axis.

**1**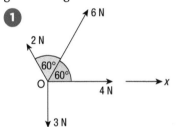

**2**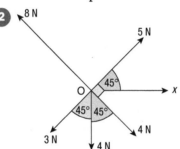

**3**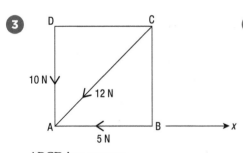

ABCD is a square.

**4**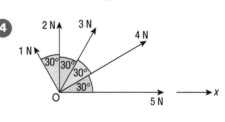

**5**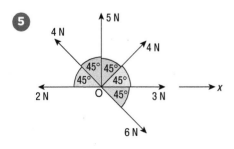

**6**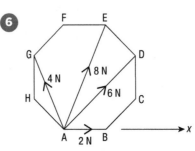

ABCDEFGH is a regular octagon.

**7** Find the magnitude of the resultant and give the angle that the resultant makes with the vector **i** of the forces, $(4\mathbf{i} - 3\mathbf{j})$ N, $(\mathbf{i} + 6\mathbf{j})$ N, $(-2\mathbf{i} + 5\mathbf{j})$ N, and $3\mathbf{i}$ N.

**8** Find the magnitude of the resultant and give the angle that the resultant makes with the vector **i** of the velocities $(4\mathbf{i} - 7\mathbf{j})$ ms$^{-1}$, $(-3\mathbf{i} + 8\mathbf{j})$ ms$^{-1}$, $(2\mathbf{i} + 3\mathbf{j})$ ms$^{-1}$, $8\mathbf{i}$ ms$^{-1}$ and $(\mathbf{i} + \mathbf{j})$ ms$^{-1}$.

**9** Find the magnitude of the resultant and give the angle that the resultant makes with the vector **i** of the displacements $(-6\mathbf{i} + \mathbf{j})$ m, $(2\mathbf{i} - 5\mathbf{j})$ m, $(\mathbf{i} + 4\mathbf{j})$ m and $(3\mathbf{i} + 2\mathbf{j})$ m.

**10** Find the magnitude of the resultant and give the angle that the resultant makes with the vector **i** of the forces $(2\mathbf{i} + 2\mathbf{j})$ N, $(\mathbf{i} - 7\mathbf{j})$ N, $(-6\mathbf{i} + \mathbf{j})$ N.

**11** ABCD is a rectangle in which AB = 4 m and BC = 3 m. A force of magnitude 3 N acts along AB towards B. Another force of magnitude 4 N acts along AC towards C and a third force, 3 N, acts along AD towards D. Find the magnitude of the resultant of these forces and find the angle the resultant makes with AD.

**12** A surveyor starts from a point O and walks 200 m due north. He then turns clockwise through 120° and walks 100 m after which he walks 300 m due west. What is his resultant displacement from O?

**13** Three boys are pulling a heavy trolley with three ropes. The boy in the middle is exerting a pull of 100 N. The other two boys, whose ropes both make an angle of 30° with the centre rope, are pulling with forces of 80 N and 140 N. What is the resultant pull on the trolley and at what angle is it inclined to the centre rope?

**14** ABC is an equilateral triangle and D is the midpoint of BC. Forces of magnitudes 8 N, 6 N and 12 N act along AB, AC and DA respectively (the order of the letters gives the direction of the force). Find the magnitude of the resultant force and the angle between the resultant and DA.

## 11.3 Concurrent forces in equilibrium

An object that is at rest, or is moving with constant velocity, is in a state of **equilibrium**.

> The acceleration of a body in equilibrium is zero in any direction therefore the resultant force in any direction is also zero.

The converse of this statement is not always true, because although forces with zero resultant cannot make an object move in a line they can, as we shall see later on, cause an object to turn. For example as shown in the diagram:

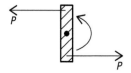

However, a set of **concurrent** forces (that is, all passing through one point) cannot cause turning.

When the resultant is zero, the sum of the components in each direction must also be zero.

Applying this fact to a concurrent system in equilibrium, in which some forces are unknown, gives a method for finding the unknown quantities.

## Example 5

A particle of weight 16 N is attached to one end of a light string whose other end is fixed. The particle is pulled aside by a horizontal force until the string is at $30°$ to the vertical. Find the magnitudes of the horizontal force and the tension in the string.

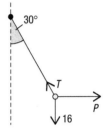

Let $P$ N and $T$ N be the magnitudes of the horizontal force and the tension respectively.

Resolving $\rightarrow$ gives $\quad P - T \sin 30° = 0$

$\Rightarrow \qquad\qquad P - T \times \dfrac{1}{2} = 0 \qquad\qquad$ [1]

Resolving $\uparrow$ gives $\quad T \cos 30° - 16 = 0$

$\Rightarrow \qquad\qquad T \times \dfrac{1}{2}\sqrt{3} - 16 = 0 \qquad\qquad$ [2]

From [2] $\qquad\qquad T = \dfrac{32}{\sqrt{3}} = \dfrac{32\sqrt{3}}{3}$

From [1] $\qquad\qquad P = \dfrac{1}{2}T = \dfrac{16\sqrt{3}}{3}$

Therefore the magnitude of the horizontal force is $\dfrac{16\sqrt{3}}{3}$ N and the magnitude of the tension is $\dfrac{32\sqrt{3}}{3}$ N.

## Example 6

A load of mass 26 kg is supported in equilibrium by two ropes inclined at $30°$ and $60°$ to the horizontal as shown in the diagram. Find in terms of $g$ the tension in each rope.

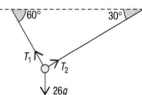

Assume that the ropes are light and that the load can be treated as a particle.

The tensions in the ropes act in perpendicular directions so resolve in these directions.

Let the tensions in the ropes be $T_1$ and $T_2$ newtons.

Resolving $\nwarrow \qquad T_1 - 26g \cos 30° = 0$

$\Rightarrow \qquad\qquad T_1 = 26g\left(\dfrac{\sqrt{3}}{2}\right) = 13g\sqrt{3}$

Resolving $\nearrow \qquad T_2 - 26g \sin 30° = 0$

$\Rightarrow \qquad\qquad T_2 = 26g\left(\dfrac{1}{2}\right) = 13g$

The tensions in the ropes are $13g\sqrt{3}$ N and $13g$ N.

## Example 7

**Question**

Forces $2\mathbf{i} - 3\mathbf{j}$, $7\mathbf{i} + 4\mathbf{j}$, $-5\mathbf{i} - 9\mathbf{j}$, $P\mathbf{i} + 2\mathbf{j}$, and $\mathbf{i} - Q\mathbf{j}$ are in equilibrium. Find the values of $P$ and $Q$.

**Answer**

The forces are given in the form $a\mathbf{i} + b\mathbf{j}$ so because $a$ and $b$ are the magnitudes of components in the directions of $\mathbf{i}$ and $\mathbf{j}$, the sum of the coefficients of $\mathbf{i}$ is zero and similarly for $\mathbf{j}$.

$$(2\mathbf{i} - 3\mathbf{j}) + (7\mathbf{i} + 4\mathbf{j}) + (-5\mathbf{i} - 9\mathbf{j}) + (P\mathbf{i} + 2\mathbf{j}) + (\mathbf{i} - Q\mathbf{j})$$
$$= (5 + P)\mathbf{i} + (-6 - Q)\mathbf{j}$$
$$= 0\mathbf{i} + 0\mathbf{j}$$

Therefore $5 + P = 0$ and $-6 - Q = 0$ $\Rightarrow$ $P = -5$ and $Q = -6$

## Friction

Friction was introduced in M1 Chapter 18.

Here is a reminder of that work.

▶ When the surfaces of two objects are in rough contact and have a tendency to move relative to each other, equal and opposite frictional forces act, one on each of the objects, so as to oppose the potential movement.

▶ Until it reaches its limiting value, the magnitude of the frictional force $F$ is just sufficient to prevent motion.

▶ When the limiting value is reached and the object moves, $F = \mu R$, where $R$ is the normal reaction between the surfaces and $\mu$ is the **coefficient of friction** for those two surfaces.

▶ For all rough contacts $0 < F \leq \mu R$.

▶ When contact is smooth $\mu = 0$.

## Example 8

**Question**

A small block of weight 32 N is lying in rough contact on a horizontal plane. A horizontal force of $P$ newtons is applied to the block until it is just about to move the block.

**a** When $P = 8$ find the coefficient of friction $\mu$ between the block and the plane.

**b** When $\mu = 0.4$, find the value of $P$.

**Answer**

**a** Resolving $\rightarrow$ $\quad 8 - \mu R = 0 \Rightarrow \mu R = 8$

Resolving $\uparrow$ $\quad R - 32 = 0 \Rightarrow R = 32$

$\Rightarrow \qquad\qquad \mu = \dfrac{\mu R}{R} = \dfrac{8}{32}$

The coefficient of friction is $\dfrac{1}{4}$.

**b** Resolving $\rightarrow$ $\quad P - 0.4 \times R = 0$

Resolving $\uparrow$ $\qquad\qquad R = 32$

$\therefore \qquad\qquad P - 0.4 \times 32 = 0$

$\qquad\qquad\qquad\quad P = 12.8$

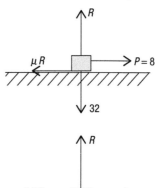

## Example 9

A small block of weight 24 N rests in rough contact with a horizontal plane. A light string is attached to the block and is inclined at $30°$ to the plane. The block is just about to slip when the tension in the string is 12 N. Find the coefficient of friction between the block and the plane.

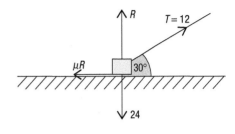

Friction is limiting so $F = \mu R$.

Resolving $\rightarrow$ $\quad 12 \cos 30° - \mu R = 0$

$\Rightarrow \quad\quad\quad 12 \times \dfrac{\sqrt{3}}{2} - \mu R = 0 \quad \Rightarrow \quad \mu R = 6\sqrt{3}$

Resolving $\uparrow$ $\quad 12 \sin 30° + R - 24 = 0 \quad \Rightarrow \quad R = 18$

$$\dfrac{\mu R}{R} = \dfrac{6\sqrt{3}}{18} \Rightarrow \mu = \dfrac{\sqrt{3}}{3}$$

The coefficient of friction is $\dfrac{\sqrt{3}}{3}$.

## Example 10

A particle of weight 8 N is resting in rough contact with a plane inclined at an angle $\alpha$ to the horizontal where $\tan \alpha = \dfrac{3}{4}$. The coefficient of friction between the particle and the plane is $\mu$. A horizontal force of $P$ newtons is applied to the particle. When $P = 16$ the particle is on the point of slipping up the plane.

a Find $\mu$.

b Find the value of $P$ such that the particle is just prevented from slipping down the plane.

a Resolving parallel and perpendicular to the plane involves $\mu$ in only one equation.

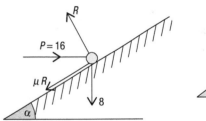

Resolving $\nearrow$ $\quad 16 \cos \alpha - \mu R - 8 \sin \alpha = 0$

$\Rightarrow \quad\quad 16 \times \dfrac{4}{5} - \mu R - 8 \times \dfrac{3}{5} = 0$

$\Rightarrow \quad\quad\quad\quad\quad\quad \mu R = 8$

Resolving $\nwarrow$ $\quad R - 16 \sin \alpha - 8 \cos \alpha = 0$

$\Rightarrow \quad\quad R - 16 \times \dfrac{3}{5} - 8 \times \dfrac{4}{5} = 0$

$\Rightarrow \quad\quad\quad\quad\quad\quad R = 16$

$$\dfrac{\mu R}{R} = \dfrac{8}{16} \Rightarrow \mu = \dfrac{1}{2}$$

b This time resolving horizontally and vertically uses $P$ in only one equation.

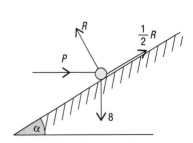

Resolving $\uparrow$ $\quad R \cos \alpha + \dfrac{1}{2} R \sin \alpha - 8 = 0$

$\Rightarrow \quad\quad\quad\quad\quad\quad R = \dfrac{80}{11}$

Resolving $\rightarrow$ $\quad P + \dfrac{1}{2} R \cos \alpha - R \sin \alpha = 0$

$\Rightarrow \quad\quad\quad\quad\quad\quad P = \dfrac{R}{5} = \dfrac{16}{11}$

# Exercise 3

In this exercise all forces are measured in newtons.

In questions 1 to 3, the forces shown in the diagram are in equilibrium.

Find the values of $P$, $Q$ and where appropriate, the value of $\theta$.

**1 a**
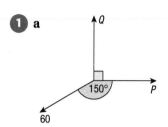

**b**
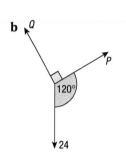

**c**
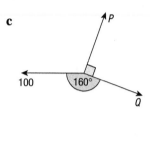

**2 a**
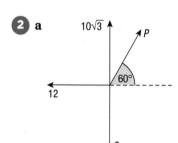

**b**
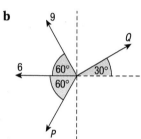

**c**
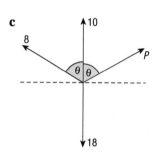

**3 a**
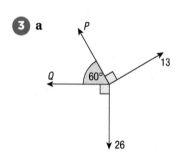

**b**
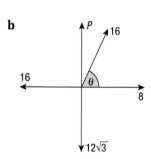

**c**
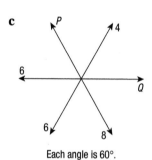

Each angle is 60°.

**4** A light inextensible string is of length 50 cm. It is fixed to a wall at one end A and a particle of mass 4 kg is attached to the other end B.

A horizontal force applied to the end B holds the particle in equilibrium at a distance of 30 cm from the wall. Find, in terms of $g$, the tension in the string.

**5** A small block of weight 20 N is attached to two light inelastic strings.

The other ends of the strings are fixed to two fixed points on the same level, 1 m apart.

The lengths of the strings are 0.6 m and 0.8 m.

State the angle between the strings.

By resolving in the directions of the strings, find the tension in each string.

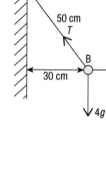

**6** A small block of weight 30 N rests on a smooth plane inclined at 30° to the horizontal and is held in equilibrium by a light string inclined at 30° to the plane.

Find the tension in the string.

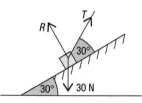

**7** A block of weight 100 N rests in equilibrium on a rough plane inclined at 30° to the horizontal.

Find the magnitude of the frictional force.

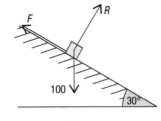

**8** Write down, in the form $a\mathbf{i} + b\mathbf{j}$, each of the forces shown in the diagram.

Given that the forces are in equilibrium, find $P$ and $Q$.

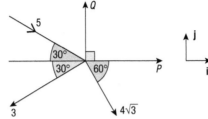

**9** Four forces act on a particle keeping it in equilibrium. Find the values of $p$ and $q$ when the forces are

**a**  $2\mathbf{i} + 7\mathbf{j}, 5\mathbf{i} - p\mathbf{j}, 9\mathbf{i} + 4\mathbf{j}$ and $q\mathbf{i} - 11\mathbf{j}$

**b**  $\mathbf{i} - 6\mathbf{j}, -8\mathbf{i} + 3\mathbf{j}, p\mathbf{i} + q\mathbf{j}$ and $3\mathbf{i} + 10\mathbf{j}$

**10** The resultant of the forces $7\mathbf{i} - 2\mathbf{j}, -6\mathbf{i} + 5\mathbf{j}, 3\mathbf{i} + 6\mathbf{j}$ and $a\mathbf{i} + b\mathbf{j}$ is $11\mathbf{i} - 2\mathbf{j}$.

**a**  Find the values of $a$ and $b$.

**b**  When a fifth force is added to the given forces, equilibrium is established. Write down in the form $a\mathbf{i} + b\mathbf{j}$ the force that is added.

In questions 11 to 17, the particle is of weight 24 N and has rough contact with the given surface; $\mu$ is the coefficient of friction between the particle and the surface.

**11** The particle is just about to slip down a plane inclined at 30° to the horizontal.

Find the value of $\mu$.

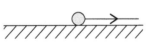

**12** The particle is on a horizontal plane and is being pulled by a horizontal string.

It is just on the point of moving when the tension in the string is 8 N, find the value of $\mu$.

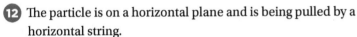

**13** The particle is just about to slip up a plane inclined at 30° to the horizontal, when being pushed by a force parallel to the plane. Given $\mu = \dfrac{1}{2}$, find the magnitude of the force.

**14** The particle is on a horizontal plane and is being pulled by a string inclined at 60° to the horizontal. It is just on the point of moving when the tension in the string is 16 N. Find the value of $\mu$.

**15** The particle is supported in limiting equilibrium on a plane inclined at 30° to the horizontal, by a string parallel to the plane.

When $\mu = \dfrac{1}{5}$, find the tension in the string.

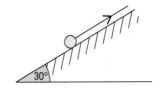

**16** The particle is resting on a plane inclined at an angle $\alpha$ to the horizontal, where $\tan \alpha = \dfrac{4}{3}$. A force of 12 N parallel to the plane is just able to prevent the particle from slipping down the plane. Find the value of $\mu$.

**17** The particle is held in limiting equilibrium on a plane inclined at 30° to the horizontal, by a string inclined at 30° to the plane as shown.

Given that the value of $\mu$ is $\dfrac{1}{4}$, find the tension in the string when the particle is on the point of moving

**a** up the plane

**b** down the plane.

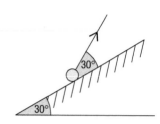

# 11.4 Moments and rigid objects in equilibrium

## Moments

Up to this point equilibrium problems have involved objects that can be modelled as a particle so the forces acting on them are concurrent (all passing through one point).

When forces act on a rigid body of significant size there is no reason why these forces should be concurrent. Therefore these forces are capable of producing rotation.

To find out how to measure turning effect, look at a rod pivoted at its midpoint P.

If it is perfectly uniform, the rod can hang in a horizontal position.

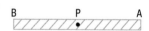

When a downward force $F$ is applied at one end A the rod rotates clockwise as shown.

The force has not made the rod move bodily downwards, it has caused the rod to turn about the pivot.

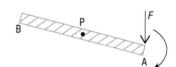

It can be shown experimentally that an additional force $2F$, applied downwards halfway along PB, will maintain the rod in its original position.

Each force exerts a turning effect on the rod and together they restore the balance of the rod.

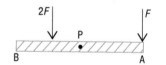

Experiments show that the turning effect of a force is given by magnitude of force × perpendicular distance from pivot

The turning effect of a force is called the **moment** of the force (also called torque).

Rotation may also occur about a hinge. In all cases the rotation is at 90° to an axis called the **axis of rotation**.

The line (axis) is perpendicular to the plane in which the forces act.

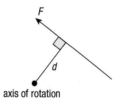

> The magnitude of the moment of a force $F$, acting at a perpendicular distance $d$ from the axis of rotation, is given by $F \times d$.

The unit in which moment is measured is the **newton metre**, Nm.

To give a full description of the turning effect of a force we must also give the sense of rotation, that is clockwise or anticlockwise. To do this when dealing with a system of moments, choose a positive sense of rotation. For example, if anticlockwise is chosen as the positive sense then an anticlockwise moment has a + sign while a clockwise moment has a − sign.

> **Note**
> The positive sense does not have to be anticlockwise; a choice can be made for each problem.

> **The resultant moment of a number of forces is then the algebraic sum of the separate moments.**

When a force passes through the axis of rotation, its distance from that axis is zero. Therefore the moment of the force about that axis is zero.

There are some people who cannot see immediately from a diagram the sense of rotation that a particular force would cause. If you have this problem, stick a pin into the point on the diagram about which turning will take place and pull the page (gently) in the direction of the force. You will then *see* the rotation happening.

## Example 11

ABCD is a square **lamina** subjected to the forces shown in the diagram. (A lamina is a flat object which is modelled as having no thickness.)

Find the clockwise moment of each of the forces about an axis through

a   B          b   A.

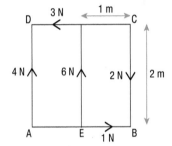

a

| Magnitude of force | 1 N | 2 N | 3 N | 4 N | 6 N |
|---|---|---|---|---|---|
| ⊥ distance from B | 0 | 0 | 2 m | 2 m | 1 m |
| clockwise moment about B | 0 | 0 | −6 Nm | 8 Nm | 6 Nm |

b

| Magnitude of force | 1 N | 2 N | 3 N | 4 N | 6 N |
|---|---|---|---|---|---|
| ⊥ distance from A | 0 | 2 m | 2 m | 0 | 1 m |
| clockwise moment about A | 0 | 4 Nm | −6 Nm | 0 | −6 Nm |

## Example 12

The diagram shows a rod AB, free to rotate about the end A.

a   Taking the anticlockwise sense as positive, find the moment about A of each of the forces acting on the rod.

b   Hence find the resultant (total) moment of the forces about A.

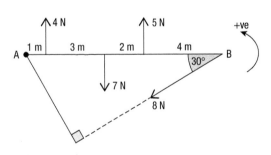

a   The moment of the 4 N force is $(4 \times 1)$ Nm = 4 Nm

The moment of the 7 N force is $-(7 \times 4)$Nm = −28 Nm

The moment of the 5 N force is $(5 \times 6)$ Nm = 30 Nm

The perpendicular distance from A to the force of 8 N is AB sin 30° = 10 sin 30°.

The moment of the 8 N force is $-(8 \times 10 \sin 30°)$ Nm = −40 Nm

b   The resultant moment is $4 + (-28) + 30 + (-40)$ Nm = −34 Nm

The resultant moment about A is 34 Nm clockwise.

When the collected moments of a number of forces are required about an axis through a point A, we say we are taking moments about A. This is denoted by the symbol A↺; the sense of the curved arrow indicates the positive sense of rotation, so A↺ means taking anticlockwise moments about A.

## Exercise 4

**1** Find the magnitude and sense of the moment of the given force about the point O.

**a**    **b**    **c**

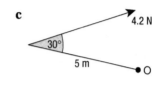

**2** Find, in magnitude and sense, the resultant moment of the given forces about the point A.

**a**    **b**

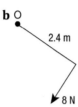
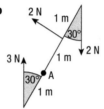

**c**    **d**

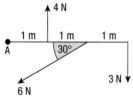
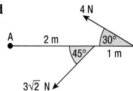

**e**    **f**

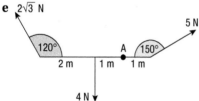
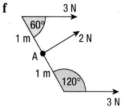

**3** ABCD is a square of side 1 m. Find the magnitude and sense of the resultant moment of the given forces about

**a**   A

**b**   D.

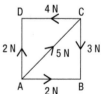

**4** The diagram shows an equilateral triangle of side 2 m.

Find the resultant anticlockwise moment of the forces shown about

**a**   A

**b**   C.

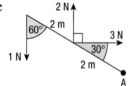

# Equilibrium under parallel forces

The object shown in the diagram is in equilibrium under the action of the set of parallel forces.

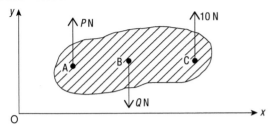

There are no components of force in the direction O$x$, so there can be no change in motion in that direction. Therefore, since the object is in equilibrium,
the resultant force in the direction O$y$ is zero
the resultant moment about A (or any other point) is zero.

As there are only two unknown forces, only two equations are needed to find them. So we can

either        collect the parallel forces and take moments about one axis

or        take moments about each of two axes.

Note that, because there is a choice of method, you can use the method not chosen first as a check.

Note also that if only one unknown force has to be found, just one of the above equations may give enough information.

## Example 13

The diagram shows a uniform beam of length 3 m and weight 40 N, suspended in equilibrium in a horizontal position by two vertical ropes, one attached at the end A and the other at C, 1 m from the other end B.

Find the tension in each rope. (The weight of a uniform beam acts through the midpoint.)

There are no horizontal forces so no information is given by resolving horizontally, but as there are only two unknowns only two equations are needed. We will take moments about both A and C.

A↺        $T_2 \times 2 - 40 \times 1.5 = 0$        [1]

C↺        $T_1 \times 2 - 40 \times 0.5 = 0$        [2]

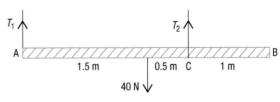

From [1]        $T_2 = 30$

From [2]        $T_1 = 10$

The tensions in the two ropes are 30 N and 10 N.

(Check: Resolving ↑ gives $30 + 10 - 40 = 0$.)

## Example 14

A uniform plank of weight 30 N and length 12 m rests on two supports at points B and C as shown in the diagram. The plank carries a load 60 N at the end A and a load 90 N at the end D. Find the force exerted by each support.

| B↻ gives | $60 \times 2 + F_2 \times 6 - 30 \times 4 - 90 \times 10 = 0$ | [1] |
|---|---|---|
| ↑ gives | $F_1 + F_2 - 60 - 30 - 90 = 0$ | [2] |
| From [1] | $F_2 = 150$ | |
| From [2] | $F_1 + F_2 - 180 = 0$, so $F_1 = 30$ | |

The supporting forces at B and C are 30 N and 150 N respectively.

Check: C↻ gives $\qquad 60 \times 8 - F_1 \times 6 + 30 \times 2 - 90 \times 4 = 0$

## Example 15

A scaffold board of weight 50 N and length 4 m lies partly on a flat roof and projects 2 m over the edge. A load of weight 30 N is carried on the overhanging end B and the board is prevented from tipping over the edge by a force applied at the other end A. The weight of the scaffold board acts through a point 1 m from the end A. Find the least force needed to stop the board tipping.

The least force will *just* prevent the board from tipping when it is about to lose contact with the roof except at the edge, so the reaction between the board and the roof acts at the edge of the roof.

In this problem only the magnitude of the force at A is required and not $R$.

So taking moments about M, in order to avoid introducing $R$, means only one equation is needed.

M↻ gives $30 \times 2 - 50 \times 1 - F \times 2 = 0$

$\Rightarrow \qquad F = 5$

So the least force needed is 5 N.

## Exercise 5

In questions 1 to 4, a light beam (the weight is negligible) rests in a horizontal position on two supports, one at A and the other at B, and carries loads as shown. Find the force exerted at each support.

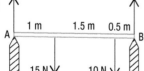

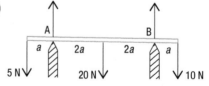

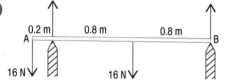

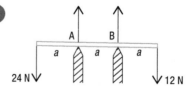

In questions 5 to 8, a horizontal uniform beam (the weight acts through the midpoint) is supported by vertical ropes. Find the tensions in the ropes.

**5**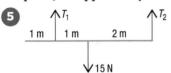

**6**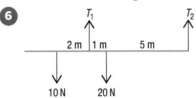

**7**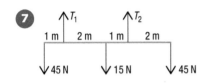

**8**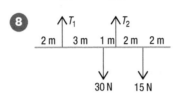

In questions 9 to 11, state any assumptions that are made.

**9** A non-uniform plank of wood 3 m long is being carried by two men, one at each end of the plank.

Mick is taking a load of 42 N and Tom at the other end is supporting 22 N.

Find the distance from Mick's end of the point through which the weight of the plank acts.

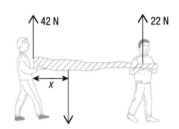

**10** A boy builds a bridge over a stream by supporting a uniform plank of wood symmetrically on two small brick piers, one on each bank. The piers are 2.6 m apart, the plank weighs 300 N and the boy weighs 420 N.

Find the force exerted by each pier when the boy stands

**a** over one of the piers

**b** 1 m from one pier

**c** in the centre of the bridge.

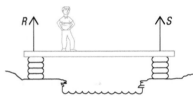

**11** A uniform plank AB of mass 200 kg and length 5 m overhangs a flat roof by 2 m. A load of mass 12 kg is placed on the end A. A boy can just stand 0.6 m from the end B without the plank tipping.

**a** What is the mass of the boy?

**b** Find the smallest extra load that must be placed at the end A to enable the boy to walk right to the end B.

**12** A uniform beam AB of weight 50 N and 2.5 m long has weights of 20 N and 30 N hanging from the ends A and B respectively. Find the distance from A of the point on the beam where a support should be placed so that the beam will rest horizontally.

## Equilibrium of a rigid object in a plane

When the forces shown in the diagram act on an object, that object will be in equilibrium only if both the resultant force and the resultant moment are zero.

The forces acting on the object are not parallel so the resultant force has components in the directions of both O$x$ and O$y$.

Therefore, when the object is in equilibrium, the sum of the components in each of the directions O$x$ and O$y$ must be zero.

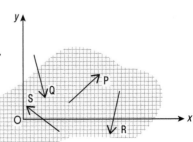

So the general conditions necessary for an object to be in equilibrium under the action of a set of non-parallel coplanar forces are

The resultant force in the direction $Ox$ is zero.
The resultant force in the direction $Oy$ is zero.
The resultant moment about any axis is zero.

Applying these conditions to a particular problem gives three equations, so three unknown quantities can be found.

In some problems it is more convenient to use an alternative set of three independent equations, that is

The resultant in the direction $Ox$ (or $Oy$ but not both) is zero.
The resultant moment about a particular axis is zero.
The resultant moment about a different axis is also zero.

Examples 16 to 19 and Exercise 6 illustrate the use of both of these methods applied to a variety of problems.

> **Note**
>
> Another way to produce three independent equations is to take moments about each of three axes, provided that these axes are not in line. (If they are in line, the third resultant moment would be a combination of the first two and not an independent fact.)

## Example 16

**Question**

A ladder of length 4 m and weight 50 N rests in equilibrium with its foot A on horizontal ground and resting against a vertical wall at the top B. The ladder is uniform; contact with the wall is smooth but contact with the ground is rough and the coefficient of friction is $\frac{1}{3}$. Find the angle $\theta$ between the ladder and the wall when the ladder is on the point of slipping.

**Answer**

The ladder is about to slip so friction is limiting and $F = \mu R$.

Resolving horizontally and vertically gives

$$\rightarrow \qquad S - \frac{1}{3}R = 0$$

$$\uparrow \qquad R - 50 = 0$$

Hence $R = 50$ and $S = \frac{50}{3}$.

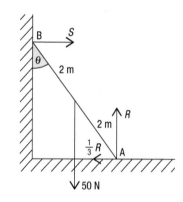

The third equation is given by taking moments about *any* axis. The best choice of axis is through A because both $F$ and $R$ pass through A and therefore have zero moment.

$$A\circlearrowleft \ S \times 4 \cos\theta - 50 \times 2 \sin\theta = 0 \quad \Rightarrow \quad S = 25 \tan\theta$$

$$\frac{50}{3} = 25 \tan\theta \Rightarrow \tan\theta = \frac{2}{3}$$

The angle between the ladder and the wall is $34°$ (nearest degree).

In examples where slipping could occur, it is important not to assume that friction is limiting. In such cases remember that $F < \mu R$.

## Example 17

**Question**

A uniform ladder of length 4 metres and weight 60 N is in equilibrium with its foot A resting on horizontal ground and resting at an angle of $10°$ against a smooth vertical wall.

*(continued)*

*(continued)*

a   Find the magnitude of the reaction and frictional force at A.

b   The angle at which the ladder is inclined to the vertical wall is increased to 30°. Find the least value of the coefficient of friction between the foot of the ladder and the ground that prevents the ladder slipping.

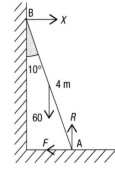

a   The rod is in equilibrium but we are not told that it is on the point of slipping so $F \neq \mu R$

Resolving vertically gives $R = 60$

Taking clockwise moments about B gives

$60 \times 2 \sin 10° + F \times 4 \cos 10° = R \times 4 \sin 10°$

$\Rightarrow F = \dfrac{60 \sin 10° - 30 \sin 10°}{\cos 10°} = 30 \tan 10° = 5.28$   [1]

Therefore the reaction at A is 60 N and the frictional force is 5.28 N

b   Resolving vertically again gives $R = 60$.

Using [1] and changing the angle to 30° gives

$F = 30 \tan 30° = 17.32$

When the ladder is on the point of slipping $F = \mu R$

therefore the least value of $\mu$ is given by $17.32 = 60\mu$

$\Rightarrow$                           $\mu = 0.289$

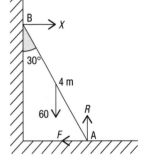

## Example 18

The end A of a uniform rod AB of length $2a$ and weight 5 N is smoothly pivoted to a fixed point on a wall. The end B carries a load of weight 10 N. The rod is held in a horizontal position by a light string joining the midpoint G of the rod to a point C on the wall, vertically above A. The string is inclined at 60° to the wall.

a   Find the tension in the string.

b   Find the horizontal and vertical components of the force exerted by the pivot on the rod.

Represent the components of the force at the pivot by $X$ and $Y$ as shown.

There are three unknown quantities, $X$, $Y$ and $T$, so three equations are needed.

a   Resolving $\rightarrow$       $X - T \sin 60° = 0$       [1]

Resolving $\uparrow$       $Y + T \cos 60° - 5 - 10 = 0$       [2]

Take moments about A so that $X$ and $Y$ are not involved.

A$\circlearrowleft$   $5 \times a + 10 \times 2a - T \times a \sin 30° = 0$       [3]

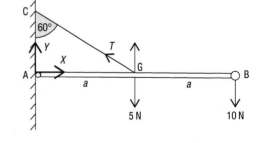

From [3],       $Ta\left(\dfrac{1}{2}\right) = 25a \Rightarrow T = 50$

Therefore the tension in the string is 50 N.

b   Using this value of $T$ in [2] gives

$Y + 50\left(\dfrac{1}{2}\right) - 15 = 0 \Rightarrow Y = -10$

and [1] gives $X - 50\left(\dfrac{\sqrt{3}}{2}\right) = 0 \Rightarrow X = 25\sqrt{3}$

The vertical component of the pivot force is 10 N downwards.

The horizontal component of the pivot force is $25\sqrt{3}$ N acting away from the wall.

## Example 19

A wooden plank AB of weight 600 N and length 4 m rests with A on rough horizontal ground where the coefficient of friction is $\frac{1}{2}$. The plank rests in rough contact with the top C of a rail of height 1.5 m, where AC = 3 m. The plank is just about to slip.

**a** Find the normal contact forces at A and at C.

**b** Find the coefficient of friction at C.

State any assumptions that have been made.

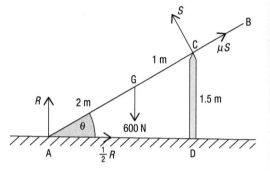

In triangle ACD, $\sin\theta = \dfrac{CD}{AC} = \dfrac{1.5}{3} = \dfrac{1}{2} \quad \Rightarrow \quad \theta = 30°$

As the plank is about to slip, friction is limiting at both A and C.

$R$ and $S$ can be found separately by taking moments about A and C.

**a** A↺ $\quad S \times 3 - 600 \times 2 \cos 30° = 0$

$\Rightarrow \quad S = 400 \cos 30° = 346.4...$

C↺ $\quad 600 \times 1 \times \cos 30° - R \times 3 \cos 30° + \dfrac{1}{2}R \times 3 \sin 30° = 0$

$\Rightarrow \quad R(\cos 30° - \dfrac{1}{2} \sin 30°) = 200 \cos 30° \quad \Rightarrow \quad R = 281.1...$

The normal reaction at A is 281 N and that at C is 346 N.

**b** Resolving ← $\quad S \sin 30° - \mu S \cos 30° - \dfrac{1}{2}R = 0$

$\Rightarrow \quad\quad\quad \mu S \cos 30° = S \sin 30° - \dfrac{1}{2}R$

Therefore $\quad \mu = \dfrac{173 - 140}{346 \times 0.866} = 0.110$

The coefficient of friction at C is 0.110 (3 significant figures).

It has been assumed that: the plank is uniform and straight; the top of the rail is small enough to be treated as a point.

## Exercise 6

In questions 1 to 6, a uniform ladder of mass 200 N and length 2 m rests with one end A on level ground and the other end B resting against a vertical wall. When the ladder is in limiting equilibrium (that is just about to slip) the angle between the ladder and the wall is $\theta$. Give angles to the nearest degree.

**1** Contact with the wall is smooth; contact with the ground is rough and the coefficient of friction is $\frac{1}{3}$. Find $\theta$.

**2** Contact with the wall is smooth; contact with the ground is rough, the coefficient of friction is $\mu$ and $\theta = 45°$.

   **a** Find the normal reaction with the ground.

   **b** Find the frictional force.

   **c** Find the value of $\mu$.

**3** Contact is rough both with the wall and with the ground; the coefficient of friction in both cases is $\frac{1}{4}$. Find $\theta$. (Remember that the ladder will not slip until friction is limiting at both points of contact.)

**4** Contact with the wall is smooth and contact with the ground is rough. When $\theta = 60°$ a workman of mass 80 kg can climb one quarter of the way up the ladder before limiting equilibrium is reached. Find the reaction at the wall and the coefficient of friction. Model the workman as a point-load.

**5** If, in question 4, $\theta$ is reduced to 30°, find how far up the ladder the man can now climb.

**6** If, in question 4, contact is rough at both ends of the ladder and $\mu = \dfrac{1}{3}$ in each case, find $\theta$ if the workman can just climb to the top of the ladder.

In questions 7 to 10, a uniform rod PQ of length 2 m and weight 24 N is hinged at the end P to a fixed point and is in equilibrium. R is a point vertically above P.

**7** PQ is kept horizontal by a support at Q. Find the magnitude of the supporting force and the magnitude and direction of the force exerted on the rod by the hinge.

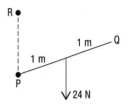

**8** PQ is held at an angle of 60° to the upward vertical through P, by a light string joining Q to R and QR = 2 m.

   **a** Find the tension in the string.

   **b** Find the magnitude of the reaction at the hinge.

**9** PQ is held at an angle of 60° to the downward vertical through P by a horizontal force of $F$ newtons applied at the end Q. Find the value of $F$.

**10** PQ is horizontal. A light string of length 2.5 m connects Q to R and a load of 20 N is applied to the rod at the end Q.

   **a** Find the tension in the string.

   **b** Find the magnitude and direction of the force acting on the rod at the hinge.

**11** A uniform rod is 6 m long and weighs 30 N. The rod rests with its end A on rough ground and against a smooth peg at C where AC is 4 m. The rod rests at an angle $\theta$ to the ground. The coefficient of friction between A and the ground is 0.2.

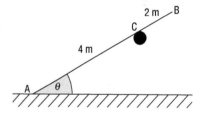

   **a** Show that the rod will not slip when $\theta = 20°$.

   **b** Show that the rod cannot rest in equilibrium when $\theta = 45°$.

**12** A uniform rod AB of length $4a$ and weight 20 N rests with the end A in rough contact with level ground where the coefficient of friction is $\dfrac{1}{2}$. A point C on the rod, at a distance of $3a$ from A, rests against a smooth peg. The rod is in limiting equilibrium when it is at 30° to the ground.

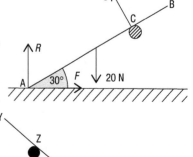

   **a** Find the reaction at the peg.

   **b** Find the frictional force.

**13** A uniform rod XY whose midpoint is M, is in equilibrium in a vertical plane as shown in the diagram.

The rod rests on a rough peg at Z and a force $F$ acts at X as shown.

Given YZ = ZM and $\tan \alpha = \dfrac{4}{3}$, find the coefficient of friction at Z and the force $F$.

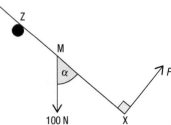

# Summary

The resultant of a set of coplanar forces can be found by resolving each force into components in two perpendicular directions. The sum of these components can then be used to find the resultant using Pythagoras' theorem and trigonometry in a right-angled triangle.

When there is friction between two surfaces, $0 < F \leq \mu R$. When friction is limiting, $F = \mu R$.

The moment of a force $F$, acting at a perpendicular distance $d$ from the axis of rotation, is given by $F \times d$ together with the sense of rotation.

When an object is in equilibrium under the action of a set of concurrent coplanar forces:

▶ The resultant force in the direction O$x$ is zero.
▶ The resultant force in the direction O$y$ is zero.

When an object is in equilibrium under the action of a set of parallel coplanar forces:

▶ The resultant force in the direction of the forces is zero.

When an object is in equilibrium under the action of a set of non-concurrent and non-parallel coplanar forces:

▶ The resultant force in the direction O$x$ is zero.
▶ The resultant force in the direction O$y$ is zero.
▶ The resultant moment about any axis is zero.

# Review

1. A horizontal force of $P$ newtons is applied to an object of weight 80 N, standing in rough contact on a horizontal table. The coefficient of friction between the object and the table is $\frac{1}{2}$. Find the magnitude of the frictional force when

   **a** $P = 10$    **b** $P = 40$    **c** $P = 50$

   State in each case whether or not the body moves.

2. A light plank of length 4 m is balanced on a pivot at the centre. A load of weight 220 N is on one end.

   **a** How far from the other end should another load, of weight 280 N, be placed if the plank is to be balanced?

   **b** What force is exerted on the plank by the pivot?

3. A uniform ladder of weight 100 N and of length 4 m stands on rough horizontal ground, with coefficient of friction 0.25 and its upper end rests against a smooth vertical wall. The ladder is inclined at 65° to the horizontal. A man, of weight 800 N, stands on the top of the ladder and another man, of weight $W$ N, stands on the bottom of the ladder. Find the least value of $W$ which will prevent the ladder from slipping. State any assumptions made and comment on how reasonable they are.

4. A uniform rod XY, whose weight is 80 N, is in equilibrium in a vertical plane. The midpoint of the rod is at M. The rod is supported on a plane inclined at 30° to the horizontal, by a string attached to the end X and held vertically. The rod is inclined to the plane at 30° as shown in the diagram.

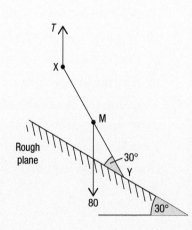

**a** Find the tension in the string.

**b** Show that the rod will not slip if the coefficient of friction between the rod and the plane is greater than $\dfrac{\sqrt{3}}{3}$.

## Assessment

**1** A small block of weight 20 N is placed on a plane inclined at an angle $\theta$ to the horizontal. The coefficient of friction between the block and the plane is $\mu$.

   **a** When $\theta = 30°$ the block is on the point of slipping. Show that $\mu = \dfrac{\sqrt{3}}{3}$.

   **b** Given $\mu = \dfrac{1}{5}$ and $\tan\theta = \dfrac{3}{4}$ find the magnitude of the horizontal force needed to prevent the block from slipping down the plane.

   **c** Given $\mu = \dfrac{2}{5}$ and $\tan\theta = \sqrt{3}$ find the magnitude of the horizontal force that will be on the point of making the block slide up the plane.

**2**

$$\tan\alpha = \tfrac{5}{12} \quad\Rightarrow\quad$$

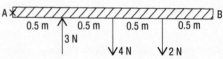

A warehouse porter is trying to push a trolley, of weight 240 N, up a plane inclined at an angle $\alpha$ to the horizontal, where $\tan\alpha = \dfrac{5}{12}$. He finds that the trolley is just on the point of moving when the horizontal force he is exerting on the handles reaches 200 N. Stating any assumptions that are necessary, use a suitable model to find the value of the coefficient of friction between the trolley and the plane.

**3 a** Find the resultant clockwise moment about a horizontal axis through A, of the forces acting on the beam AB shown in the diagram.

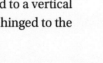

   **b** When a force $F$ newtons is applied at B, perpendicular to the beam, the resultant moment is zero. Find the value of $F$.

**4** A rod of length 1.2 m is placed on a table top with part of the rod protruding over the edge. The weight of the rod is 10 N. The rod can just hold a particle of weight 6 N, placed on the overhanging end A, without toppling over the edge.

   **a** If the rod is uniform, what length of rod is on the table top?

   **b** If, instead, the weight of the rod acts at a point G, and 0.5 m of the rod protrudes over the edge of the table, find the length of AG.

**5** In the diagram, AB is a uniform rod of weight 49 N, hinged to a vertical wall at A. The rod is supported by a light rod CD, which is hinged to the wall at D.

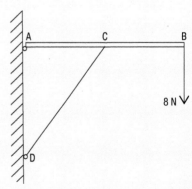

AC = CB = 0.3 m and AD = 0.4 m.

There is a load of 8 N vertically downwards at B.

   **a** Find the thrust exerted on rod AB by the rod CD.

   **b** Find the magnitude and direction of the reaction at the hinge A.

**6** A skip, of mass 800 kg, is at rest on a rough horizontal surface. The coefficient of friction between the skip and the ground is 0.4 . A rope is attached to the skip and then the rope is pulled by a van so that the rope is horizontal while it is taut, as shown in the diagram.

The mass of the van is 1700 kg. A constant horizontal forward driving force of magnitude $P$ newtons acts on the van. The skip and the van accelerate at $0.05 \text{ ms}^{-1}$.

Model both the van and the skip as particles connected by a light inextensible rope. Assume that there is no air resistance acting on the skip or on the van.

**a** Find the speed of the van and the skip when they have moved 6 metres.

**b** Draw a diagram to show the forces acting on the skip while it is accelerating.

**c** Draw a diagram to show the forces acting on the van while it is accelerating. State one advantage of modelling the van as a particle when considering the vertical forces.

**d** Find the magnitude of the friction force acting on the skip.

**e** Find the tension in the rope.

**f** Find $P$.

AQA MM1B June 2014

**7** A uniform rod, $PQ$, of length $2a$, rests with one end, $P$, on rough horizontal ground and a point $T$ resting on a rough fixed prism of semicircular cross-section of radius $a$, as shown in the diagram. The rod is in a vertical plane which is parallel to the prism's cross-section.

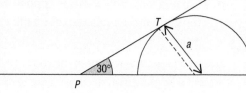

The coefficient of friction at both $P$ and $T$ is $\mu$. The rod is on the point of slipping when it is inclined at an angle of 30° to the horizontal.

Find the value of $\mu$.

AQA MM2 June 2015

# 12 Centres of Mass

## Introduction
The centre of gravity of a body is a familiar idea. This chapter shows that for bodies of normal size on the surface of the earth, the centre of mass is the same as the centre of gravity and it looks at methods for finding the centre of mass.

## Recap
You will need to remember...
► The moment of a force about an axis is equal to the magnitude of the force times the perpendicular distance of the force from the axis.

## Applications
A sculpture can be hung from a single point on the body. The angle at which it rests in equilibrium can affect the appearance of the sculpture. It is important to find the point from which it looks best.

## Objectives
By the end of this chapter, you should know how to...
► Find the centre of mass of a system of particles in a plane.
► Find the centre of mass of a uniform symmetrical lamina or solid.
► Find the centre of mass of a compound lamina whose parts have known centres of mass.
► Find the position of a body in equilibrium that is suspended from a point.

## 12.1 Definition of centre of mass

A rod of weight $Mg$ is made up of a large number of very short lengths of material, each with its own weight.

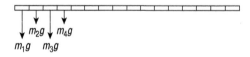

For a problem involving a rod, a single weight is marked acting at a particular point on the rod which is called 'the point through which the weight of the rod acts'.

To replace all the components of weight by a single weight we must make sure that it has exactly the same effect on the rod as the separate components have, so the total weight is the sum of all the component weights and the single weight acts through a point such that the moment of the single weight about any axis is equal to the resultant moment of the component weights.

The point through which the resultant weight of a body acts is called the **centre of gravity** of the body and is denoted by G. Its coordinates are denoted by $(\bar{X}, \bar{Y})$.

The position of the centre of gravity of any object can be found by equating the resultant moment of all the parts to the moment about the same axis of the total weight acting through the centre of gravity.

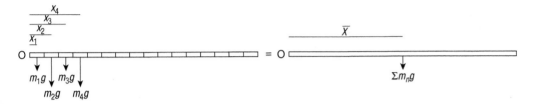

This approach can be extended to any number of weights, $m_1g$, $m_2g$, $m_3g$, ... at distances $x_1$, $x_2$, $x_3$, ... from the chosen axis.

The weight of the object is then $\Sigma m_n g$. ($\Sigma$ means 'the sum of terms of this form when $n$ takes values 1, 2, 3 ...')

Taking moments about about O gives

O↺ $\qquad m_1g \times x_1 + m_2g \times x_2 + m_3g \times x_3 + ... = \bar{X} \times \Sigma m_n g$

⇒ $\qquad\qquad \Sigma m_n g x_n = \bar{X} \Sigma m_n g$

Taking the value of $g$ as constant and cancelling it from each term in the moment equation above gives $\Sigma m_n x_n = \bar{X} \Sigma m_n$.

The solution of this equation is the location of the point G which we have so far called the centre of gravity, that is, the point about which the *weight* of an object is evenly distributed.

However, the equation above shows that the mass is also evenly distributed about G.

Therefore G can also be called the **centre of mass**.

This argument depends on the assumption that the value of $g$ is the same for all the masses (and so can be cancelled). Provided that we are dealing with normal sized bodies for example, on the earth this assumption is reasonable, so

> for an object of normal size, the centre of mass coincides with the centre of gravity.

Therefore, to find the centre of mass of a particular object, form an equation in which one side has the sum of terms like $m_1 \times x_1$, and the other side has the sum of all the masses multiplied by $\bar{X}$. This equation can be written

$$\Sigma m_n x_n = \bar{X} \Sigma m_n$$

# 12.2 The centre of mass of a system of particles

The equation $\Sigma m_n x_n = \bar{X} \Sigma m_n$ can be used to find the centre of mass of a set of particles.

## Example 1

Question

The diagram shows a set of three particles of masses 5 kg, 2 kg and 4 kg attached to a light rod at the given positions. Find the distance from O of the centre of mass, G, of the particles.

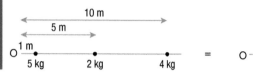

Answer

Using $\Sigma m_n x_n = \bar{X} \Sigma m_n$ gives

$5 \times 1 + 2 \times 5 + 4 \times 10 = \bar{X} \times (5 + 2 + 4)$

$\Rightarrow \qquad \bar{X} = 5$

The distance of G from O is 5 m.

In Example 1, the particles are in a straight line. When the particles are situated anywhere in a plane the position of G can be located by coordinates in an $xy$ plane.

To locate G, both its $x$ and $y$ coordinates are required.

Think of the $xy$ plane as being horizontal so that the weights act vertically.

So taking moments about the $y$-axis gives

$\Sigma m_n g x_n = \bar{X} \Sigma m_n g \quad \Rightarrow \quad \Sigma m_n x_n = \bar{X} \Sigma m_n$

and taking moments about the $x$-axis gives

$\Sigma m_n g y_n = \bar{Y} \Sigma m_n g \quad \Rightarrow \quad \Sigma m_n y_n = \bar{Y} \Sigma m_n$

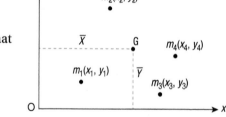

Therefore      the $x$-coordinate, $\bar{X}$, is given by $\Sigma m_n x_n = \bar{X} \Sigma m_n$

and            the $y$-coordinate, $\bar{Y}$, is given by $\Sigma m_n y_n = \bar{Y} \Sigma m_n$

## Example 2

Question

Particles A, B and C of masses 4 kg, 7 kg and 5 kg are placed respectively at points with coordinates (4, 2), (0, 6) and (1, 5). Find the centre of mass of the particles.

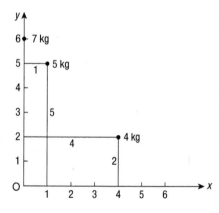

Answer

Using $\Sigma m_n x_n = \bar{X} \Sigma m_n$ gives

$\qquad 4 \times 4 + 7 \times 0 + 5 \times 1 = \bar{X}(4 + 7 + 5) \quad \Rightarrow \quad \bar{X} = 21 \div 16$

Using $\Sigma m_n y_n = \bar{Y} \Sigma m_n$ gives

$\qquad 4 \times 2 + 7 \times 6 + 5 \times 5 = 16 \bar{Y} \quad \Rightarrow \quad \bar{Y} = 75 \div 16$

The centre of mass is at the point $\left( \dfrac{21}{16}, \dfrac{75}{16} \right)$.

## Example 3

**Question**

Particles of masses 4 kg, 2 kg, 5 kg and 6 kg are placed respectively at the vertices A, B, C and D of a light square lamina of side 2 m. Find the distances of the centre of mass of the particles from AB and AD.

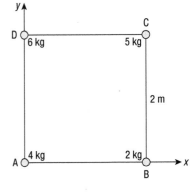

**Answer**

Using $\quad \Sigma m_n x_n = \bar{X} \Sigma m_n$

$$4 \times 0 + 2 \times 2 + 5 \times 2 + 6 \times 0 = \bar{X} \times (4 + 2 + 5 + 6)$$

$\Rightarrow \qquad \bar{X} = 14 \div 17$

Using $\quad \Sigma m_n y_n = \bar{Y} \Sigma m_n$

$$4 \times 0 + 2 \times 0 + 5 \times 2 + 6 \times 2 = \bar{Y} \times (4 + 2 + 5 + 6)$$

$\Rightarrow \qquad \bar{Y} = 22 \div 17$

The distances of the centre of mass are $\dfrac{14}{17}$ m from AD and $\dfrac{22}{17}$ m from AB.

Note that from AB (a *horizontal* line) the distance measured is *vertical*, that is $\bar{Y}$. When giving the answer to Example 3 it is easy to reverse the lines from which the distances are given.

## Example 4

**Question**

Five particles whose masses are $2m$, $3m$, $2m$, $4m$ and $m$, are placed at points with position vectors $3\mathbf{i} + \mathbf{j}$, $\mathbf{i} - 4\mathbf{j}$, $5\mathbf{i} + 6\mathbf{j}$, $-\mathbf{i} + 2\mathbf{j}$, and $3\mathbf{i}$ respectively. Find the position vector of their centre of mass.

**Answer**

The position vector of the centre of mass is $\bar{X}\mathbf{i} + \bar{Y}\mathbf{j}$.

Using $\quad \Sigma m_n x_n = \bar{X} \Sigma m_n$

$$2m \times 3 + 3m \times 1 + 2m \times 5 + 4m \times (-1) + m \times 3 = 12m \times \bar{X}$$

$\Rightarrow \qquad \bar{X} = 18 \div 12 = \dfrac{3}{2}$

Using $\quad \Sigma m_n y_n = \bar{Y} \Sigma m_n$

$$2m \times 1 + 3m \times (-4) + 2m \times 6 + 4m \times 2 + m \times 0 = 12m \times \bar{Y}$$

$\Rightarrow \qquad \bar{Y} = 10 \div 12 = \dfrac{5}{6}$

The position vector of the centre of mass is $\dfrac{3}{2}\mathbf{i} + \dfrac{5}{6}\mathbf{j}$.

When the particles are located by their position vectors, the separate $x$ and $y$ equations can be combined in the form

$$\Sigma m_n \mathbf{r}_n = \bar{\mathbf{r}} \Sigma m_n \text{ where } \bar{\mathbf{r}} = \bar{X}\mathbf{i} + \bar{Y}\mathbf{j}$$

The solution can then be given as

$$2m(3\mathbf{i} + \mathbf{j}) + 3m(\mathbf{i} - 4\mathbf{j}) + 2m(5\mathbf{i} + 6\mathbf{j}) + 4m(-\mathbf{i} + \mathbf{j}) + m(3\mathbf{i}) = 12m\bar{\mathbf{r}}$$

$\Rightarrow \qquad (18\mathbf{i} + 10\mathbf{j})m = 12m\bar{\mathbf{r}}$

$\Rightarrow \qquad \bar{\mathbf{r}} = \dfrac{3}{2}\mathbf{i} + \dfrac{5}{6}\mathbf{j}$

## Exercise 1

In questions 1 to 6, find the distance from A of the centre of mass of the given set of particles.

**1** A ——— $3a$ ——○——○—— $3a$ ——○—— $a$ —— B
$\qquad\qquad\qquad 2m \; 5m \qquad\quad 2m$

**2** A ——— $a$ —○— $a$ —○— $a$ —○— $2a$ —○ B
$\qquad\qquad 2m \; m \; 2m \qquad m$

**3**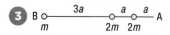
B o———3a———o—a—o—a—A
m        2m  2m

**4**  A ———2a———o———3a———o—a— B
         4m          m

**5**  A o——a——o———2a———o B
m      m            2m

**6**  B o———4a———o—a—o A
6m           4m  m

In questions 7 to 10, find the coordinates of the centre of mass of the given set of particles.

**7**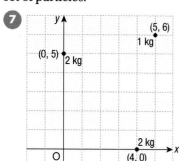

**8**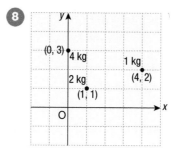

**9**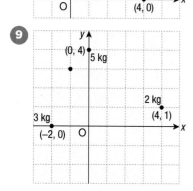

**10**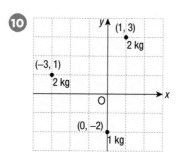

**11** The position vectors of the vertices of a light square framework are $4\mathbf{i} + \mathbf{j}$, $9\mathbf{i} + 5\mathbf{j}$, $3\mathbf{i} + 8\mathbf{j}$ and $4\mathbf{j}$. Particles of masses 2 kg, 4 kg, 2 kg and 2 kg respectively are placed at the vertices. Find the position vector of their centre of mass.

# The centre of mass of a uniform lamina

The number of particles that make up a **lamina** is extremely large. When dealing with a uniform lamina, the mass per unit area of a uniform lamina is constant throughout. When a uniform lamina has a symmetrical shape, the mass is equally distributed about the line of symmetry. Therefore, the centre of mass must lie somewhere on the line of symmetry. If there is more than one axis of symmetry, it follows that G is located at the point of intersection of these axes.

## The centre of mass of a uniform rod

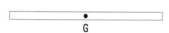

The midpoint of a uniform rod is clearly its centre of mass, as the masses of the two halves are equal and equally distributed.

# The centre of mass of a uniform square lamina

From symmetry, the centre of mass of the square lies somewhere on the line AB that bisects the square because the distribution of the mass is the same on both sides of this line. Similarly the centre of mass lies on CD and therefore is at the midpoint of each of these lines, which is the geometric centre of the square.

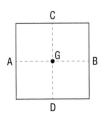

# The centre of mass of a uniform rectangular lamina

Using symmetry again, the centre of mass is at the point of intersection of AB and CD, the point where the bisectors of the sides meet.

(This is also the point of intersection of the diagonals of the rectangle.)

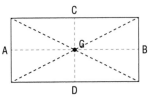

# The centre of mass of a uniform circular lamina

A circle is symmetric about any diameter. Therefore the centre of mass is the point where they meet. This is the geometric centre of the circle.

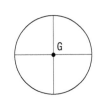

# The centre of mass of a uniform triangular lamina

There is no line of symmetry, so divide the triangle into strips parallel to one side, BC say. Then each strip can be treated as a rod.

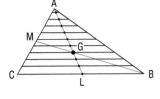

The centre of mass of each strip is at its midpoint, so the centre of mass of the triangle, G, lies on the line passing through all these midpoints, that is, on the **median** AL.

Using strips parallel to AC shows that G also lies on the median BM.

So G is at the point of intersection of the medians of the triangle.

A geometric property of a triangle is that its medians intersect at a point which is $\frac{1}{3}$ of the way from base to vertex on any median.

For example, in a right-angled triangle, G is $\frac{1}{3}$ of the way from the right-angle along each of the perpendicular sides.

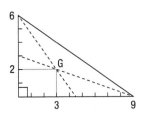

# 12.3 The centre of mass of a composite lamina

The centres of mass of a number of common laminas can be used to find the centre of mass of a uniform lamina that is a combination of these shapes.

For each part of the lamina, the mass can be expressed as the product of area and mass per unit area,

that is, mass = area × density.

Therefore the mass of each part, and hence the mass of the whole, is a multiple of $\rho$, where $\rho$ (pronounced ro) is the symbol for **density**.

## Example 5

**Question**

The uniform lamina OABCDEF is a rectangle measuring 4 cm by 6 cm with a square section of side 2 cm removed as shown in the diagram. Find the centre of mass of the lamina.

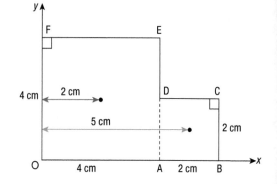

**Answer**

Let $\rho$ be the mass per unit area of the lamina and let $G(\bar{X}, \bar{Y})$ be the centre of mass of the whole lamina.
The mass of OAEF is $16\rho$ and its centre of mass is at $(2, 2)$.
The mass of ABCD is $4\rho$ and its centre of mass is at $(5, 1)$.
The mass of OABCDEF is $16\rho + 4\rho = 20\rho$.

Keeping track of the working is easier when a table is used, as shown.

| Portion | Mass | Coordinates of G | | $mx$ | $my$ |
|---|---|---|---|---|---|
| | | $x$ | $y$ | | |
| + OAEF | $16\rho$ | 2 | 2 | $16\rho \times 2$ | $16\rho \times 2$ |
| + ABCD | $4\rho$ | 5 | 1 | $4\rho \times 5$ | $4\rho \times 1$ |
| OABCDEF | $20\rho$ | $\bar{X}$ | $\bar{Y}$ | $20\rho \times \bar{X}$ | $20\rho \times \bar{Y}$ |

(The plus signs are a reminder that the two parts are *added* to give the whole.)
Working down the 5th column using $\Sigma m_n x_n = \bar{X}\Sigma m_n$ gives

$$16\rho \times 2 + 4\rho \times 5 = 20\rho \times \bar{X} \quad \Rightarrow \quad \bar{X} = \frac{13}{5}$$

Working down the 6th column using $\Sigma m_n y_n = \bar{Y}\Sigma m_n$ gives

$$16\rho \times 2 + 4\rho \times 1 = 20\rho \times \bar{Y} \quad \Rightarrow \quad \bar{Y} = \frac{9}{5}$$

The centre of mass of the whole lamina is the point $\left(\dfrac{13}{5}, \dfrac{9}{5}\right)$.

## Example 6

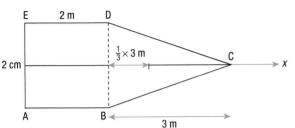

The uniform lamina ABCDE consists of a square ABDE and an isosceles triangle BCD as shown in the diagram. Find the centre of mass of the lamina.

The shape is symmetrical about the line through C that bisects AE so G lies on this line.

Use this line as the $x$-axis.

Only the $x$-coordinate of the centre of mass is unknown.

| Portion | Mass | $x$-coord. of G | $m_n x_n$ |
|---------|------|-----------------|-----------|
| + ABDE | $4\rho$ | 1 | $4\rho \times 1$ |
| + BCD | $3\rho$ | $2+1$ | $3\rho \times 3$ |
| ABCDE | $7\rho$ | $\bar{X}$ | $7\rho \times \bar{X}$ |

Working down the last column using $\Sigma m_n x_n = \bar{X} \Sigma m_n$ gives

$4\rho \times 1 + 3\rho \times 3 = 7\rho \times \bar{X}$

$\Rightarrow \quad \bar{X} = \dfrac{13}{7}$

The centre of mass of the lamina is on the line that bisects AE and DB and distant $\dfrac{13}{7}$ m from AE.

> **Note**
>
> Note that when the position of G has been found it is a good idea to mark it on the diagram, to see whether it looks about right.

## Example 7

The diagram shows a uniform rectangular lamina ABCD of mass 5 kg. A uniform circle of mass 2 kg is attached to B and a uniform circle of mass 3 kg to C. Find the distances from AB and AD of the centre of mass of the lamina with the circles attached.

|  |  | Coords of G | |  |  |
|--|--|--|--|--|--|
| Portion | Mass | $x$ | $y$ | $mx$ | $my$ |
| + ABCD | 5 kg | 1 | 2 | 5 | 10 |
| + Particle B | 2 kg | 2 | 0 | 4 | 0 |
| + Particle C | 3 kg | 2 | 4 | 6 | 12 |
| Loaded lamina | 10 kg | $\bar{X}$ | $\bar{Y}$ | $10\bar{X}$ | $10\bar{Y}$ |

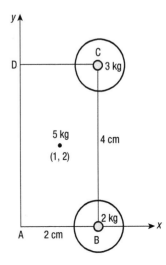

Using $\Sigma m_n x_n = \bar{X} \Sigma m_n$ gives

$5 + 4 + 6 = 10\bar{X} \quad \Rightarrow \quad \bar{X} = 1.5$

Using $\Sigma m_n y_n = \bar{Y} \Sigma m_n$ gives

$10 + 0 + 12 = 10\bar{Y} \quad \Rightarrow \quad \bar{Y} = 2.2$

The centre of mass is distant 1.5 cm from AD and 2.2 cm from AB.

## Example 8

The diagram shows a uniform square lamina ABCD of side 2 m, whose density is 1.2 kg per square metre. A circular hole of radius 0.5 m is cut from it. The centre of the hole is distant 0.8 m from both AB and AD. A thin uniform strengthening strip is attached to the edge BC. Given that the mass of the strip is 1.5 kg, find the distances of the centre of mass of the combined lamina and strip from AB and from AD.

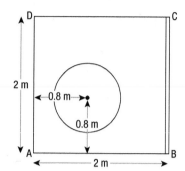

The mass of ABCD is $2^2 \times 1.2$ kg, i.e. 4.8 kg, and the mass removed for the hole is $\pi(0.5)^2 \times 1.2$ kg, i.e. 0.9425 kg.

|  |  | Distance of C of M | | | |
|---|---|---|---|---|---|
| Portion | Mass | $x$ from AD | $y$ from AB | Mass $\times x$ | Mass $\times y$ |
| + ABCD | 4.8 | 1 | 1 | 4.8 | 4.8 |
| − Hole | 0.9425 | 0.8 | 0.8 | 0.754 | 0.754 |
| + Strip | 1.5 | 2 | 1 | 3 | 1.5 |
| Whole | 5.358 | $\overline{X}$ | $\overline{Y}$ | $5.358\overline{X}$ | $5.358\overline{Y}$ |

Note. The circle is removed from the square, so we subtract the mass of the circle from the square and strip.

Using $\sum mx = \overline{x}\sum m$ gives $\qquad (4.8 - 0.754 + 3) = 5.358\ \overline{x}$

$\Rightarrow \qquad\qquad \overline{x} = 1.315$

Using $\sum my = \overline{y}\sum m$ gives $\qquad (4.8 - 0.754 + 1.5) = 5.358\ \overline{y}$

$\Rightarrow \qquad\qquad \overline{y} = 1.035$

The centre of mass is 1.32 m from AD and 1.04 m from AB (3 sf).

## Exercise 2

In this exercise all laminas and rods are uniform.

**1** State the coordinates of the centre of mass of each lamina.

a

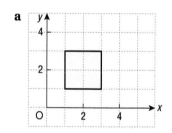

b

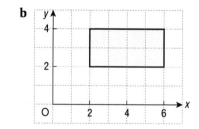

c

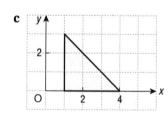

**2** State the coordinates of the centre of mass of each section of the given shape.

**a**

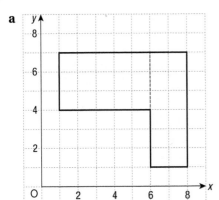

**b**

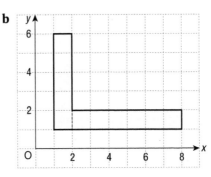

Keep your solutions to questions 3 to 5 because you will need them in Exercise 3.

In questions 3 to 5, choose axes based on symmetry and find the coordinates of the centre of mass of each lamina. Take the side of one square as 1 cm.

**3**

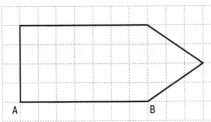

**4**

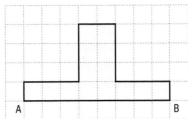

**5**

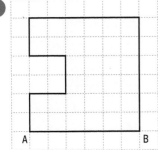

Find the centre of mass of each lamina.

**6**

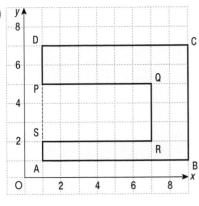

**7**

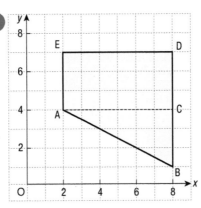

**8** The diagram shows a uniform right-angled triangular lamina of mass 3 kg, with a uniform circular lamina of mass 2 kg attached at B. Find the coordinates of the centre of mass of the system.

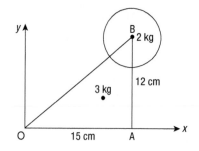

**9** Particles of masses 1 kg, 3 kg and 4 kg are attached to the vertices A, C and D respectively of a uniform square of side 2 cm and mass 6 kg. Find the distances of the centre of mass of the composite lamina from AB and AD.

**10** A uniform square lamina of mass 2 kg has a uniform rod of mass 3 kg attached along the edge AB as shown in the diagram.

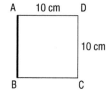

 **a** State the distance of the centre of mass of the composite shape from BC.

 **b** Find the distance of the centre of mass of the composite shape from AB.

**11** A uniform rectangular lamina of mass 30 kg has two uniform rectangular blocks attached at C and D as shown in the diagram. The mass of each block is 10 kg.

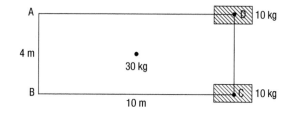

 **a** State the distance of the centre of mass of the composite shape from BC.

 **b** Find the distance of the centre of mass of the composite shape from AB.

**12** A uniform square lamina ABCD of side 6 cm has a mass of 2 kg. The point E is the midpoint of DC. A uniform triangular lamina ABE of mass 1.5 kg is attached to the square as shown in the diagram. Find the distance of the centre of mass from AB.

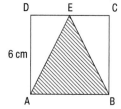

**13** The mass of the Earth is $5.976 \times 10^{24}$ kg and the mass of the Moon is $7.350 \times 10^{22}$ kg. The distance between their centres at perigee (when they are closest) is 356 400 km.

 **a** Find the distance of their centre of mass from the centre of the Earth at this time.

 **b** State any assumptions that you make.

# 12.4 Suspended bodies

When a body is suspended by a string attached to one point of the body, two forces act on the body: the tension in the string vertically upwards and the weight of the object vertically downwards.

When the body hangs at rest it is in equilibrium and the two forces must therefore be in the same line, so the centre of gravity, and therefore the centre of mass, is vertically below the point of suspension.

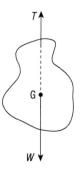

When the object is one whose centre of mass is known, a right-angled triangular lamina for example, the position in which the lamina will hang can be found by joining the point of suspension A to G.

In equilibrium, AG is vertical.

It is not always easy to see how to find the angle between the vertical and one of the sides of the lamina in its suspended position.

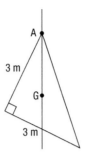

It can be more straightforward, if it is possible, to draw the lamina so that two of its sides are horizontal and vertical, and mark the line which would be vertical.

For example in the right-angled triangle shown above,

AG is vertical and G is distant 1 m from both BA and BC.

Therefore, to find the angle $\theta$ between AB and the vertical we use

$$\tan \theta = \frac{1}{2}$$

$$\Rightarrow \qquad \theta = 27° \text{ (nearest degree)}$$

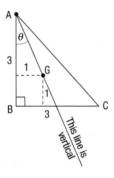

## Example 9

A uniform lamina ABCD is a trapezium of mass 1.5 kg in which AB = 4 cm, AD = 3 cm, DC = 1 cm and angle DAB is 90°.

The distance of the centre of mass, G, of the lamina, from AB and AD is $\frac{6}{5}$ cm and $\frac{7}{5}$ cm respectively.

**a** The lamina is suspended from A and hangs freely. Find the angle between AB and the vertical.

**b** The lamina is suspended from A and D by two strings with AD horizontal. Find the tensions in the strings.

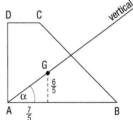

**a** When ABCD is hanging freely from A, AG is vertical, so the angle needed is angle GAB.

$$\tan \alpha = \frac{\overline{Y}}{\overline{X}} = \frac{6}{5} \div \frac{7}{5} = \frac{6}{7}$$

$\alpha = 41°$ (nearest degree)

**b** The forces acting are the tensions in the strings, $T_1$ and $T_2$ and the weight 1.5g N

They are all vertical and the lamina is in equilibrium.

Taking moments about A↺ : $T_2 \times 3 - 1.5\,g \times \frac{6}{5} = 0$

$$\Rightarrow \quad T_2 = \frac{1.5 \times 9.8 \times 1.2}{3} = 5.88$$

Vertically $\quad T_1 + T_2 = 1.5\,g \quad \Rightarrow \quad T_1 = 1.5 \times 9.8 - 5.88 = 8.82$

The tension in the string at A is 8.82 N and the tension in the string at B is 5.88 N.

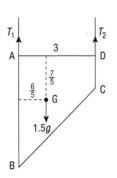

## Example 10

A sculpture is in the form of a uniform solid cylinder of radius 2 cm, height 8 cm and mass 4 kg, with a small lead bead of mass 1 kg attached to a point A on the rim of the base.

**a** Find the centre of mass of the sculpture.

**b** The sculpture is suspended from the point B, directly above A on the upper rim. Find the angle that AB makes with the vertical.

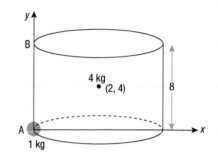

**a**

|  | | Coords of G | | | |
|---|---|---|---|---|---|
| Portion | Mass | $x$ | $y$ | $\Sigma mx$ | $\Sigma my$ |
| Cylinder | 4 | 2 | 4 | 8 | 16 |
| Lead bead | 1 | 0 | 0 | 0 | 0 |
| Sculpture | 5 | $\bar{X}$ | $\bar{Y}$ | $5\bar{X}$ | $5\bar{Y}$ |

Using $\Sigma mx = \bar{X}\Sigma m$ gives $8 = 5\bar{X} \Rightarrow \bar{X} = \dfrac{8}{5}$

Using $\Sigma my = \bar{Y}\Sigma m$ gives $16 = 5\bar{Y} \Rightarrow \bar{Y} = \dfrac{16}{5}$

Therefore the centre of mass is 1.6 cm from a vertical through A and 3.2 cm above the lower rim.

**b** A solid body suspended from a point P hangs in equilibrium with its centre of mass, G, vertically below P.

When the body has a plane of symmetry in which P is located and the position of G is known, the equilibrium position of the body can be found in the same way as if that plane of symmetry is a lamina.

There is a plane of symmetry containing G through A and B

BG is vertical therefore angle ABG is the angle needed.

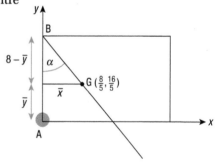

$\tan \alpha = \dfrac{\bar{X}}{8 - \bar{Y}} = \dfrac{\frac{8}{5}}{\frac{24}{5}} = \dfrac{1}{3} \Rightarrow \alpha = 18.4°$ (1 decimal place)

Therefore AB makes an angle of 18.4° with the vertical.

## Exercise 3

**1** When it is freely suspended from A, find the angle between AB and the vertical, for the lamina given in Exercise 2 question 3.

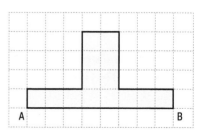

**2** When it is freely suspended from A, find the angle between AB and the vertical, for the lamina given in Exercise 2 question 4.

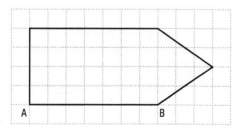

**3** When it is freely suspended from A, find the angle between AB and the vertical, for the lamina given in Exercise 2 question 5.

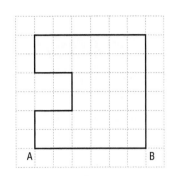

**4** The centre of mass of this uniform solid is $\dfrac{5}{3}$ cm from O. The solid is suspended from A.

Find the angle between AB and the vertical.

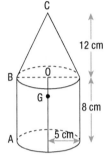

**5** The diagram shows a section through the centre of a uniform **frustum** of a cone. The centre of mass of the frustum of the cone is 4.7 cm above the base. The frustum is suspended freely from the point P.

Find the angle between the axis of symmetry and the vertical.

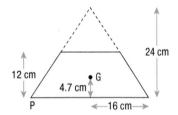

**6** The uniform solid shown in the diagram is freely suspended from C, the midpoint of AB. Find the angle between AB and the vertical given that the height above the base of the centre of mass is $\dfrac{83}{88}$ cm.

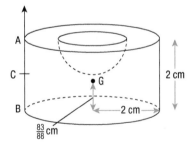

**7** This uniform solid is suspended from B and hangs freely.

The distance below O of the centre of mass is given by $\dfrac{h^2 - 24}{2(h + 4)}$.

Find $h$ given that

**a** AB is horizontal

**b** BC is horizontal.

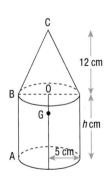

# Summary

For an object of normal size, the centre of mass coincides with the centre of gravity.

The distance of the centre of mass of a particular object from two perpendicular axes can be found from the equations $\Sigma m_n x_n = \overline{X} \Sigma m_n$ and $\Sigma m_n y_n = \overline{Y} \Sigma m_n$.

The centre of mass of a uniform lamina lies on any axis of symmetry that the lamina has.

The centre of mass of a triangle lies on the point of intersection of the medians. This is $\frac{1}{3}$ the distance from base to vertex of any median.

The centre of mass of a uniform solid lies in any plane of symmetry that the solid has.

When a body is suspended from a point, the line through that point and the centre of mass is vertical.

# Review

In questions 1 and 2, find the distance from A of the centre of mass of the given set of particles.

**1**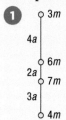

**2**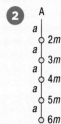

**3** A light L-shaped wire ABC is in the $xy$ plane with particles of equal mass 2 kg attached to A, B and C. The coordinates of A, B and C are $(-3, 2)$, $(2, 2)$ and $(2, -1)$ respectively. Find the centre of mass of the system.

**4** State the coordinates of the centre of mass of the uniform lamina shown in the diagram.

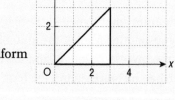

**5** State the coordinates of the centre of mass of each section of the uniform lamina shown in the diagram.

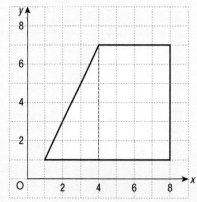

**6** A composite body is made up of a uniform square lamina ABCD of mass 1.5 kg and a uniform triangular lamina BEC of mass 1 kg attached to one side of the square as shown in the diagram.

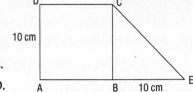

a Find the distance of the centre of mass of the composite body from AE.

b Find the distance of the centre of mass of the composite body from AD.

c The composite body is suspended in equilibrium from E.

Find the angle that AE makes with the vertical.

**7** A solid right circular cone, of height 4 cm and base radius 1 cm, is suspended freely from a point P on the circumference of the base. The centre of mass of the cone is on the line of symmetry, $\frac{3}{4}$ of the way from the vertex to the base.

Find the angle $\alpha$ between PO and the vertical, where O is the centre of the base.

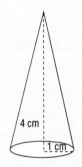

## Assessment

**1** The diagram shows a light lamina with particles A, B, C and D attached. The mass of A is 2 kg, the mass of B is 4 kg, the mass of C is 3 kg and the mass of D is 5 kg.

a Find the coordinates of G, the centre of mass of the system.

b The lamina is suspended from the point E and hanging in equilibrium.

Find the inclination of GE to the direction of the $x$-axis.

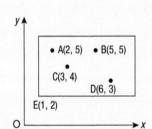

**2** The diagram shows a uniform lamina made up of three rectangles with the dimensions shown.

a Explain why the centre of mass lies on the line through the midpoint of AH and DE.

b Find the distance of the centre of mass from the side AH.

c The lamina is suspended from the point H and hangs freely in equilibrium. Find the angle made by AH with the vertical.

d The lamina is then hung from two strings attached at H and G with HG horizontal. 1 cm² has mass 1 kg. Find the tensions in the two strings.

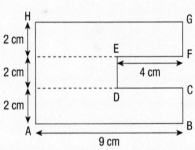

**3** The diagram shows a sculpture in the form of a sphere of mass 10 kg and radius 8 cm with a sphere of mass 2 kg and radius 3 cm attached as shown. The line AB goes through the centre of both spheres and through the point of attachment of the two spheres.

a Find the distance of the centre of mass of the sculpture from A.

b The point E on the surface of the larger sphere is on a diameter perpendicular to AB.

The sculpture is suspended from E and hangs in equilibrium.

Find the angle that AB makes with the vertical.

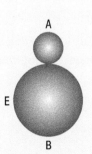

**4** The diagram shows a uniform rectangular lamina of mass 3 kg and measuring 6 cm by 4 cm. Three uniform circular laminas each of mass 1 kg are attached to the rectangle, one each with their centres at B and C and one with its centre at the midpoint of OA.

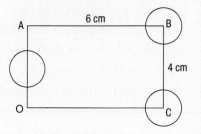

**a** Find the distance of the centre of mass of the composite body from

  **i** OC  **ii** OA

**b** The composite body is freely suspended from A.

Find the angle that AB makes with the vertical.

**5** A uniform rod *AB*, of mass 4 kg and length 6 metres, has three masses attached to it. A 3 kg mass is attached at the end *A* and a 5 kg mass is attached at the end *B*. An 8 kg mass is attached at a point *C* on the rod.

Find the distance *AC* if the centre of mass of the system is 4.3 m from point *A*.

AQA MM2 June 2015

**6** The diagram shows a uniform lamina which is in the shape of two identical rectangles *AXGH* and *YBCD* and a square *XYEF*, arranged as shown.

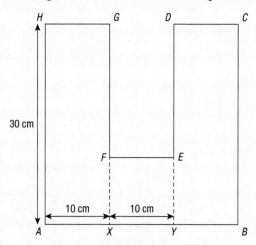

The length of *AX* is 10 cm, the length of *XY* is 10 cm and the length of *AH* is 30 cm.

**a** Explain why the centre of mass of the lamina is 15 cm from *AH*.

**b** Find the distance of the centre of mass of the lamina from *AB*.

**c** The lamina is freely suspended from the point *H*.

Find, to the nearest degree, the angle between *HG* and the horizontal when the lamina is in equilibrium.

AQA MM2 January 2013

# 13 Motion in One, Two and Three Dimensions

## Introduction

This chapter applies Newton's Law, $F = ma$, to motion in a straight line, motion in a plane with constant acceleration and to problems with variable acceleration where $F$ is a function of time.

## Recap

You will need to remember...

▶ Newton's laws of motion, $F = ma$ and 'action and reaction are equal and opposite'.

▶ The relationship between friction and normal reaction. Chapter 11

▶ Forces can be resolved into two components in perpendicular directions.

▶ The tension in a string is the same at each end.

▶ In a system of connected particles the acceleration of each particle is the same.

▶ The equations of motion for motion in a straight line with constant acceleration. Chapter 11

▶ When $s$, $v$ and $a$ are functions of time, $v = \dfrac{\mathrm{d}s}{\mathrm{d}t} \Rightarrow s = \int v \, \mathrm{d}t$ and $a = \dfrac{\mathrm{d}v}{\mathrm{d}t} \Rightarrow v = \int a \, \mathrm{d}t$.

## Objectives

By the end of this chapter, you should know how to...

▶ Solve problems involving the linear motion of a body under the action of constant forces.

▶ Solve problems, in two or three dimensions, involving a particle under the action of a variable force expressed as a function of time in vector form.

▶ Solve problems involving projectiles moving only under the action of gravitational attraction.

## Applications

When a golfer hits a ball, where it lands and whether it misses trees or other obstacles in its path depends (among other factors such as wind) on how hard the ball is hit and the angle at which it is hit. Assuming no wind and no air resistance, the initial speed and direction of the ball can be calculated to achieve the best outcome.

## 13.1 Linear motion with constant acceleration

When a body moves in a straight line, the forces acting on it are not always acting along the line. For example, a body can be pulled along horizontal ground by a rope inclined at an angle to the ground. If the surface is smooth, none of the forces acting on the block are horizontal.

A body can also move in a straight line on an inclined plane. All the forces, except friction if present, do not act along the plane.

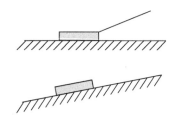

When the forces acting are constant, the resultant force perpendicular to motion is zero and the resultant force parallel to motion is used with $F = ma$ to find the acceleration.

## Example 1

A block of mass 40 kg is pulled along a horizontal plane by a rope inclined at $10°$ to the horizontal.

The coefficient of friction between the block and the ground is 0.2.

a Draw a diagram showing all the forces acting on the block.

b Find the tension in the rope that will make the block accelerate at $0.1 \text{ ms}^{-2}$.

a

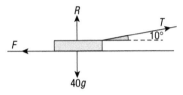

b The block is moving horizontally, therefore there is no resultant force vertically $\Rightarrow R = 40g - T \sin 10$.

$F = \mu R \quad \Rightarrow F = 0.2 \times (39.2 - T \sin 10)$

Using $F = ma$ horizontally gives $T \cos 10° - 7.84 \, g + 0.2 \, T \sin 10 = 40 \times 0.1$

Therefore $(T \cos 10 + 0.2 \sin 10) = 82.4$

The tension in the rope is 80.8 N.

## Example 2

A body of mass 3 kg is sliding down a smooth plane inclined at $30°$ to the horizontal.

a Find the acceleration of the mass in terms of g.

b Show that the normal reaction exerted by the plane on the mass is $\dfrac{3g\sqrt{3}}{2}$ N.

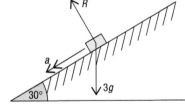 

The diagram shows all the forces that act on the body. The body can move down the plane, so the acceleration is marked in that direction. The acceleration is caused only by the components of force that act along the plane.

a The resultant force $\swarrow$ is $3g \sin 30° = 3g \left( \dfrac{1}{2} \right)$

Using $F = ma$ gives $\left( \dfrac{3}{2} \right) g = 3a \Rightarrow a = \dfrac{1}{2} g$

b As there is no acceleration perpendicular to the plane, the resultant of the forces in that direction is zero.

Resolving $\nwarrow$ gives $R - 3g \cos 30° = 0$.

$R = 3g \left( \dfrac{\sqrt{3}}{2} \right) = \dfrac{3g\sqrt{3}}{2}$

## Example 3

A cyclist is riding up a hill inclined at 20° to the horizontal. His speed at the foot of the hill is 10 ms⁻¹ but after 30 seconds it has dropped to 4 ms⁻¹. The total mass of the cyclist and his machine is 100 kg and there is a wind of strength 15 N down the slope.

**a** Find the acceleration of the cyclist assuming it is constant.

**b** Find the constant driving force exerted by the cyclist up the slope.

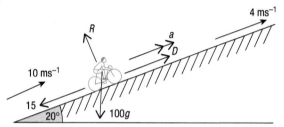

**a** For the motion up the slope:

Known $u = 10$ Required $a$

$v = 4$

$t = 30$

Using $v = u + at$ gives $4 = 10 + 30a$

$$a = -\frac{6}{30} = -\frac{1}{5}$$

The cyclist has an acceleration of $-\frac{1}{5}$ ms⁻².

**b** Resolving $\nearrow$ gives $D - 15 - 100g \sin 20°$

Using $F = ma \nearrow$ gives $D - 15 - 100g \sin 20° = 100a$

Therefore $D = 15 + 980(0.3420) + 100\left(-\frac{1}{5}\right) = 330.1...$

The cyclist's driving force is 330 N.

## Exercise 1

**1** The diagram shows a small block of mass 2 kg being pulled up a plane inclined at 30° to the horizontal.

The block has an acceleration of 0.5 ms⁻².

Find an expression in terms of $g$ for the tension in the string when

**a** the plane is smooth

**b** the plane is rough and exerts a frictional force of 4 N.

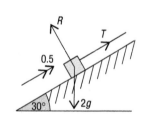

**2** A truck is being pulled along a horizontal track by two cables, against resistances totalling 1100 N, with an acceleration of 0.8 ms⁻².

One cable is horizontal and the other is inclined at 40° to the track.

The tensions in the cables are shown on the diagram.

**a** Find the mass of the truck.

**b** Find the vertical force exerted by the track on the truck.

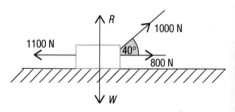

**3** The diagram shows a small block of mass 10 kg being pulled along a rough horizontal plane by a string inclined at 60° to the plane. There is a frictional force of 18 N.

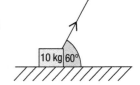

   **a** Copy the diagram and on it mark all the forces that act on the block.

   **b** The block has an acceleration of 3 ms⁻². Find the tension in the string.

   **c** Show that the normal reaction exerted by the plane on the block can be expressed as $(10g - 48\sqrt{3})$ N.

**4** A body of mass 6 kg is sliding down a smooth plane inclined at 30° to the horizontal. Its speed is controlled by a rope inclined at 10° to the plane as shown; the tension in the rope is 10 N. The body starts from rest. Find how far down the plane it travels in 5 seconds.

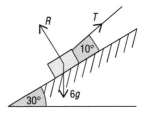

**5** A particle of mass 4 kg rests on a smooth plane inclined at 60° to the horizontal. The particle is attached to one end of a light inelastic string which passes over a fixed smooth pulley at the top of the plane and carries a particle of mass 2 kg at the other end.

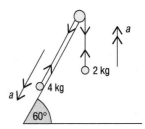

   **a** Find the acceleration of the system.

   **b** Find the tension in the string.

   **c** Find the force exerted on the pulley by the string.

**6** A block of mass 8 kg is sliding down a rough plane inclined at 20° to the horizontal. The block is accelerating at 0.2 ms⁻².

   Find the coefficient of friction between the block and the plane.

## 13.2 Newton's Law applied to motion with variable acceleration in two or three dimensions

Section 10.2 shows that when a particle is moving in a plane its motion in two perpendicular directions can be dealt with separately. When a particle is moving in three dimensions, its motion in three perpendicular directions can also be dealt with separately.

In each case, $\mathbf{F} = m\mathbf{a}$ where $\mathbf{F}$ is the force acting on the particle. When $\mathbf{F}$ is a function of $t$,

$$\mathbf{a} = \frac{d\mathbf{v}}{dt} \quad \Rightarrow \quad \mathbf{v} = \int \mathbf{a}\, dt \text{ and } \mathbf{v} = \frac{d\mathbf{s}}{dt} \quad \Rightarrow \quad \mathbf{s} = \int \mathbf{v}\, dt$$

### Example 4

A force, in newtons, at time $t$ seconds is given by $\mathbf{F} = 2\mathbf{i} + 3(t^2 - 1)\mathbf{j}$.
The force acts on a particle P, of mass 2 kg, moving in a horizontal plane.
When $t = 0$, P is at rest at the point with position vector $\mathbf{i} + \mathbf{j}$.

   **a** Find **i** the acceleration vector **ii** the position vector of P at time $t$.

   **b** Write down separate equations for the $x$ and $y$ coordinates of P at time $t$.

   **c** By eliminating $t$ from these two equations, find the Cartesian equation of the path of P.

**a  i**  Using Newton's Law, $\mathbf{F} = m\mathbf{a}$, gives  $2\mathbf{i} + 3(t^2 - 1)\mathbf{j} = 2\mathbf{a}$

$$\Rightarrow \quad \mathbf{a} = \mathbf{i} + \frac{3}{2}(t^2 - 1)\mathbf{j}$$

**ii**  First find the velocity vector.

$$\mathbf{v} = \int \mathbf{a}\ dt = \int \left[ \mathbf{i} + \frac{3}{2}(t^2 - 1)\mathbf{j} \right] dt = t\mathbf{i} + \left( \frac{1}{2}t^3 - \frac{3}{2}t \right)\mathbf{j} + \mathbf{A}$$

When $t = 0, \mathbf{v} = 0$ therefore $\mathbf{A} = 0$

therefore $\quad \mathbf{v} = t\mathbf{i} + \left( \frac{1}{2}t^3 - \frac{3}{2}t \right)\mathbf{j}$

$$\mathbf{r} = \int \mathbf{v}\ dt = \frac{1}{2}t^2\mathbf{i} + \left( \frac{1}{8}t^4 - \frac{3}{4}t^2 \right)\mathbf{j} + \mathbf{B}$$

When $t = 0, \mathbf{r} = \mathbf{i} + \mathbf{j}$ therefore $\mathbf{i} + \mathbf{j} = \mathbf{B}$

therefore $\quad \mathbf{r} = \frac{1}{2}t^2\mathbf{i} + \left( \frac{1}{8}t^4 - \frac{3}{4}t^2 \right)\mathbf{j} + \mathbf{i} + \mathbf{j} \Rightarrow \mathbf{r} = \left( \frac{1}{2}t^2 + 1 \right)\mathbf{i} + \left( \frac{1}{8}t^4 - \frac{3}{4}t^2 + 1 \right)\mathbf{j}$

**b**  $x = \frac{1}{2}t^2 + 1$ $\qquad$ [1]

$y = \frac{1}{8}t^4 - \frac{3}{4}t^2 + 1$ $\qquad$ [2]

**c**  From [1]  $t^2 = 2(x - 1)$

Substituting in [2] gives  $y = \frac{1}{8}[4(x-1)^2] - \frac{3}{4}[2(x-1)] + 1$

Multiplying throughout by 2 gives  $2y = (x-1)^2 - 3(x-1) + 2$

$$\Rightarrow \qquad 2y = x^2 - 5x + 6 = (x-2)(x-3)$$

## Example 5

A particle P of mass 2 kg is moving in space. At any time $t$ the position vector of P is $\mathbf{r} = (\cos t)\mathbf{i} + t^2\mathbf{j} - (1 - e^t)\mathbf{k}$.

Show that the magnitude of the force acting on P at time $t = \frac{1}{2}\pi$ is given by $\sqrt{16 + 4e^\pi}$.

First find the acceleration of P, then apply Newton's Law.

$\mathbf{r} = (\cos t)\mathbf{i} + t^2\mathbf{j} - (1 - e^t)\mathbf{k} \Rightarrow \mathbf{v} = \dfrac{d\mathbf{r}}{dt} = (-\sin t)\mathbf{i} + 2t\mathbf{j} + e^t\mathbf{k}$

and $\mathbf{a} = \dfrac{d\mathbf{v}}{dt} = (-\cos t)\mathbf{i} + 2\mathbf{j} + e^t\mathbf{k}$

Using $\mathbf{F} = m\mathbf{a}$ gives $\mathbf{F} = (-2\cos t)\mathbf{i} + 4\mathbf{j} + 2e^t\mathbf{k}$

When $t = \frac{1}{2}\pi$, $\qquad \mathbf{F} = 4\mathbf{j} + 2e^{\frac{\pi}{2}}\mathbf{k}$

Therefore $\qquad |\mathbf{F}| = \sqrt{16 + 4e^\pi}$

## Exercise 2

**1**  A force $\mathbf{F}$ acts on particle P of mass 2 kg. Given that $\mathbf{F} = 4t\mathbf{i} + 6\mathbf{j}$, and that P is initially at O with velocity $5\mathbf{j}$, find $\mathbf{v}$ and $\mathbf{r}$ when $t = 3$.

**2**  A force represented by $12\mathbf{j}$ is the only force acting on a particle P. Initially P, whose mass is 3 kg, is at O with a speed of 6 ms⁻¹ in the direction O$x$.

$\quad$ **a**  Find $\qquad$ **i**  the initial velocity vector $\mathbf{v}$ $\qquad$ **ii**  the acceleration vector $\mathbf{a}$

$\qquad\qquad\qquad$ **iii**  $\mathbf{v}$ and $\mathbf{r}$ after $t$ seconds.

$\quad$ **b**  Show that P moves on a parabola.

3. Two forces **P** and **Q**, of magnitudes $P$ newtons and $Q$ newtons, act on a particle A of mass 2 kg where $\mathbf{P} = 6\mathbf{i} - \mathbf{j}$ and $\mathbf{Q} = 2\mathbf{i} + 5\mathbf{j}$.

   **a** Find the resultant of **P** and **Q** and hence the acceleration of A.

   **b** The particle A is initially at rest at a point with position vector $3\mathbf{j}$. Find the position vector of A after $t$ seconds.

4. A force $\mathbf{F}_1 = 8\mathbf{i} + 6\mathbf{j}$ and a force $\mathbf{F}_2 = 6\mathbf{i} + 2t\mathbf{j}$ act on a particle P of mass 2 kg. The particle is initially at rest at the origin.

   **a** Find the resultant of $\mathbf{F}_1$ and $\mathbf{F}_2$.

   **b** Find the acceleration of P and the velocity of P after $t$ seconds.

   **c** Find the position vector of P after 2 seconds.

5. A particle P of mass 2 kg is at rest, when a force $4\mathbf{i} + 6t\mathbf{j} + \mathbf{k}$ acts on it for one second.

   Find the velocity of P at the end of this time.

6. At any time $t$ seconds, the position vector of a particle P of mass 4 kg is given by $\mathbf{r} = 6t\mathbf{i} - \mathbf{j} + (\sin t)\,\mathbf{k}$. Only one force, **F**, acts on P.

   **a** Find the velocity and acceleration vectors of P at time $t$.

   **b** Find the force **F**.

7. A force represented by $4\mathbf{i} - 6\mathbf{j} - e^t\mathbf{k}$ acts on a particle P of mass 2 kg which is initially at rest at the origin O.

   **a** Find the acceleration of P.

   **b** Find the velocity of P after $t$ seconds.

## 13.3 Projectiles

A particle that is thrown into the air is called a **projectile**. Assuming that air resistance is small enough to be ignored, the only force acting on a projectile, once it has been thrown, is its own weight.

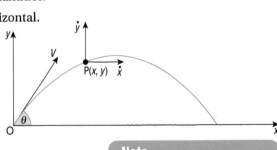

It follows that the projectile has an acceleration downward of magnitude $g$, but has constant velocity horizontally.

Therefore the horizontal motion and the vertical motion are analysed separately.

The horizontal and vertical components of velocity and displacement are involved and a new form of notation is used to denote these quantities.

A particle is projected with a velocity $V$ at an angle $\theta$ to the horizontal.

Take O as the point of projection, $Ox$ as a horizontal axis and $Oy$ as a vertical axis.

At any time during its flight, the projectile is at a point where:

▶   the horizontal displacement from O is $x$

▶   and the vertical displacement from O is $y$.

The horizontal velocity is the rate at which $x$ increases with respect to time and this is denoted by $\dot{x}$.

Similarly the vertical velocity at any time is denoted by $\dot{y}$.

The initial components of velocity are:

     $V \cos\theta$   in the direction $Ox$

     $V \sin\theta$   in the direction $Oy$.

> **Note**
>
> The dot over the $x$ and $y$ means 'the rate of increase with respect to time'.

For the horizontal motion, where the velocity is constant, at any time $t$ seconds after projection,

$$\dot{x} = V \cos \theta \qquad\qquad [1]$$

and by integration $\quad x = (V \cos \theta) \times t = Vt \cos \theta \qquad [2]$

For the vertical motion where there is an acceleration $g$ downwards,

using $v = u + at$ and $s = ut + \dfrac{1}{2}at^2$ where $u = V \sin \theta$, gives

$$\dot{y} = V \sin \theta - gt \qquad\qquad [3]$$

$$y = (V \sin \theta) \times t - \frac{1}{2}gt^2 = Vt \sin \theta - \frac{1}{2}gt^2 \qquad [4]$$

These four equations provide all the information needed to solve problems on the motion of any projectile.

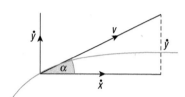

For example, to find the speed $v$ of the particle at a particular time, $\dot{x}$ and $\dot{y}$ can be found from equations [1] and [3], then $v^2 = \dot{x}^2 + \dot{y}^2$

Also, the direction of motion (the way the particle is moving), is given by the direction of the velocity. So when the direction of motion makes an angle $\alpha$ with the horizontal, then $\tan \alpha = \dfrac{\dot{y}}{\dot{x}}$.

## Example 6

A particle P is projected from a point O with a speed of 40 ms⁻¹, at 60° to the horizontal. Find, 3 seconds after projection,

a   the speed of the particle

b   the horizontal and vertical displacements of the particle from O

c   the distance of P from O.

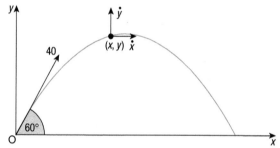

a   Use equations [1] and [3] with $V = 40$, $\theta = 60°$ and $t = 3$.

Horizontally,   $\dot{x} = 40 \cos 60° = 20$

Vertically,     $\dot{y} = 40 \sin 60° - gt = 34.64 - 29.4$

$\Rightarrow$   $\dot{y} = 5.24$

If $v$ ms⁻¹ is the speed of the particle after 3 seconds,

$v^2 = 5.24^2 + 20^2 \quad \Rightarrow \quad v = 20.67...$

The speed of the particle is 20.7 ms⁻¹.

b   Use equations [2] and [4].

Horizontally,   $x = 40 \times 3 \cos 60° = 60$

Vertically,     $y = 40 \times 3 \sin 60° - \dfrac{1}{2}gt^2 = 103.92 - 44.1$

$y = 59.82...$

The displacements of the particle from O are:

60 m horizontally and 59.8 m vertically.

c   $OP^2 = x^2 + y^2$

$= 60^2 + 59.82^2 = 7178.7...$

The distance of P from O is 84.7 m.

# Example 7

A particle P is projected from a point 5 m above the ground. The horizontal and vertical components of the velocity of projection are each 24 ms⁻¹.

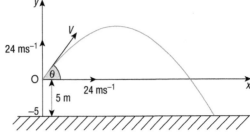

**a** Find the angle of projection.

**b** Taking $g$ as 10, find the horizontal distance of P from the point of projection when it hits the ground.

**a** When the velocity of projection is $V$ then

$V\cos\theta = 24$ and $V\sin\theta = 24$

$\tan\theta = \dfrac{V\sin\theta}{V\cos\theta} = \dfrac{24}{24} = 1$

$\theta = 45°$

**b** When the particle hits the ground it is 5 m *below* the point of projection so $y = -5$.

At any time $t$, $y = 24t - \dfrac{1}{2}gt^2$

when $y = -5$, $-5 = 24t - 5t^2 \Rightarrow 5t^2 - 24t - 5 = 0$

Hence $(5t+1)(t-5) = 0 \Rightarrow t = -\dfrac{1}{5}$ or 5

A negative time has no meaning in this problem, so $t = 5$.

At any time $t$, $x = 24t$ so, when $t = 5$, $x = 120$.

When P hits the ground its horizontal distance from O is 120 m.

# Example 8

The top of a tower is 10 m above horizontal ground. A boy fires a stone from a catapult with a speed of 12 ms⁻¹. Find how far from the foot of the tower the stone hits the ground if

**a** it is fired horizontally

**b** it is fired at 30° below the horizontal.

**a** When the initial velocity is horizontal, $\theta = 0$, therefore $\sin\theta = 0$ and $\cos\theta = 1$.

At any time $t$, $x = Vt$ and $y = -\dfrac{1}{2}gt^2$

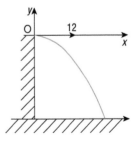

When the stone hits the ground, $y = -10$.

$-\dfrac{1}{2}gt^2 = -10 \Rightarrow t^2 = \dfrac{20}{9.8} = 2.04$

$t = 1.4$ ($t$ cannot be negative)

When $t = 1.4$, $x = Vt = 12 \times 1.4 = 16.8$

The stone hits the ground 16.8 m from the foot of the tower.

**b** The stone is fired at a downward angle so $\theta$ is a negative acute angle.

At any time $t$, $x = Vt\cos(-30°) = Vt\cos 30°$

and $y = Vt\sin(-30°) - \dfrac{1}{2}gt^2 = -Vt\sin 30° - \dfrac{1}{2}gt^2$

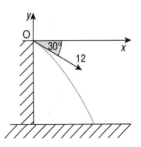

*(continued)*

*(continued)*

When the stone hits the ground, $y = -10$.

$$12t\left(-\frac{1}{2}\right) - \frac{1}{2}gt^2 = -10 \implies 4.9t^2 + 6t - 10 = 0$$

$$\implies \quad t = \frac{-6 \pm \sqrt{232}}{9.8}$$

$t = 0.9419$ ($t$ cannot be negative)

When $t = 0.9419$, $x = Vt\cos 30° = 12 \times 0.9419 \times 0.8660$

$$= 9.789$$

The stone hits the ground 9.79 m from the foot of the tower.

## Exercise 3

In questions 1 to 5, a particle P is projected from a point O on a horizontal plane with velocity $V$ at an angle $\theta$ to the horizontal. All units are based on metres and seconds.

1 Given that $V = 24.5$ and $\theta = 30°$, find the speed of P after

    **a** 1 second     **b** 2 seconds.

2 When $V = 20$ and $\theta = 60°$, find the height of P above the plane when

    **a** $t = 1$     **b** $t = 2$     **c** $t = 3$.

3 Given that $V = 30$ and $\tan\theta = \frac{3}{4}$, find the speed and the coordinates of the position of P after

    **a** 1 second

    **b** 2 seconds.

4 When $V = 10$ and $\theta = 60°$, find the time taken for P to travel a horizontal distance of 5 m and find the height of P at this time.

5 After 4 seconds P hits the plane at a point A. When $\theta = 45°$

    **a** find $V$

    **b** find the distance OA.

6 A particle P is projected from a point O with a velocity of 10 ms⁻¹ at an angle of 30° to the horizontal. Find the horizontal and vertical displacements of P from O after half a second and hence find the distance from O to P at this time.

7 A man throws a ball with a speed of 26 ms⁻¹ at an angle $\alpha$ above the horizontal, where $\tan\alpha = \frac{12}{5}$. (Take $g = 10$ in this question.)

    **a** Find the times at which the ball is 16 m above the ground.

    **b** Find the horizontal distance covered between these times.

8 A stone is thrown downwards from a point A into a quarry that is 25 m deep. Find the initial speed and the direction of projection if, after 2 seconds, the stone lands at the bottom of the quarry at a horizontal distance from A of

    **a** 30 m

    **b** 20 m.

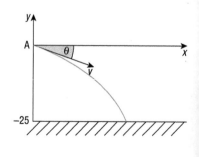

9 A particle is projected with a velocity of $20\sqrt{2}$ ms⁻¹ at an angle of 45° to the horizontal. Find how long it is before the particle reaches its highest point and what the greatest height is.

**10** A girl throws a ball from a window to her friend on the ground 24 m below. The ball is thrown with speed 10 ms⁻¹ at an angle $\alpha$ above the horizontal where $\tan \alpha = \dfrac{5}{12}$.

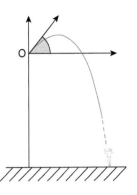

**a** Find how long it takes for the ball to reach ground level.

**b** If the friend is standing 8 m horizontally from the window, will she be able to catch the ball without moving?

**11** A particle P is projected from a point O with a velocity of 25 ms⁻¹ at an angle of 60° to the horizontal. For how long is P at least 15 m above the level of O?

**12** A stone is thrown from the top of a cliff at an angle $\alpha$ to the horizontal and with a speed of 19.5 ms⁻¹.

The stone falls into the water 37.5 m from the foot of the cliff.

Given that $\tan \alpha = \dfrac{12}{5}$, find the height of the cliff when $\alpha$ is

**a** above the horizontal

**b** below the horizontal.

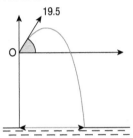

## The equation of the path

Using $x$ and $y$ axes through the point of projection, the coordinates of the position of the projectile at any time can be expressed as

$x = Vt \cos \theta$ and $y = Vt \sin \theta - \dfrac{1}{2}gt^2$. As these two equations give $x$ and

$y$ each in terms of the variable $t$, they are the parametric equations of the path of the projectile (which is also called the **trajectory**).

From the first equation, $t = \dfrac{x}{V \cos \theta}$.

Substituting in the second equation gives

$$y = \frac{Vx \sin \theta}{V \cos \theta} - \frac{1}{2}g\left(\frac{x}{V \cos \theta}\right)^2$$

$$\Rightarrow \qquad y = x \tan \theta - \left(\frac{gx^2}{2V^2}\right)\sec^2\theta$$

> **Note**
>
> You cannot quote this equation but you need to know how to derive it. Remember that it is derived by eliminating $t$ from the equations $x = Vt \cos \theta$ and $y = Vt \sin \theta - \dfrac{1}{2}gt^2$

For any particular projectile, $g$, $V$ and $\theta$ are constants, so $y$ is a quadratic function of $x$, showing that the path of a projectile is a parabola with a vertical axis of symmetry.

## The greatest height reached

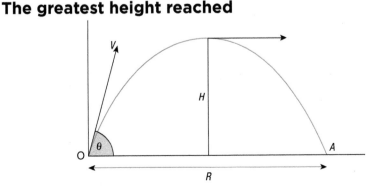

When P is at its greatest height it is momentarily travelling horizontally so $\dot{y} = 0$.

Using $\qquad v^2 = u^2 + 2as$

gives $\qquad 0 = V^2 \sin^2 \theta - 2gH$

$$\Rightarrow \qquad \boldsymbol{H = \dfrac{V^2\sin^2\theta}{2g}.}$$

## The range on a horizontal plane

The range of a projectile on a horizontal plane is the distance between the point of projection O and the point where the projectile returns to the level of O. The range is usually denoted by $R$ as shown in the previous diagram.

When the projectile reaches A, $y = 0$,

therefore $\qquad Vt \sin \theta - \dfrac{1}{2} gt^2 = 0 \qquad \Rightarrow \qquad t = \dfrac{2V \sin \theta}{g}$

This is the time taken for the whole of the journey; it is called the **time of flight** (and it is twice the time taken to reach the highest point on the path).

Now $R$ is the value of $x$ (= $Vt \cos \theta$) at this time,

so $\qquad R = V \cos \theta \left( \dfrac{2V \sin \theta}{g} \right) = \dfrac{V^2}{g} (2 \sin \theta \cos \theta)$

Therefore $\qquad R = \dfrac{2V^2 \sin \theta \cos \theta}{g} = \dfrac{V^2 \sin 2\theta}{g}$

## The maximum horizontal range

The formula for the horizontal range of a projectile shows that the value of $R$ depends upon the initial velocity *and* the angle of projection.

Therefore, for a given value of $V$, the range varies only with $\theta$ and the maximum range occurs when $\sin 2\theta$ is maximum.

The greatest value of $\sin 2\theta$ is 1 and this is when $2\theta = 90° \Rightarrow \theta = 45°$

Therefore the *maximum* horizontal range, $R_{\max}$, is $\dfrac{V^2}{g}$ and it is achieved when $\theta = 45°$.

## Example 9

A gun is fired with a rubber bullet at a speed of 100 ms$^{-1}$ in a tunnel whose roof is 4 m above the point of projection.

a   Find the greatest possible angle of projection if the bullet is not to hit the roof.

b   Find the range of the gun with this angle of projection.

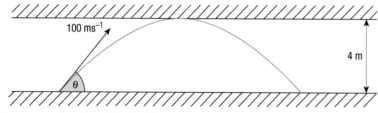

a   The greatest height of the bullet is 4 m above O.

*(continued)*

*(continued)*

Using $v^2 = u^2 + 2as$ gives

$$0 = 100^2 \sin^2 \theta - (2)(9.8)(4)$$

therefore $\sin^2 \theta = \dfrac{78.4}{10\,000} = 0.00784$

$\Rightarrow \quad \sin \theta = 0.08854 \quad \Rightarrow \quad \theta = 5.1°$

**b** The range is the value of $x$ when $y = 0$, that is when $100t(0.08854) - 4.9t^2 = 0$

$\Rightarrow \quad t = 0$ or $t = \dfrac{8.854}{4.9} = 1.8069...$

When $t = 0$, $x = 0$, when $t = 1.8069$, $x = 100 \times 1.8069...$

Therefore the range is 181 m.

## Example 10

A stone catapulted from a point O, with velocity 40 ms⁻¹ at an angle $\alpha$ to the horizontal, passes through a point distant 32 m horizontally and 45 m vertically from O.

Show that there are two possible angles of projection and give their values.

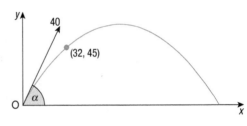

The coordinates of a point on the path of the stone are given and the time when the point is reached does not matter; so use

$$x = Vt \cos \theta \quad \text{and} \quad y = Vt \sin \theta - \frac{1}{2} gt^2$$

$\Rightarrow \quad 32 = 40t \cos \alpha \quad \text{and} \quad 45 = 40t \sin \alpha - 4.9t^2$

Eliminating $t$ gives $45 = 32 \tan \alpha - \dfrac{9.8 \times (32)^2}{2 \times 1600 \times \cos^2 \alpha} = 32 \tan \alpha - \dfrac{9.8 \times (32)^2 \times (\tan^2 \alpha + 1)}{2 \times 1600}$

(Using $\dfrac{1}{\cos^2 \alpha} = \sec^2 \alpha = \tan^2 \alpha + 1$)

Using $\tan \alpha = T$, this equation simplifies to

$$45 = 32T - 3.136(T^2 + 1)$$

$\Rightarrow \quad 3.136T^2 - 32T + 48.136 = 0$

This is a quadratic equation for $T$ in which $a = 3.136$, $b = -32$, $c = 48.136$.

Solving the equation by using the formula gives

$$T = \tan \alpha = \frac{32 \pm \sqrt{420.18...}}{6.272} = \frac{32 \pm 20.49...}{6.272} = 8.3689 \text{ or } 1.8351$$

Therefore the two values of $\alpha$ are 61.4° and 83.2°.

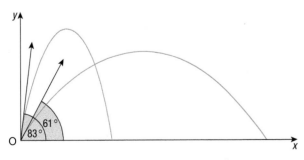

# Exercise 4

In questions 1 to 5, a particle P is projected from a point O on a horizontal plane, with speed $V$ and angle of elevation $\theta$. The greatest height reached is $H$ and the range on the plane is $R$. Do not use the formulae for $H$ and $R$, derive them.

1. $V = 24 \text{ ms}^{-1}$ and $\theta = 30°$, find $H$ and $R$.

2. $V = 20 \text{ ms}^{-1}$ and $R = 28$ m, find $\theta$.

3. $\theta = 45°$ and $H = 10$ m, find $V$ and $R$.

4. $R = 60$ m and $\theta = 60°$, find $V$ and $H$.

5. $H = 16$ m and $V = 40 \text{ ms}^{-1}$, find $\theta$ and $R$.

6. A stone is thrown with speed $26 \text{ ms}^{-1}$ at an angle $\alpha$ above the horizontal where $\tan \alpha = \dfrac{5}{12}$. Find how far it has travelled horizontally when

   a  it is at the same level as the point of projection

   b  it is 2 m below the point of projection.

7. A particle is projected from a point O with speed $50 \text{ ms}^{-1}$ at an angle of elevation of $40°$.

   a  Taking O$x$ as the horizontal axis and O$y$ vertically upward, find the equation of the flight path of the particle.

   b  Hence find the height of the particle when it is distant 20 m horizontally from O.

8. A particle is projected from a point on a horizontal plane with a speed of $12 \text{ ms}^{-1}$. The time of flight is half a second.

   a  Find the angle of projection.

   b  Find the range.

9. Use a horizontal $x$-axis and a vertical $y$-axis to find the equation of the path of a projectile whose initial velocity is

   a  $5 \text{ ms}^{-1}$ at an angle $\theta$ above the horizontal where $\tan \theta = \dfrac{4}{3}$

   b  $30 \text{ ms}^{-1}$ at an angle of $45°$ below the horizontal.

   Take $g = 10$ in this question.

10. A ball is thrown with a speed of $16 \text{ ms}^{-1}$ from a point on horizontal ground. The angle of projection is $\alpha$ where $\tan \alpha = \dfrac{3}{4}$.

    a  Find the time for which the ball is in the air.

    b  Find the horizontal distance it travels in this time.

11. A stone is thrown from ground level with a speed of $15 \text{ ms}^{-1}$, so that when it is travelling horizontally it just passes over a tree of height 3 m. Find the angle of projection.

**12** A ball is fired at 80 ms⁻¹ at an angle $\alpha$ to the horizontal. The ball must pass over an obstruction that is 20 m high and 120 m away in the line of flight. Find the smallest possible value of $\alpha$. Take g as 10 in this question.

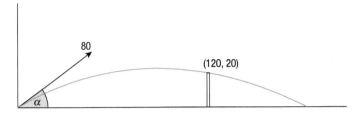

# Summary

## Linear motion with constant acceleration

When a body moves in a straight line with constant acceleration under the action of forces that are not collinear

▶ the resultant force perpendicular to the direction of motion is zero

▶ the resultant force in the direction of motion is equal to mass times acceleration.

## Motion with variable acceleration in two or three dimensions

When a particle is moving in two or three dimensions, its motion in two or three perpendicular directions can be dealt with separately.

In each case, $\mathbf{F} = m\mathbf{a}$ where $\mathbf{F}$ is the force acting on the particle and when $\mathbf{F}$ is a function of $t$,

$$\mathbf{a} = \frac{\mathrm{d}\mathbf{v}}{\mathrm{d}t} \Rightarrow \mathbf{v} = \int \mathbf{a}\,\mathrm{d}t \text{ and } \mathbf{v} = \frac{\mathrm{d}\mathbf{s}}{\mathrm{d}t} \Rightarrow \mathbf{s} = \int \mathbf{v}\,\mathrm{d}t$$

## Projectiles

Problems on the motion of a projectile are solved using the equations of motion horizontally and vertically where

$\dot{x} = V\cos\theta$ and $x = Vt\cos\theta$

$\dot{y} = V\sin\theta - gt$ and $y = Vt\sin\theta - \frac{1}{2}gt^2$

The equation of the path of a projectile is derived by eliminating $t$ from the equations $x = Vt\cos\theta$ and $y = Vt\sin\theta - \frac{1}{2}gt^2$.

The greatest height reached by a projectile can be found by using $v^2 = u^2 + 2as$ when $v$, and so $\dot{y}$ is zero.

The range of a projectile is found by finding the value of $x$ at the time when $y = 0$.

# Review

1. A small block of mass 3 kg is being pulled by a rope along a rough horizontal plane.

   The coefficient of friction between the block and the plane is 0.5.

   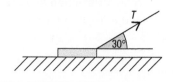

   **a** Copy the diagram and show all the forces acting on the block.

   **b** Find an expression in terms of $T$ for the normal reaction.

   **c** The tension in the rope is 40 N. Find the acceleration of the block.

2. A particle of mass 3 kg moves in three dimensions under the action of a single force **F** N. At time $t$ seconds the postion vector of the particle from the point O is given by $\mathbf{r} = 2t\mathbf{i} + 3e^{-t}\mathbf{j} + t^3\mathbf{k}$.

   **a** Find the velocity vector at time $t$.

   **b** Find the acceleration vector at time $t$.

   **c** Hence find the vector **F**.

3. An arrow is shot from the top of a building 26 m high. The initial speed of the arrow is 30 ms$^{-1}$ and it is fired at an angle of 20° above the horizontal. Find how long the arrow is in the air.

4. A golf ball is hit at 25 ms$^{-1}$ from a point O towards the green which is 2 m below the level of projection. The ball is hit at an angle $\theta$ to the horizontal where $\tan\theta = \dfrac{7}{24}$ and lands in the hole. Find the horizontal distance from O to the hole.

   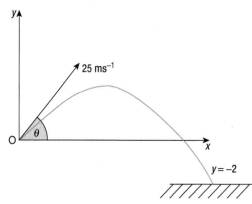

5. A ball is hit with speed 36 ms$^{-1}$ at a height of 0.5 m above level ground. The ball is hit at an angle $\theta$ to the ground.

   **a** Find the equation of the path of the ball in terms of $\theta$.

   **b** The ball is caught 70 m away at a height of 2.2 m. Find the angle at which the ball was hit.

   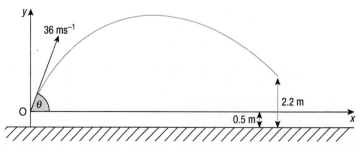

# Assessment

**1** A crate of mass 100 kg is being pulled by a rope up a slope inclined at 10° to the horizontal . The coefficient of friction between the crate and the slope is 0.2 and the rope attached to the crate is parallel to the slope.

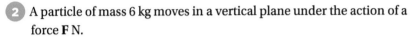

  **a** Copy the diagram and show all the forces acting on the crate.

  **b** The crate is accelerating at 0.1 ms⁻². Find the tension in the rope.

**2** A particle of mass 6 kg moves in a vertical plane under the action of a force **F** N.

At time $t$ seconds, the velocity, **v** ms⁻¹, of the particle is given by $\mathbf{v} = 3t\mathbf{i} - \cos t\,\mathbf{j}$ where **i** and **j** are unit vectors in the horizontal and vertical directions respectively.

  **a** Find the acceleration **a** ms⁻² of the particle at time $t$.

  **b** Find the force **F** N acting on the particle at time $t$.

  **c** Find the magnitude of **F** when $t = \pi$.

  **d** Find, in the interval $0 \le t \le 2\pi$, the values of $t$ when **F** is acting horizontally.

  **e** When $t = 0$, the position vector of the particle is 2**i**. Find the position vector of the particle at time $t$.

**3** A golf ball is struck with a speed of 26 ms⁻¹ at an angle $\alpha$ to the horizontal where $\tan \alpha = 2$.

  **a** Find the time taken for the ball to travel 20 m horizontally.

  **b** Find the speed of the ball when it has travelled 20 m horizontally.

  **c** Find the equation of the path of the ball.

  **d** A tree is 22 m high and is 20 m from the point of projection and in the path of the ball.

  Determine whether the ball will clear the tree.

**4** The speed of a tennis player's serve is 45 ms⁻¹. At the moment when the racquet strikes, the ball is exactly over the base line and at a height of 2.8 m. The height of the net is 0.9 m and the distance from the base line to the point where the ball reaches the net is 12 m. The ball is served at an angle of 8° below the horizontal.

  **a** Find the time it takes for the ball to travel 12 m towards the net.

  **b** Find the direction in which the ball is moving when it reaches the net.

  **c** Determine whether the ball goes over the net.

**5** A small block A of mass 6 kg rests on a horizontal table and is connected by a light inextensible string that passes over a smooth pulley, fixed on the edge of the table, to another small block B of mass 5 kg which is hanging freely.

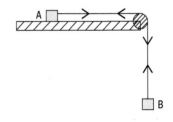

Find the acceleration of the system and the tension in the string if

  **a** the table is smooth

  **b** the table is rough and exerts a frictional force of 20 N.

**6** A particle, of mass 50 kg, moves on a smooth horizontal plane.
A single horizontal force

$$[(300t - 60t^2)\mathbf{i} + 100e^{-2t}\mathbf{j}] \text{ newtons}$$

acts on the particle at time $t$ seconds.

The vectors $\mathbf{i}$ and $\mathbf{j}$ are perpendicular unit vectors.

**a** Find the acceleration of the particle at time $t$.

**b** When $t = 0$, the velocity of the particle is $(7\mathbf{i} - 4\mathbf{j})$ ms$^{-1}$.

Find the velocity of the particle at time $t$.

**c** Calculate the speed of the particle when $t = 1$.

AQA MM2 January 2012

**7** In a scene from an action movie, a car is driven off the edge of a cliff and
lands on the deck of a boat in the sea, as shown in the diagram.

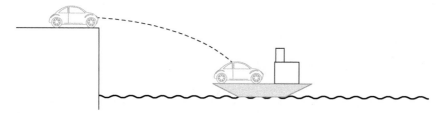

To land on the boat, the car must move 20 metres horizontally from the
cliff. The level of the deck of the boat is 8 metres below the top of the cliff.
Assume that the car is a particle which is travelling horizontally when it
leaves the top of the cliff and that the car is not affected by air resistance as
it moves.

**a** Find the time that it takes for the car to reach the deck of the boat.

**b** Find the speed at which the car is travelling when it leaves the top of
the cliff.

**c** Find the speed of the car when it hits the deck of the boat.

AQA MM1 June 2013

# 14 Work, Power and Energy

## Introduction

Work, energy and power are all closely related. There are different forms of energy such as light, heat, sound, electrical energy and chemical energy. These can often be converted from one form to another, for example electrical energy can be used to give heat or light energy.

This chapter looks at these concepts in relation to forces acting on bodies.

## Recap

You will need to remember...

▶ The component of a force $F$ at an angle $\theta$ to the direction of $F$ is $F\cos\theta$.

▶ $\sin^{-1} x$ means the angle whose sine is $x$.

## Applications

Anyone pushing a heavy crate across a floor will probably think that it is hard work. Making an effort to move an object is a common concept of work. That effort also uses energy and power. The power of a car is often quoted in the manufacturer's specification because it gives an indication of how fast the car can go.

## Objectives

By the end of this chapter, you should know how to...

▶ Define the work done by a force.

▶ Use the definition of the power of a moving vehicle and use it to solve problems.

▶ Find kinetic energy and potential energy.

▶ Understand and use the work-energy principle.

▶ Understand and use conservation of mechanical energy.

## 14.1 Work

When an object moves under the action of a constant force $F$, the amount of **work** done by the force is given by:

the component of $F$ in the direction of motion × distance moved.

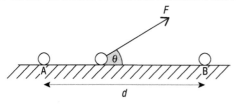

When a constant force $F$ moves an object from A to B, the amount of work done by $F$ is

$(F\cos\theta) \times d = Fd\cos\theta$

When the force is measured in newtons and the distance in metres, the work done is measured in **joules** (J).

For example, if a force of 12 N acts on a body and moves it a distance of 3 m in the direction of the force, the amount of work done by the force is 36 J.

When several forces act on one body, the work done by each force can be found independently of the others.

## Work done against a particular force

Sometimes the work that is needed to *overcome* an opposing force is needed.

For example, an object is pulled at constant speed from A to B, a distance *s* along a rough surface.

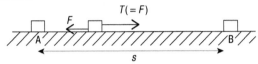

A frictional force of magnitude *F* acts on the object, opposing the motion.

Because there is no acceleration, the force causing the displacement is equal and opposite to the frictional force (that is, it is of magnitude *F*) and therefore the amount of work it does is given by *Fs*.

This work is done to overcome friction and is called *the work done against friction*.

> The work done *against* a force is given by
> the magnitude of that force × the distance moved in the opposite direction.

Look at a body of mass *m*, raised vertically through a height *h*.

The weight, *mg*, acts vertically downward and has to be overcome by an upward force in order to raise the body.

The force needed to raise the body vertically at constant speed is *mg* upwards.

The work done by this force is $mg \times h$.

This amount of work is needed to overcome the opposing force of gravity, therefore the work done against gravity is *mgh*.

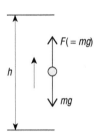

> The work done *against* gravity is given by
> the magnitude of the gravitational force (*mg* down) × the distance moved in the opposite direction (*h* up).

Alternatively, work done against a force can be taken as being negative work done *by* that force. Therefore, in the situation above, the work done *by* gravity is −*mgh*.

## Work done by a moving vehicle

Several forces can act on a moving vehicle, including friction, air resistance, the weight of the vehicle, reaction with the ground and so on, but most important is the driving force which is produced by the engine.

> *The work done by the driving force is* D × s

## Example 1

A body resting in smooth contact with a horizontal plane moves 2.6 m along the plane under the action of a force of 20 N. Find the work done by the force when it is applied

**a** horizontally

**b** at 60° to the plane.

**a** The whole of the force acts in the direction of motion.

Work done = $20 \times 2.6$ J = 52 J

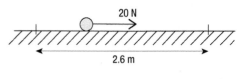

**b** The component of the force that acts in the direction of motion is $F \cos 60°$

Work done = $(20 \cos 60° \times 2.6)$ J = $\left(20 \times \dfrac{1}{2} \times 2.6\right)$ J = 26 J

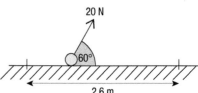

## Example 2

Sixteen crates, each of mass 250 kg, are raised 3 m by a hoist, to be placed on a platform. Find the work done against gravity by the hoist.

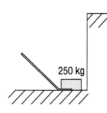
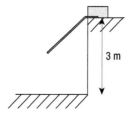

The work done against gravity in raising one crate is $(250 \times g \times 3)$ J = $750g$ J

The work done against gravity in raising 16 crates is $(16 \times 750g)$ J = $12\,000g$ J = 118 kJ

> **Note**
>
> 1000 J is 1 kilojoule, 1 kJ.

## Example 3

A small block of mass 2 kg slides, at constant speed, $\dfrac{4}{5}$ m down the face of a plane inclined at 30° to the horizontal. Contact between the block and the plane is rough.

**a** Find the work done by

**i** the weight of the block

**ii** the reaction between the block and the plane.

**b** Find the work done against friction.

**a** The direction of motion is down the plane.

**i** The component of weight down the plane is $2g \sin 30°$ N = 9.8 N

Therefore the work done by the weight is $9.8 \times \dfrac{4}{5}$ J = 7.84 *J*

**ii** The reaction has no component parallel to the plane.

Therefore no work is done by the reaction.

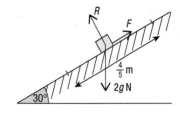

*(continued)*

*(continued)*

**b** The acceleration of the block is zero so the resultant force along the plane is zero.

Resolving ↙ gives $\qquad 2g \sin 30° - F = 0 \Rightarrow F = g$

Therefore the frictional force is 9.8 N up the plane.

So the work done against friction is $\left(9.8 \times \dfrac{4}{5}\right)$ J $= 7.84$ J

## Example 4

A tractor climbs, at a steady speed of 5 ms⁻¹, up a slope inclined at an angle $\alpha$ to the horizontal. The mass of the tractor is 1400 kg and $\sin\alpha = \dfrac{3}{25}$.

**a** By modelling the tractor as a particle, find the work done by the tractor against gravity per minute.

**b** The total work done by the tractor in this time is 780 kJ.

Find the resistance to motion.

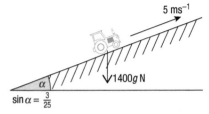

**a** In one minute:

the tractor climbs up the slope a distance $5 \times 60$ m $= 300$ m

the vertical distance climbed is $\qquad 300\,(\sin\alpha)$ m $= 36$ m

Therefore the work done against gravity is $(1400 \times 9.8 \times 36)$ J $= 494$ kJ

**b** The speed of the tractor is constant so the acceleration is zero.

Resolving ↗ gives $\quad D - R - mg \sin\alpha = 0$

$D = R + 1400 \times 9.8 \times \dfrac{3}{25} = R + 1646.4$

The work done by the tractor = the work done by the driving force

$$= (D \times 300)\text{ J}$$
$$= (R + 1646.4) \times 300\text{ J}$$

This is given as 780 kJ. Therefore $780\,000 = 300(R + 1646.4)$

$\Rightarrow R = 2600 - 1646.4 = 953.6$

The resistance to motion is 954 N.

> **Note**
>
> In Example 3, the work done is found from component of the force in the direction of the distance × distance.
> In Example 4, the work done is found from component of displacement in the direction of the force × force.
> You can use either method.

## Exercise 1

In questions 1 to 3, a small object moves from A to B under the action of the forces shown in each diagram.

Find the work done by each force.

**1** AB = 2 m

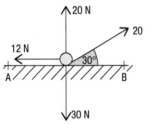

**2** AB = 3 m

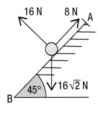

**3** AB = 4 m

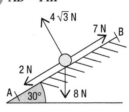

In questions 4 to 8, find the work done against gravity.

**4** A block of mass 3 kg is raised vertically through 2.1 m.

**5** A workman of mass 87 kg climbs a vertical ladder of length 7 m.

**6** Eight crates of beer, each of mass 24 kg, are lifted from the ground on to a shelf that is 1.8 m high.

**7** A forklift truck loads a tiger in a cage into the hold of an aircraft. The combined mass of tiger and cage is 340 kg and the floor of the hold is 7.2 m above the ground.

**8** A crane lifts a one-tonne block of stone out of a quarry that is 11 m deep.

**9** A block of mass 14 kg is pulled a distance of 6 m up a plane inclined at 20° to the horizontal. The contact is rough and the magnitude of the frictional force is 30 N. Find the work done against

   **a** friction

   **b** gravity.

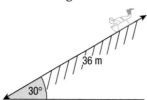

**10** There is an average resistance of 480 N to the motion of a train as it travels 5.8 km between two stations. Find the work done against the resistance.

**11**

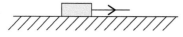

A boy, whose mass is 40 kg, has a skate board of mass 6 kg. He pulls the skate board 36 m up a slope inclined at 30° to the horizontal.

   **a** Find the amount of work that the boy does against gravity to pull the skate board up.

   The boy then sits on the skate board and rolls back to the foot of the slope.

   **b** Find the work done by gravity during the descent.

In questions 12 to 15, a box of mass 6 kg is pulled by a rope along a horizontal surface at a constant speed.

**12** The speed is 4 ms⁻¹. Find the work done by the rope in 20 seconds when the tension in it is 18 N.

**13** The work done by the rope in moving the box 8 m is 200 J. Find the tension in the rope.

**14** The coefficient of friction between the box and the surface is $\frac{1}{3}$. Friction is the only resistance to the motion of the box. Find the work done by the rope when it pulls the box 5 m.

**15** The work done by the rope against friction when it pulls the box a distance of 12 m is 180 J. Find the coefficient of friction between the box and the surface.

**16** A girl pushes her bicycle 150 m up a hill inclined at an angle $\alpha$ to the horizontal where $\sin \alpha = \frac{1}{10}$. The combined weight of the girl and her bicycle is 700 N.

**a** Find the work she does against gravity.

**b** There is an average resistance to motion of 20 N. Find the total work done by the girl.

## 14.2 Power

The rate at which the work is being done is called **power**.

One unit of power is produced when work is done at the rate of 1 joule per second. This unit is called the **watt** (W).

A machine working at the rate of 1 joule per second has a power of 1 watt.

When 1000 joules of work are done per second, the power is 1000 watts or 1 kilowatt (1 kW).

When the total work done in a certain time is known, the *average power* can be found.

For example, a force that does 45 joules of work in 9 seconds is working at an average rate of 5 joules per second, so the average power of the force is 5 watts.

### The power of a moving vehicle

The power of an engine is defined as the rate at which the driving force is working.

A vehicle that has a speed of $v$ ms$^{-1}$ is moved $v$ metres in 1 second by the driving force, $D$ newtons.

Therefore the work done in 1 second by the driving force is $Dv$ joules,

so the power is $Dv$ joules/second which is $Dv$ watts.

Therefore, if $P$ watts is the power of an engine,

$P = Dv$

When the speed of the vehicle is constant, both $D$ and $v$ are constant and therefore the power is constant. If the speed is not constant, the value of $Dv$ gives the power *at the instant* when the speed is $v$ ms$^{-1}$.

Note that when the vehicle is stationary, the power is zero.

There is a maximum value of the power a particular engine can generate. When the speed of a vehicle is the maximum possible in a given situation, the power of the engine is also at its maximum. Therefore no acceleration is possible, so the resultant force acting on the vehicle is zero.

An engine can use less power than the maximum available when, for example, a lower speed is wanted or the resistance falls.

When solving problems involving the power of the engine of a moving vehicle, it can help to express the driving force as $\dfrac{P}{v}$.

Doing this can shorten the solution.

## Example 5

A man can carry a load of bricks up a vertical five-metre ladder in 46 seconds. Find the average power required given the combined mass of the man and his bricks is 92 kg. What assumptions have been made?

The work done against gravity is $92g \times 5$ J $= 4508$ J

This work is done in 46 s so the work is done at an average rate of

$\dfrac{4508}{46}$ joules per second $\Rightarrow 98$ W

Therefore the average power is 98 W.

It is assumed that the man and his bricks can be modelled as a particle which rises exactly 5 m.

## Example 6

On a level track a train has a maximum speed of 50 ms$^{-1}$.

The total resistance to motion is 28 kN.

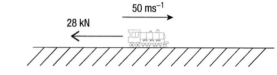

a  Find the maximum power of the engine.

b  The resistance is reduced and the power needed to maintain the same speed as before is 1250 kW.

Find the lower resistance.

a  At maximum speed there is no acceleration so the resultant force in the direction of motion is zero.

Model the train as a particle.

Resolving $\rightarrow$ $D - 28\,000 = 0$ $\Rightarrow$ $D = 28\,000$

To achieve maximum speed, maximum power is needed.

Maximum power = driving force $\times$ maximum velocity

$\qquad\qquad\quad = 28\,000 \times 50$ W

$\qquad\qquad\quad = 1400$ kW

b  As the velocity is constant the acceleration is zero, so the resultant force is zero.

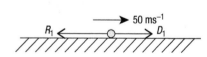

$P = Dv$ $\Rightarrow$ $1\,250\,000 = D_1 \times 50$ $\Rightarrow$ $D_1 = 25\,000$

Resolving $\rightarrow$ $25\,000 - R_1 = 0$

The reduced resistance is 25 kN.

## Example 7

When a car of mass 1200 kg is driving up a hill inclined at $\alpha$ to the horizontal, with the engine working at 32 kW, the maximum speed is 25 ms$^{-1}$. Given that $\sin \alpha = \dfrac{1}{16}$, and assuming that the car can be modelled as a particle, find the resistance to motion.

Speed is maximum so acceleration is zero and the resultant force up the hill is zero.

$$\text{Driving force} = \frac{\text{power}}{\text{velocity}}$$

The driving force is $\dfrac{32\,000}{25}$ N = 1280 N

Resolving ↗ $D - R - W\sin\alpha = 0$

$$\Rightarrow \quad 1280 - R - 1200g \times \frac{1}{16} = 0$$

$$\therefore \quad R = 1280 - \frac{1200 \times 9.8}{16} = 545$$

The resistance to motion is 545 N.

When a vehicle uses a driving force that is larger than all the forces opposing motion, there is a resultant force in the direction of motion. Therefore the vehicle has an acceleration which can be found by applying Newton's Law.

## Example 8

A car of mass 1200 kg has a maximum speed of 144 kmh$^{-1}$ on a level road when there is a resistive force of 56 N. Find the acceleration of the car at the instant when its speed is 81 kmh$^{-1}$ and the engine is working at maximum power.

$144 \text{ kmh}^{-1} = 144 \times \dfrac{5}{18} \text{ ms}^{-1} = 40 \text{ ms}^{-1}$; and $81 \text{ kmh}^{-1} = 22.5 \text{ ms}^{-1}$

Modelling the car as a particle, at maximum speed,

resolving → $\qquad\qquad\qquad\qquad D - 56 = 0 \Rightarrow D = 56$

The maximum power $P$ is given by $\qquad P = Dv = 56 \times 40 = 2240$

The maximum power is 2240 W.

At the lower speed,

the resultant force in the direction of motion is $D - R$.

Applying $F = ma$ in this direction gives

$$\frac{2240}{22.5} - 56 = 1200a \Rightarrow a = \frac{43.56}{1200} = 0.0363$$

Therefore the acceleration is 0.0363 ms$^{-2}$.

## Exercise 2

When the speed of a vehicle is given in kmh$^{-1}$, the unit must be converted to ms$^{-1}$ to be consistent with the other units being used. This conversion can be done by using the fact that $1 \text{ kmh}^{-1} = \dfrac{5}{18} \text{ ms}^{-1}$.

In questions 1 to 5, find the average rate at which work is done.

1 A mass of 60 kg is lifted vertically through 4 m in 9 seconds.

2 A mass of 40 kg is lifted vertically at a constant speed of 5 ms$^{-1}$.

3 A cat weighing 24 N climbs up a 3 metre high wall in 2 seconds.

4 A worker stacks 36 bottles in a minute. Each bottle weighs 5 N and is lifted up 1.6 m.

5 An elevator lifts boxes, each with a mass of 23 kg, up to a height of 2.1 m above ground level. On average, 64 boxes are lifted per hour.

**6** A car driving at a constant speed $v$, against a constant resistance $R$, is working at a rate $P$.

**a** When $v = 25$ ms$^{-1}$ and $R = 960$ N, find $P$.

**b** When $P = 60$ kW and $v = 120$ kmh$^{-1}$, find $R$.

**c** When $R = 1300$ N and $P = 26$ kW, find $v$.

**7** A train has a maximum speed of 90 kmh$^{-1}$ on a level track when the resistive force is a constant 40 kN.

Find the maximum power of the engine.

**8** The maximum power that a van of mass 900 kg can exert is 36 kW. The resistance to the motion of the van is a constant 1500 N. Find the maximum speed that the van can reach on a slope of inclination 1 in 15 $(\sin \alpha = \dfrac{1}{15})$ when driving　　**a** up the slope　　**b** down the slope.

**9** A lorry of mass 2000 kg moves against a constant resistance. The maximum speed of the lorry down a slope of 1 in 10 $(\sin \theta = 0.1)$ is 24 ms$^{-1}$ and the maximum speed up the same slope is 12 ms$^{-1}$.

**a** Find the maximum power of the engine.

**b** Find the constant resistance.

**c** Find the maximum speed of the lorry on a level road.

**10** The resistive forces opposing the motion of a car of mass 2000 kg total 5000 N. The engine is working at 70 kW. Find the acceleration at the instant when the speed is 40 kmh$^{-1}$.

**11** A car of mass 1000 kg has a maximum power of 50 kW. The car is travelling up a hill with a gradient of $\sin^{-1} 0.125$ against resistance to motion of 3000 N. Find the acceleration at the instant when the speed is 30 kmh$^{-1}$.

**12** A train of mass 400 tonnes is moving down an incline of $\sin^{-1} \dfrac{1}{15}$ using a power output of 50 kW. The resistance to motion is 30 kN. Find the acceleration at the instant when the speed is 20 ms$^{-1}$.

## 14.3 Energy

Anything that has the capacity to do work has energy. This energy can be used up in doing work.

Conversely, in order to give energy to an object, work must be done to it, therefore

work and energy are interchangeable and so are measured in the same unit, the joule.

This section deals with **mechanical energy**, which is the capacity to do work as a result of motion or position.

## Kinetic energy

A body moving with speed $v$ has **kinetic energy** (KE). The value of the kinetic energy is equal to the amount of work needed to bring that body from rest to the speed $v$ and an expression for its value is found as follows.

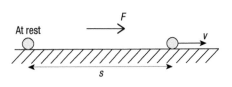

A body of mass $m$ starts from rest and reaches a speed $v$ after moving through a distance $s$ under the action of a constant force $F$.

The acceleration, $a$, is given by $\qquad v^2 - u^2 = 2as \Rightarrow a = \dfrac{v^2 - 0}{2s}$

Then Newton's Law, $F = ma$, gives $\qquad F = \dfrac{mv^2}{2s} \Rightarrow Fs = \dfrac{1}{2}mv^2$

The work done in producing the kinetic energy is $Fs$, therefore $\dfrac{1}{2}mv^2$ is the value of the kinetic energy.

Therefore, for a body of mass $m$ moving with speed $v$,

$$KE = \frac{1}{2}mv^2$$

Note that both $m$ and $v^2$ are always positive quantities showing that KE is always positive and does not depend upon the direction of motion, so kinetic energy is a scalar quantity.

## Potential energy

**Potential energy** (PE) is a property of position. When a body is in a position that, if it could be released, it would begin to move, it has potential energy.

For example, a body that is held at a height $h$ above a fixed level will, if released, begin to fall, so it will begin to possess kinetic energy. Therefore before it is released it has the *potential* to move, hence the name for energy due to position.

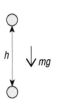

The value of the potential energy is equal to the work needed to raise the body through a vertical distance $h$.

The work done in raising a body of mass $m$ is the work done against gravity, that is $mg \times h$.

$PE = mgh$

When the body falls from rest and reaches a speed $v$ at the bottom then, using $v^2 - u^2 = 2as$ gives $v^2 = 2gh, \Rightarrow gh = \dfrac{1}{2}v^2$.

Therefore $mgh = \dfrac{1}{2}mv^2$ confirming that potential energy is converted into kinetic energy.

There is no absolute value for the potential energy of an object, as the height $h$ is measured from some chosen fixed level. When a different level is chosen the potential energy is changed without the body itself moving. Therefore in every problem the level from which height is measured must be clearly stated. As the potential energy of an object that is *on* the chosen level is zero, we identify this level by marking it 'PE = 0'.

When an object is *below* the level chosen, the value of $h$ is negative ($h$ is the height *above* that level), so the object has *negative potential energy*.

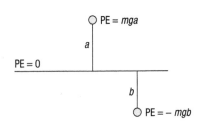

## Example 9

A window cleaner of mass 72 kg climbs up a ladder to a second-floor window, 5 m above ground level. Assuming that the window cleaner can be treated as a particle, find his potential energy relative to the ground.

He then descends 3 m to clean a first-floor window. Find how much potential energy he has lost.

Is the assumption reasonable?

At the second-floor window, A,

$m = 72, g = 9.8, h = 5$

Therefore $\qquad PE = mgh = 72 \times 5 \times 9.8 \, J = 3530 \, J$

In descending to the lower window, B, the reduction in height is 3 m.

Loss in PE $= mg \times$ (reduction in $h$)

$\qquad = 72 \times 9.8 \times 3 \, J = 2120 \, J$

(Alternatively find the values of PE at the two windows and subtract.)

The height of the point where the weight of the man acts may be quite different from the height of the window, so treating him as a particle that rises *exactly* 5 m is not a reasonable assumption, but gives a rough approximation.

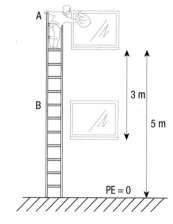

## Example 10

A particle of mass 6 kg has a velocity $v$ metres per second.

Find is its kinetic energy when

**a** $v = 7$       **b** $v = 4\mathbf{i} - 3\mathbf{j}$

**a** $KE = \dfrac{1}{2}mv^2 = \dfrac{1}{2} \times 6 \times 7^2 \, J = 147 \, J$

**b** The speed of the particle is $|\mathbf{v}|$

When $\mathbf{v} = 4\mathbf{i} - 3\mathbf{j}$, $|\mathbf{v}| = \sqrt{4^2 + (-3)^2} = 5$

$KE = \left( \dfrac{1}{2} \times 6 \times 5^2 \right) J = 75 \, J$

> **Note**
>
> Total mechanical energy =
> KE + PE

## Example 11

A bird of mass 0.6 kg, flying at 9 ms$^{-1}$, skims over the top of a tree 6.2 metres high.

What is the total mechanical energy of the bird as it clears the tree?

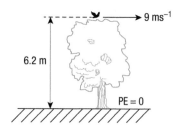

Modelling the bird as a particle,

KE of the bird is $\dfrac{1}{2}mv^2 = \dfrac{1}{2} \times 0.6 \times 81 \, J = 24.3 \, J$

PE of the bird is $mgh = 0.6 \times 9.8 \times 6.2 \, J = 36.5 \, J$

The total mechanical energy is KE + PE = 60.8 J

## Exercise 3

State any assumptions that you make.

1. The potential energy of a particle of mass $m$ kilograms, which is at a height $h$ metres above a given line, is $N$ joules.

   **a** When $m = 4$ and $h = 11$, find $N$.      **b** When $N = 48$ and $m = 6$, find $h$.

2. Find, in joules, the kinetic energy of

   **a** a block of mass 8 kg moving at 9 ms$^{-1}$

   **b** a car of mass 1200 kg travelling at 36 kmh$^{-1}$

   **c** a bullet of mass 16 g moving at 500 ms$^{-1}$

   **d** a body of mass 10 kg with a velocity $\mathbf{v}$ ms$^{-1}$ where $\mathbf{v} = 3\mathbf{i} + 2\mathbf{j}$.

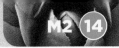

**3** On an assault course, a woman of mass 54 kg starts from ground level at A, climbs 9 m up to B, drops from B to C, 3.6 m below B, then runs up to D at a height of 8 m above the ground.

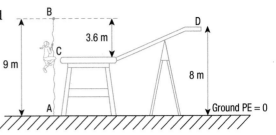

   **a** Find her potential energy relative to the ground at

     **i** A          **ii** D

   **b** Find the gain in potential energy between A and B.

   **c** Find the loss in potential energy between B and C.

   **d** Find the gain in potential energy between C and D.

   **e** Using the answers to parts **b**, **c** and **d**, find the PE at D.

**4**  **a** Find the gain in kinetic energy when the speed of a body of mass 4 kg increases from $7 \text{ ms}^{-1}$ to $11 \text{ ms}^{-1}$.

   **b** Find the kinetic energy lost when the speed of the same body falls from $18 \text{ ms}^{-1}$ to $5 \text{ ms}^{-1}$.

**5** A car of mass 900 kg is travelling at $72 \text{ kmh}^{-1}$.

   **a** Find the kinetic energy lost when the speed falls to $54 \text{ kmh}^{-1}$.

   **b** The kinetic energy rises to 281.25 kJ. Find the speed at which the car is travelling.

**6** A particle of mass 3 kg is at rest. It begins to move with constant acceleration and five seconds later it has kinetic energy of 150 J.

   **a** Find the speed at the end of the 5 seconds.

   **b** Find how far the particle has travelled in this time.

## The work-energy principle

Mechanical energy is defined as the capacity of the forces acting on a body to do work. The link between energy and work done can be expressed more precisely.

> The work done by external forces acting on a body is equal to the change in the mechanical energy of the body.

> When the external forces act in a direction that helps promote the motion of the body the mechanical energy increases, whereas opposing external forces cause a decrease in mechanical energy.

The weight of an object is not counted as an external force in this context because work done by weight is potential energy and is already accounted for.

In solving problems involving this principle, use it in the form

> Final ME ~ Initial ME = Work Done (~ means *the difference between*)

This avoids confusion in cases where one type of energy increases and the other decreases.

## Example 12

A force acting on a body of mass 6 kg, moving horizontally, causes the speed to increase from $3 \text{ ms}^{-1}$ to $8 \text{ ms}^{-1}$.

**a** Find the work done by the force.

**b** The magnitude of the force is 11 N. Find how far the body moves during this speed change.

**a** The body is moving horizontally so there is no change in PE. Only KE changes.

Work done = Final ME − Initial ME

$$= \frac{1}{2}(6)(8)^2 - \frac{1}{2}(6)(3^2) \, J = 165 \, J$$

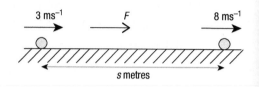

**b** Work done by force = $Fs$

therefore $165 = 11s \Rightarrow s = 15$

The body moves 15 m.

## Example 13

A small block of mass 3 kg is moving on a horizontal plane against a constant resistance of $R$ newtons. The speed of the block falls from 12 ms$^{-1}$ to 7 ms$^{-1}$ as the block moves 5 m.

Find the magnitude of the resistance.

There is no change in PE and, as the speed is reducing, the final KE is less than the initial KE

Initial ME − Final ME $= \frac{1}{2}(3)(12^2) - \frac{1}{2}(3)(7^2)$

$$= 216 - 73.5 \, J = 142.5 \, J$$

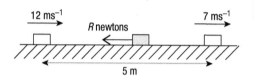

Work done by resistance = Change in ME

therefore $R \times 5 = 142.5 \Rightarrow R = 28.5$

The magnitude of the resistance force is 28.5 N.

## Example 14

A stone falls down through a tank of oil. The speed of the stone as it enters the oil is 2 ms$^{-1}$ and at the bottom of the tank it is 3 ms$^{-1}$. The oil is 2.4 m deep.

Find the resistance, $F$ newtons, that the oil exerts on the stone whose mass is 4 kg.

The resistance is an opposing force so the work it does reduces the ME of the stone. Both KE and PE change.

$$\text{Initial ME} = \left[\frac{1}{2}(4)(2^2) + (4)(9.8)(2.4)\right] J = 8 + 94.08 \, J = 102.08 \, J$$

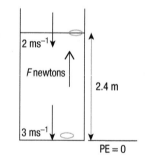

$$\text{Final ME} = \frac{1}{2}(4)(3^2) + 0 = 18 \, J$$

Work done by the resistance = Change in ME

Therefore $F \times 2.4 = 102.08 - 18$

$\Rightarrow \qquad F = 35.03...$

The resistance is 35.0 N.

## Example 15

A car of mass 1000 kg drives up a slope of length 750 m and inclination 1 in 25. Resistance forces are negligible. Find the driving force of the engine given that the speed at the foot of the incline is 25 ms$^{-1}$ and the speed at the top is 20 ms$^{-1}$. (Remember that an inclination of 1 in 25 means the angle is $\sin^{-1}\frac{1}{25}$.)

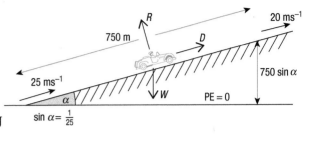

The driving force, $D$ newtons, acts for 750 m

therefore work done by driving force $= 750D$ J

Both kinetic energy and potential energy change.

Final KE $= \dfrac{1}{2}(1000)(20^2)$ J $= 200$ kJ

Final PE $= (1000)(9.8)(750 \sin \alpha) = 9800\left(750 \times \dfrac{1}{25}\right) = 294$ kJ

Therefore    Final ME $= 494$ kJ

Initial KE $= \dfrac{1}{2}(1000)(25^2)$ J $= 312.5$ kJ

Initial PE $= 0$

Therefore    Initial ME $= 312.5$ kJ

Work done by driving force = Final ME − Initial ME

Therefore    $750D = 494\,000 - 312\,500 \Rightarrow D = 242$

The driving force is 242 N.

## Exercise 4

Use the work-energy principle for each question. Model each large object as a particle, ignore air resistance unless it is specifically mentioned and state any *other* assumptions made.

1. A mass of 6 kg is pulled by a string across a smooth horizontal plane.

   As the block moves through a distance of 4.2 m, the speed increases from 2 ms$^{-1}$ to 6 ms$^{-1}$. Find the tension in the string.

   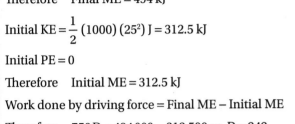

2. A body of mass 8 kg, travelling on a rough horizontal plane at 12 ms$^{-1}$, is brought to rest by friction. Find the work done by the frictional force.

3. A ball of mass 0.4 kg is thrown vertically upwards with a speed of 10 ms$^{-1}$. It comes instantaneously to rest at a height of 3.6 m above the point of projection, P.

   a   Find the resistance to its motion.

   b   The ball then falls back to P. The resistance is unchanged. Find the speed at P.

   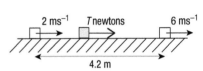

4. A body of mass 0.5 kg is lifted by a vertical force from rest at a point A to a point B that is 1.7 m vertically above A. The body has a speed of 3 ms$^{-1}$ when it reaches B.

   a   Find the work done by the force.

   b   Hence find the magnitude of the force.

5. A car of mass 750 kg is travelling along a level road at 10 ms$^{-1}$ against a constant resistance of 200 N.

   Using a driving force of 1200 N, the driver accelerates for 20 m.

   Find the speed of the car at the end of the 20 m.

   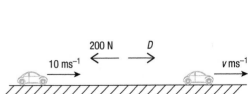

**6** A block of mass 5 kg lies on a horizontal plane. It is pulled from rest through a distance of 8 m by a horizontal force of 12 N. Find the speed of the block at the end of the 8 m when the contact between block and plane is

    **a**   smooth            **b**   rough, with a coefficient of friction of $\dfrac{1}{10}$.

**7** A particle of mass 7 kg is pulled by a force of $F$ newtons, 4 metres up a smooth plane inclined at 30° to the horizontal. Find the work done by the pulling force if

    **a**   the particle is pulled at a constant speed

    **b**   the speed changes from 1 ms$^{-1}$ initially to 2 ms$^{-1}$ at the end.

**8** A block of mass 3 kg slides down a plane inclined at $\alpha$ to the horizontal where $\sin\alpha = \dfrac{1}{4}$. The block starts from rest and there is a constant frictional force of 4 N.

    **a**   Find how far has the block travelled when its speed reaches 6 ms$^{-1}$.

    **b**   When the block has moved 4 m down the plane, what is its speed?

**9** A stone of mass 2 kg is pushed horizontally off a wall 5 m high. The stone hits the ground at 9 ms$^{-1}$. Find the air resistance.

**10** Two boys are kicking a ball about. It has a mass of 0.5 kg. The ball comes towards one boy at 8 ms$^{-1}$ and he passes it back at 12 ms$^{-1}$.

    **a**   Find the work he does in bringing the ball to instantaneous rest.

    **b**   Find the work he does in giving it a speed of 12 ms$^{-1}$.

## Conservation of mechanical energy

The work-energy principle states that the total change in the mechanical energy of a body is equal to the work done on the body.

Therefore it follows that:

> When the total work done by the external forces acting on a body is zero there is no change in the total mechanical energy of the body, so energy is conserved i.e KE + PE is constant.
> This is the *principle of conservation of mechanical energy.*

The weight of a body is not an external force in this context as the work done by the weight is already included as potential energy.

## Example 16

A particle is projected with speed 8 ms$^{-1}$. Find its speed after it has moved a vertical distance of 2 m

    **a**   upwards           **b**   downwards.

Let the mass of the particle be $m$ kg and the final speed be $v$ ms⁻¹.

At A $\quad$ PE $= 0$ and KE $= \dfrac{1}{2}m \times 8^2$ J

Therefore $\quad$ Total ME $= 32m$ J

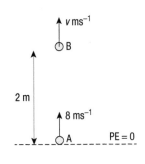

a $\;$ At B $\;$ PE $= mgh = 19.6m$ J

$\quad$ and $\quad$ KE $= \dfrac{1}{2}mv^2$ J

$\quad$ Therefore $\quad$ Total ME $= \left(19.6\,m + \dfrac{1}{2}v^2 m\right)$ J $\qquad$ 2 m

$\quad$ Total ME at A = Total ME at B

$\quad$ Therefore $\quad 32m = 19.6\,m + \dfrac{1}{2}v^2 m$

$\quad \Rightarrow \quad v^2 = 24.8 \Rightarrow v = 4.98$

The speed at B is 4.98 ms⁻¹.

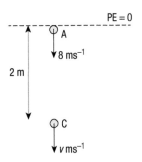

b $\;$ At C PE $= mg(-2)$ J and KE $= \dfrac{1}{2}mv^2$ J

$\quad$ Therefore $\;$ Total ME $= \left(\dfrac{1}{2}mv^2 - 19.6\,m\right)$ J

$\quad$ Total ME at A = Total ME at C

$\quad$ Therefore $\quad 32m = \dfrac{1}{2}mv^2 - 19.6m$

$\quad \Rightarrow \quad v^2 = 103.2 \Rightarrow v = 10.2$

The speed at C is 10.2 ms⁻¹.

Note that in Example 16 the value of the mass, $m$, was not given and was not needed as it cancelled in each conservation of energy equation. This will always be the case as $m$ is a factor of both PE and KE.

Note also that the direction in which the particle is projected is not given. The diagrams assume that this direction is vertical, but the particle can be projected in any direction because kinetic energy does not depend on the direction of the velocity.

## Example 17

A bead is threaded on to a circular ring of radius 0.5 m and centre O, which is fixed in a vertical plane. The bead is projected from the lowest point of the ring, A, with a speed of 4 ms⁻¹, and first comes to instantaneous rest at a point B. Contact between the ring and the bead is smooth and there is no other resistance to motion. Find the height of B above A.

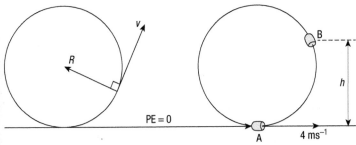

The normal reaction $R$ is always perpendicular to the direction of motion of the bead therefore no work is done by $R$. No other external force is acting so energy is conserved.

(continued)

**Answer**

At A $\quad$ PE = 0 and KE = $\frac{1}{2} mv^2 = \frac{1}{2} m(4)^2$ J

$\qquad$ Total ME = $(0 + 8m)$ J

At B $\quad$ PE = $mgh = 9.8mh$ J and KE = 0

$\qquad$ Total ME = $(9.8mh + 0)$ J

$\qquad$ Total ME at A = Total ME at B

Therefore $\qquad$ $8m = 9.8mh \Rightarrow h = \dfrac{8}{9.8}$

B is 0.816 m above A.

> **Note**
>
> The letter m is used both for mass and metre. It is easy to see the difference between the two m's in type because, for mass, $m$ is italic. When hand-writing a solution this cannot be done so it is sensible to write out the word 'metre' in full.

## Exercise 5

Use the Conservation of Mechanical Energy to solve these problems.

1. A particle of mass $m$ kilograms is projected upwards with speed 6 ms$^{-1}$. Find the height it reaches before first coming to rest.

2. A stone is thrown downwards with speed 3 ms$^{-1}$. Find its speed after it has fallen 4.5 m.

3. A ball is thrown upwards, from a point A, with speed 8 ms$^{-1}$. The point A is 1 m above ground.

    a Find the greatest height reached by the ball.

    b Find the speed of the ball when its height is 2.4 m.

    c Find the speed of the ball as it hits the ground.

Questions 4 to 6 concern a particle P, moving on a plane inclined to the horizontal at 30° and A and B are two points on the plane. Contact between the particle and the plane is smooth.

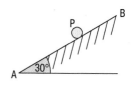

4. The particle P is projected up the plane from A with speed 4 ms$^{-1}$ and comes to instantaneous rest at B.

    Find the distance AB.

5. The particle P is released from rest at B. Find its speed as it passes through A given that AB = 2.4 m.

6. The particle P is moving down the plane and passes through B with speed $v$ ms$^{-1}$. The particle passes through A with speed 6 ms$^{-1}$ and AB = 2.1 m. Find $v$.

In questions 7 to 9, a smooth bead is threaded on to a smooth circular wire with centre O and radius $a$ metres. The wire is fixed in a vertical plane.

7. The bead is released from rest at a point level with O. Given $a = 0.5$, find the speed of the bead as it passes through the lowest point.

8. The bead is projected from the lowest point on the wire with speed 4.2 ms$^{-1}$. Given $a = 0.6$, find the height above O at which the bead first comes to rest.

9. The bead is projected from the lowest point and just reaches the highest point. Given that $a = 0.8$, find the speed of projection.

10 A girl wants to swing across a stream on a rope.

The rope is attached to an overhanging tree at point A.

The bank on the opposite side of the stream is 1.2 m higher than the bank on which she is standing.

Find the speed she must push off with in order just to get there.

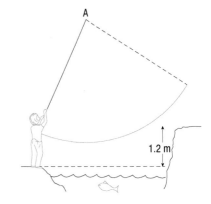

11 Peter Pan 'flies' across the stage on a harness, which slides along a smooth wire AB.

The end A of the wire is fixed at a height of 6 m and end B at a height of 5.5 m.

The lowest point of the wire as he crosses the stage is at a height of 4 m.

He starts from rest at A.

a Find his maximum speed.

b Find his speed at B.

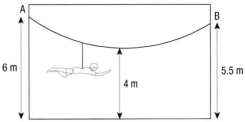

# Summary

## Work

When a constant force acts on an object the amount of work done by the force is given by

► component of force in the direction of motion × distance moved by object

The unit of work is the joule (J), which is the amount of work done when a force of 1 newton moves an object through 1 m.

The work done by a vehicle moving at constant speed is

► driving force × distance moved

## Power

Power is the rate at which work is done and is measured in watts where 1 watt (W) is 1 joule per second.

The power of a moving vehicle is the rate at which the driving force is working and this is given by driving force × velocity

## Energy

Energy is the ability to do work.

Energy and Work are interchangeable so energy is measured in joules.

The Kinetic Energy (KE) of a moving object is given by $\frac{1}{2}mv^2$; it is a scalar and it can never be negative.

Potential Energy (PE) is equivalent to work done by gravity and is given by $mgh$ where $h$ is the height of an object above a chosen level.

PE is negative for a body below the chosen level.

## Work-energy principle

The work done by external forces acting on a body is equal to the change in the mechanical energy of the body.

## Conservation of energy

When the total work done by the external forces acting on a body is zero the total mechanical energy of the body remains constant. Work done by gravity is accounted for as potential energy so weight is not included as an external force.

## Review

1. A wardrobe is lowered by a rope at a steady speed from the balcony of a fifth-floor flat to the ground, 12 m below. Given that the mass of the wardrobe is 37 kg, find the work done by the rope during the descent.

2. A car has a maximum speed of 140 kmh$^{-1}$ on a level road with the engine working at 54 kW. Find the resistance to motion.

3. A block is pulled 5 m at constant speed along a rough surface by a rope against a frictional force of 10 N. Find the average work done against friction.

4. A car has a maximum power of 30 kW and encounters resistive forces totalling 1050 N.

   Find its maximum speed on level ground.

5. A tractor working at 20 kW is travelling at a constant 5 ms$^{-1}$. Find the resistance to motion.

6. Find the potential energy of a particle of mass 3 kilograms, which is 10 metres above a given line.

7. Find the kinetic energy of a block of mass 7 kg moving at 5 ms$^{-1}$.

8. A bullet of mass 0.02 kg is fired horizontally at a speed of 360 ms$^{-1}$ into a fixed block of wood. The bullet is embedded 0.06 m into the block. Find the average resisting force exerted by the wood.

9. A particle slides down a smooth inclined plane on a line AB.

   At A its velocity down the plane is 3 ms$^{-1}$. The height of A above B is 16 m.

   Find the velocity of the particle when at B.

## Assessment

1. A crate of mass 40 kg is pulled by a rope at a constant 1.5 ms$^{-1}$ down a slope inclined at 15° to the horizontal. Contact is rough and the coefficient of friction is 0.7.
   Find

   a  the frictional force

   b  the tension in the rope

   c  the work done by the rope per second

   d  the work done by gravity while the crate moves down the slope for 6 seconds.

**2** A car of mass 1100 kg has a maximum power output of 44 kW. The resistive forces are constant at 1400 N. Find the maximum speed of the car

   **a**  on the level

   **b**  up an incline with gradient $\sin^{-1} 0.05$

   **c**  down the same incline when using half the maximum power.

**3** The constant resistance to the motion of a car of mass 1200 kg is 960 N.

  The car is driving along a level road and has an acceleration of $0.2$ ms$^{-2}$ at the instant when the speed is 25 ms$^{-1}$.

  Find the power exerted by the engine.

**4** A boy slides on level icy ground. He runs up and starts the slide at a speed of 5 ms$^{-1}$. After sliding for 6 m his speed is 4 ms$^{-1}$. His mass is 45 kg.

   **a**  Find the loss of kinetic energy.

   **b**  Find the resistance to motion.

   **c**  Assuming that air resistance is negligible, find the coefficient of friction.

**5** The seat of a swing is 0.4 m above the ground when it is stationary. A child is swinging so that she passes through the lowest point with speed 5.4 ms$^{-1}$. Find the height of the seat above ground when she first comes to rest. State what assumptions you have made and comment on their suitability.

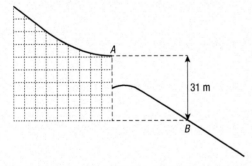

**6** Alan, of mass 76 kg, performed a ski jump. He took off at the point $A$ at the end of the ski run with a speed of 28 ms$^{-1}$ and landed at the point $B$.

  The level of the point $B$ is 31 metres vertically below the level of the point $A$, as shown in the diagram.

  Assume that his weight is the only force that acted on Alan during the jump.

   **a**  Calculate the kinetic energy of Alan when he was at the point $A$.

   **b**  Calculate the potential energy lost by Alan during the jump as he moved from the point $A$ to the point $B$.

   **c**  **i**  Find the kinetic energy of Alan when he reached the point $B$.

      **ii**  Hence find the speed of Alan when he reached the point $B$.

AQA MM2 June 2012

## Introduction

Most people think of centrifugal force as a force trying to push a body out of a circular path. This force is a reaction, it is not the force making a body move in a circle. This chapter explains the force necessary to make a body move in a circle, and it is not outwards.

## Recap

You will need to remember...

▶ One radian is equal to $\dfrac{180}{\pi}$ degrees.

▶ The relationship between limiting friction and normal reaction is $F = \mu R$ where $\mu$ is the coefficient of friction.

## Applications

Satellites orbit the Earth in approximately circular paths at constant speed. Several calculations are needed to successfully launch a satellite in the correct orbit above the Earth. One of these is the speed needed to keep the satellite in the orbit.

## Objectives

By the end of this chapter, you should know how to...

▶ Use the relationship between angular speed and linear speed of a particle moving in a horizontal circle.

▶ Use Newton's Law, $F = ma$, to find the relationship between the acceleration and the forces acting on a particle moving in a horizontal circle with constant speed.

· · · · · · · · · · · · · · · · · · · · · · · · · · · · · ·

## 15.1 Angular velocity and acceleration

A particle moves round a circle with centre O. When the particle moves from a point P on the circle to a point Q on the circle and the angle POQ is $\theta$ radians, then the rate at which $\theta$ is increasing with respect to time is $\dfrac{\mathrm{d}\theta}{\mathrm{d}t}$.

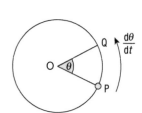

This is the **angular speed** of the particle and is denoted by the symbol $\omega$.

When the direction of rotation (clockwise or anticlockwise) is stated this gives the **angular velocity** of the particle.

A positive sign is used to denote the anticlockwise direction of rotation and a negative sign for clockwise rotation.

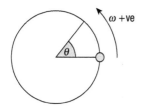
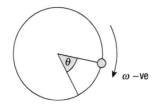

For example, the second hand of a clock rotates through 1 revolution in 1 minute so:

its angular speed is 1 rev min$^{-1}$

and its angular velocity is 1 rev min$^{-1}$ clockwise, or −1 rev min$^{-1}$.

Angular velocity can be measured in revolutions per second, but is more usually given in radians per second. Either of these units can be converted to the other using 1 revolution = $2\pi$ radians.

A particle that describes a circle at constant speed covers equal arcs in equal times so its angular speed is also constant.

Suppose that the radius of the circle is $r$, the speed round the circumference is $v$ ms$^{-1}$ and the angular speed is $\omega$ rad s$^{-1}$.

In 1 second the particle travels round an arc of length $v$ and this arc length is also given by $r\omega$, therefore

$$v = r\omega$$

## Example 1

Question

Express  **a**  3 revolutions per minute in radians per second

**b**  0.005 radians per second in revolutions per hour.

Answer

**a**  3 rev/minute $= \dfrac{3}{60}$ rev/second

$= \dfrac{1}{20} \times 2\pi$ rad s$^{-1}$ $= \dfrac{\pi}{10}$ rad s$^{-1}$

**b**  0.005 rad s$^{-1}$ $= 0.005 \times 3600$ rad/hour $= 18$ rad/hour

$= 18 \div 2\pi$ rev/hour $= \dfrac{9}{\pi}$ rev/hour

## Example 2

Question

A point on the circumference of a disc is rotating at a constant speed of 3 ms$^{-1}$. The radius of the disc is 0.24 m. Find, in rad s$^{-1}$, the rate at which the disc is rotating.

Answer

Using $v = r\omega$, with $v = 3$ and $r = 0.24$ gives

$3 = 0.24\omega \quad \Rightarrow \quad \omega = 12.5$

Therefore the disc rotates at 12.5 rad s$^{-1}$.

## Acceleration

When the velocity of a moving object is changing, the object has an acceleration. As velocity is a vector, it can change in magnitude or in direction or both. It is easy to accept that changing *speed* involves acceleration but it is not so easy to see that, for example, a car going round a corner at *constant* speed is accelerating because its direction is changing.

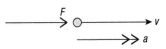

A change in speed is caused by a force that acts *in the direction* of motion of the object to which it is applied.

A force of this type cannot produce a change in the direction of the velocity so, if no other force is acting, the object continues to move in a straight line. The acceleration produced is a change in speed, that is, a change in the magnitude of the velocity.

A force that is perpendicular to the direction of motion of an object will push or pull the object off its previous line of motion but cannot alter the speed.

The acceleration in this case is a change in the *direction* of the velocity and the object therefore moves in a curve; the actual curve described depends on the particular force acting.

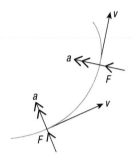

A force that is neither parallel nor perpendicular to the direction of motion of an object has a component in each of these directions. Therefore it causes a change both in the speed and in the direction of motion of the object and the object moves with varying speed on a curved path.

## Motion in a circle with constant speed

The direction of motion of a particle moving in a circle is constantly changing, so there must be a force acting perpendicular to the direction of motion of the particle at any instant.

When the particle is moving with constant speed, there is no force acting in the direction of motion.

At any point on its path the particle is moving in the direction of the tangent at that point.

A force that is perpendicular to this direction acts along the radius at that point. Because it is moving the particle from the tangent on to the circumference, the force must act *inwards* along the radius.

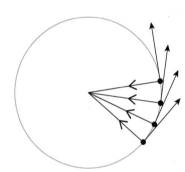

> **The force that produces circular motion with constant speed is at any instant acting along the radius towards the centre of the circle.**

## The magnitude of the radial acceleration

A particle is moving at constant speed, $v$, round a circle of centre O and radius $r$. In a time $\delta t$, the particle moves from a point P to a nearby point Q, through a small angle $\delta\theta$ measured in radians.

The length of the arc PQ is $r\delta\theta$ and, as this arc is covered in time $\delta t$ at speed $v$, its length is also $v\delta t$.

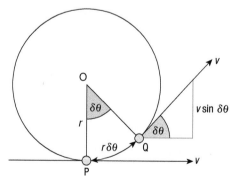

Therefore    $r\delta\theta = v\delta t$

$\Rightarrow$    $\dfrac{v}{r} = \dfrac{\delta\theta}{\delta t}$

When the particle is at P it has no velocity component along PO, but when it reaches Q the velocity component parallel to PO is $v\sin\delta\theta$.

Therefore the average acceleration from P to Q, in the direction of PO, is given by $\dfrac{v\sin\delta\theta}{\delta t}$.

The closer Q is to P, the nearer this expression is to the actual radial acceleration at P and, at the same time, both $\delta\theta$ and $\delta t$ approach zero.

For any angle $\alpha$ measured in radians, it can be shown that as $\alpha \to 0$, $\sin\alpha \to \alpha$

Therefore as $\delta\theta \to 0 \qquad \sin\delta\theta \to \delta\theta$

$$\Rightarrow \qquad \frac{v\sin\delta\theta}{\delta t} \to v\frac{\delta\theta}{\delta t} \to v\frac{d\theta}{dt}$$

$\dfrac{d\theta}{dt}$ is the rate at which $\theta$ increases with respect to time, so $\dfrac{d\theta}{dt}$ is the angular speed, $\omega$.

Hence the radial acceleration of the particle is $v\omega$.

Using $v = r\omega$, this acceleration can be expressed as either $r\omega^2$ or $\dfrac{v^2}{r}$.

> **Note**
>
> A proof of this property is not given here but comparing the values of small angles (in radians) and their sine ratios using a calculator shows that this appears to be correct.

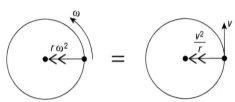

The acceleration of a particle travelling in a circle of radius $r$, at constant speed $v$ (or constant angular speed $\omega$), is directed *towards* the centre of the circle and is of magnitude $r\omega^2$ or $\dfrac{v^2}{r}$.

Therefore a particle can describe a circle with constant speed only under the action of a force of constant magnitude directed *towards the centre of the circle*.

## Example 3

A particle P is travelling round a circle of radius 0.8 m at a constant speed of 2 ms$^{-1}$. Find the acceleration of the particle, giving its magnitude and direction.

The acceleration is given by $\dfrac{v^2}{r}$, where $v = 2$ and $r = 0.8$.

The magnitude of the acceleration is $\dfrac{2^2}{0.8} = 5$ ms$^{-2}$,

and it is directed towards the centre of the circle.

## Example 4

At any time $t$ the velocity vector of a particle P is $\mathbf{v}$ ms$^{-1}$ where $\mathbf{v} = (2\sin t)\mathbf{i} - (2\cos t)\mathbf{j}$. When $t = 0$, the position vector of P is $\mathbf{r} = -2\mathbf{i}$.

a  Find the speed of P at time $t$.

b  Find the position vector of P at time $t$.

c  Hence show that P is moving in a circle with constant speed.

d  Find the acceleration of P at time $t$.

e  Hence show that the acceleration is perpendicular to the path of P.

a    $\mathbf{v} = (2\sin t)\mathbf{i} - (2\cos t)\mathbf{j} \implies |\mathbf{v}| = \sqrt{(2\sin t)^2 + (2\cos t)^2} = \sqrt{4} = 2$

The speed of P is 2 ms$^{-1}$.

b    $\mathbf{r} = \int \mathbf{v}\,dt = -(2\cos t)\mathbf{i} - (2\sin t)\mathbf{j} + \mathbf{A}$

When $t = 0$, $-2\mathbf{i} = -2\mathbf{i} + \mathbf{A}$, so $\mathbf{A} = \mathbf{0}$

Therefore $\mathbf{r} = -(2\cos t)\mathbf{i} - (2\sin t)\mathbf{j}$

c    $x = -2\cos t$ and $y = -2\sin t$

Eliminating $t$ gives $x^2 + y^2 = 4$.

Therefore P is moving in a circle of radius 4.

d    $\mathbf{a} = \dfrac{d\mathbf{v}}{dt}$, therefore $\mathbf{a} = (2\cos t)\mathbf{i} + (2\sin t)\mathbf{j}$

e    The vector $\mathbf{v}$ is in the direction of the path.

$\mathbf{a}$ and $\mathbf{v}$ are perpendicular if $\mathbf{a.v} = 0$

$\mathbf{a.v} = [(2\cos t)\mathbf{i} + (2\sin t)\mathbf{j}].[(2\sin t)\mathbf{i} - (2\cos t)\mathbf{j}]$

$\qquad = 4\cos t \sin t - 4\sin t \cos t = 0$

Therefore the acceleration is perpendicular to the path of P.

> **Note**
>
> This can also be shown by showing that the position vector of P and the acceleration vector of P are in opposite directions.
>
>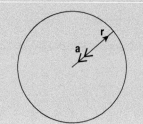

## Exercise 1

**1**   Express

   **a**   0.2 radians per minute in revolutions per hour

   **b**   100 revolutions per minute in radians per second.

**2**   Find the angular velocity, in radians per second, of the minute hand of a clock.

**3**   Find the angular speed of the Earth about its axis

   **a**   in revolutions per minute      **b**   in radians per second.

**4**   A disc is rotating about its centre with angular velocity $\omega$ rad s$^{-1}$. Point P is on the disc at a distance of $d$ metres from the centre and has speed $v$ metres per second.

   **a**   $\omega = 6$, $d = 0.2$; find $v$.

   **b**   $v = 5$, $d = 0.4$; find $\omega$.

   **c**   $v = 10$, $\omega = 2.5$; find $d$.

**5**   A roundabout has a diameter of 3 m. The speed of the edge of the roundabout is 2.4 m s$^{-1}$.

A child sits at a distance of 1 m from the centre.

   **a**   Find the angular speed in rad s$^{-1}$.

   **b**   Find the speed with which the child is moving.

In questions 6 to 10, a particle is travelling round a circle, centre O, radius $r$ metres, at a constant speed of $v$ m s$^{-1}$ and angular velocity $\omega$ rad s$^{-1}$. Its acceleration is of magnitude $a$ ms$^{-2}$.

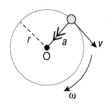

**6**   $v = 16$, $r = 5$; find $a$.

**7**   $\omega = 12$, $r = 3$; find $a$.

**8** In what direction is the acceleration

    **a** when the particle has reached a point P on the circle

    **b** when the particle has reached a point Q on the circle?

**9** $a = 75$, $r = 12$; find $v$.

**10** $a = 500$, $\omega = 6$; find $r$.

**11** The position vector of a particle P at any time $t$ is $\mathbf{r} = (\cos t)\mathbf{i} + (\sin t)\mathbf{j}$.

    **a** Show that P is moving in a circle with constant speed.

    **b** Find the magnitude of the acceleration of P.

**12** Assume the Earth is a sphere of radius 6400 km. Calculate the acceleration, in ms$^{-2}$, due to the Earth's rotation of

    **a** a person A who is standing on the equator

    **b** a person B who is at a point which is 3200 km from the Earth's axis of rotation.

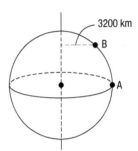

**13** An aircraft is circling at a constant height, at 500 kmh$^{-1}$. The passengers experience an acceleration of 4 ms$^{-2}$.

Find the radius of the circle.

**14** An astronaut in training is strapped into a chair which is then rotated in a horizontal circle of radius 5 m with an acceleration of $9g$ ms$^{-2}$. Find the angular velocity.

## 15.2 Motion in a horizontal circle

When a circle is described at constant speed, there is an acceleration of constant magnitude $\dfrac{v^2}{r}$ or $r\omega^2$ towards the centre of the circle and no acceleration along the tangent, so

► there is no tangential force acting

► there must be a force of constant magnitude acting towards the centre.

These conditions are satisfied when the circle is in a horizontal plane because the weight of the particle is a vertical force so has no component in the direction of motion of the particle.

There are many ways in which the force towards the centre can be provided, for example by a rotating string with one end fixed at the centre or by friction with the road surface as a car turns round a bend.

Some of the possibilities are illustrated in Examples 5, 6 and 7.

### Example 5

One end of a light inelastic string of length 1 m is fixed to a point O on a horizontal table. A particle P of mass 0.3 kg is attached to the other end of the string. The particle is given a blow which sets it moving in a circle on the table, with constant angular velocity 10 rad s$^{-1}$.

    **a** Find the tension in the string and the force exerted on P by the table.

    **b** Explain an assumption that has been made in the question.

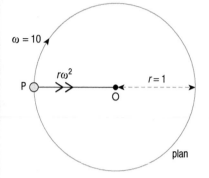

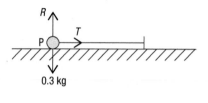

plan                vertical section

Vertically there is no acceleration; horizontally there is an acceleration of $(1)(10)^2$ ms$^{-2}$ towards O.

a   Using Newton's Law $F = ma \rightarrow$      $T = (0.3)(1)(10)^2 = 30$

    Resolving $\uparrow$                  $R - (0.3)g = 0 \Rightarrow R = 2.94$

    The tension is 30 N and the reaction exerted by the table is 2.94 N.

b   As P travels with constant speed it is assumed that there is no force along the tangent, so there is no friction between the particle and the table.

## Example 6

A small block A, of mass $m$ kg, lies on a horizontal disc which is rotating about its centre B at 3 rad s$^{-1}$ and A is 0.8 m from B. The block does not move relative to the disc.

Find the least possible value of $\mu$, the coefficient of friction between the block and the disc.

The frictional force $F$ acing on A is towards the centre of the circle because A has an acceleration towards the centre in order to travel in a circle.

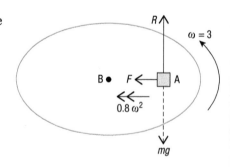

There is no friction along the tangent because A has no tendency to move in that direction.

Resolving $\uparrow$     $R - mg = 0 \Rightarrow R = mg$

Using Newton's Law $\leftarrow$    $F = mr\omega^2 \Rightarrow F = m(0.8)(3)^2 = 7.2m$

Now $F \leq \mu R \Rightarrow 7.2m \leq \mu mg$

so      $\mu \geq \dfrac{7.2}{g}$

Therefore the least value of $\mu$ is $\dfrac{7.2}{9.8} = 0.735$.

## Example 7

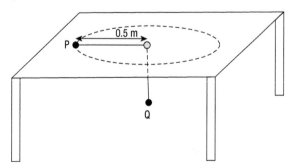

*(continued)*

*(continued)*

Two particles P and Q of masses 2 kg and 5 kg respectively are connected by a light inextensible string.

Particle P moves in a circle of radius 0.5 m with constant speed on a smooth table. The string passes through a smooth hole with particle Q hanging in equilibrium at the other end of the string.

Find the speed of the particle P.

The particle Q is in equilibrium,

therefore $\uparrow$   $T = 5g$

The tension in the string is the same throughout its length,

therefore the radial force acting on P is $5g$,

Using the acceleration in the form $\dfrac{v^2}{r}$ and $F = ma$ gives   $5g = (2)\dfrac{v^2}{0.5}$

$\Rightarrow$                                    $v^2 = \dfrac{5 \times 9.8}{4}$

Therefore   $v = 3.5$

The speed of P is 3.5 ms$^{-1}$.

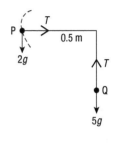

## Exercise 2

In questions 1 and 2, one end A of an inelastic string AB is fixed to a point on a smooth table.

A particle P is attached to the other end B, and moves on the table in a horizontal circle with centre A.

1. The particle is of mass 1.5 kg and its speed is 4 ms$^{-1}$. The length of the string is 2.4 m. Find the tension in the string.

2. The mass of the particle is 8 kg and it is moving with speed 5 ms$^{-1}$. Find the length of the string given that the tension in it is 12 N.

3. A circular tray of radius 0.2 m has a smooth vertical rim round the edge. The tray is fixed on a horizontal table and a small ball of mass 0.1 kg is set moving round the inside of the rim of the tray with speed 4 ms$^{-1}$.

   a   Find the acceleration of the ball.

   b   Find the horizontal force exerted on the ball by the rim of the tray.

4. A particle of mass 0.4 kg is attached to one end of a light inextensible string of length 0.6 m. The other end is fixed to a point A on a smooth horizontal table. The particle is set moving in a circular path.

   a   The speed of the particle is 8 ms$^{-1}$. Find the tension in the string.

   b   The string will snap when the tension in it exceeds 50 N. Find the greatest angular velocity at which the particle can travel.

In questions 5 and 6 choose a model. This should be clearly described and all assumptions stated.

5. An aircraft is flying at 700 kmh$^{-1}$ in a horizontal circle of radius 2 km.

   Find the horizontal component of the thrust exerted by the seat, on a passenger of mass 60 kg.

**6** A satellite, of mass 500 kilograms, is orbiting the Earth on a circular path at a height of 100 km above the surface. At this height the acceleration due to gravity is 9.5 ms$^{-2}$. Take the radius of the Earth as 6400 km. For the satellite, find

   **a**  the force exerted on it towards the centre of the Earth

   **b**  its speed in ms$^{-1}$

   **c**  its angular speed

   **d**  the time it takes to perform one orbit.

**7** Two particles A and B of masses 1 kg and 3 kg respectively are connected by a light inextensible string.

Particle A moves in a circle of radius 0.2 m with constant speed on a smooth table. The string passes through a smooth hole with particle B hanging in equilibrium at the other end of the string.

   **a**  Find the tension in the string.

   **b**  Find the angular velocity of the particle A.

   **c**  Find the speed of the particle A.

**8** A particle P has a mass of 2 kg. The position vector of P at any time $t$ is given by $\mathbf{r} = (3 \sin t)\mathbf{i} + (3 \cos t)\mathbf{j}$.

   **a**  Show that P is moving in a circle in the $xy$ plane.

   **b**  Assuming that the $xy$ plane is horizontal, find the force $\mathbf{F}$ acting on P.

# 15.3 The conical pendulum

A light inelastic string fixed at one end, A, has a particle hanging freely at the other end.

When the particle, which is not resting on a surface, is set moving in a horizontal circle, the plane of that circle will be below the level of A.

As the particle and the string rotate, they trace out the surface of a cone and the system is called a **conical pendulum**.

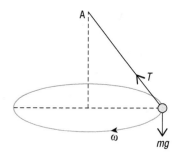

The tension in the string has two functions – the horizontal component gives the central force needed to keep the particle rotating and the vertical component balances the weight of the particle. It follows that the string can never be horizontal, as the tension in it *must* have a vertical component.

There is a similar situation when a particle moves on the inner surface of a smooth sphere, in a horizontal circle below the level of the centre of the sphere.

The vertical component of the normal reaction $R$ (which acts through the centre of the sphere) balances the weight of the particle, while the horizontal component of $R$ acts on the particle towards the centre of the circle being described.

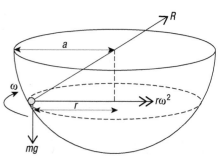

## Example 8

A particle P, of mass $m$, is attached to one end of a light inextensible string of length $a$ and describes a horizontal circle, centre O, with constant angular speed $\omega$. The other end of the string is fixed to a point Q and, as P rotates, the string makes an angle $\theta$ with the vertical. Show that

a   the tension in the string is always greater than the weight of the particle

b   the depth of O below Q is independent of the length of the string.

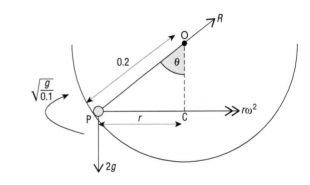

The string cannot be vertical therefore $\theta > 0$.

Resolving $\uparrow$ $\qquad\qquad T\cos\theta - mg = 0$ $\qquad\qquad$ [1]

Using Newton's Law $\leftarrow$ $\qquad T\sin\theta = m(a\sin\theta)\omega^2$

$\qquad\qquad\qquad\qquad\qquad T = ma\omega^2$ $\qquad\qquad$ [2]

a   From [1], $\qquad T = \dfrac{mg}{\cos\theta}$ $\qquad$ and $\qquad \cos\theta < 1$

$\qquad$ Therefore $\qquad T > mg$

b   In triangle POQ, $\qquad QO = a\cos\theta$ $\qquad\qquad$ [3]

$\qquad$ From [1] and [3] $\qquad QO = \dfrac{amg}{T}$

$\qquad$ Then from [2] $\qquad QO = \dfrac{amg}{ma\omega^2} = \dfrac{g}{\omega^2}$

Therefore the depth of O below Q is independent of the length of the string.

## Example 9

A particle P of mass 2 kg is moving on the inner surface of a smooth hemispherical bowl with centre O and radius 0.2 m. The particle is describing a horizontal circle, centre C, with angular speed $\sqrt{\dfrac{g}{0.1}}$.

a   Find the magnitude of the force exerted on P by the surface of the bowl.

b   Find the depth of C below O.

a   For the forces acting on P,

$\qquad$ Resolving $\uparrow$ gives

$\qquad R\cos\theta - 2g = 0$ $\qquad\qquad$ [1]

$\qquad$ Newton's Law $\rightarrow$ gives

$\qquad R\sin\theta = 2r\omega^2$

$\qquad\qquad\quad = 2(0.2\sin\theta)\left(\dfrac{g}{0.1}\right)$

$\qquad \Rightarrow \quad R = 4g$ $\qquad\qquad$ [2]

$\qquad$ The force exerted on P is 4 $g$.

b   In triangle OPC, OC $= 0.2\cos\theta$

$\qquad$ From [1] and [2], $\cos\theta = \dfrac{1}{2}$

$\qquad$ Therefore $\qquad$ OC $= 0.1$

$\qquad$ The depth of C below O is 0.1 m.

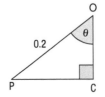

# Exercise 3

Questions 1 to 3 are about a conical pendulum which consists of an inextensible string AB with a particle P attached at B. Point A is fixed and B moves in a circle in a horizontal plane below A.

**1** The length of the string AB is 1.5 m and the mass of the particle P is 3 kg.

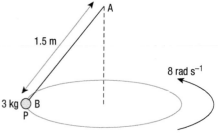

The particle P is rotating in a circle with an angular speed of 8 rad s$^{-1}$.

**a** Find the tension in the string.

**b** Find the angle between the string and the vertical.

**2** The mass of the particle P is 2 kg and P is rotating in a circle of radius 0.3 m.

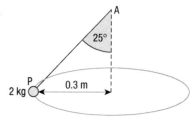

The string is inclined at 25° to the vertical.

**a** Find the tension in the string.

**b** Find the angular speed of **P**.

**3** The particle P is rotating in a circle with an angular speed of 5 rad s$^{-1}$. Find the depth below A of the plane of this circle.

**4** A small bead B of mass 0.2 kg is rotating in a horizontal circle on the inner surface of a smooth hemisphere of radius 0.3 m and centre O. The centre of the horizontal circle is 0.1 m below O.

**a** Find the magnitude of the force exerted by the bowl on the bead.

**b** Find the speed of the bead.

**5** A girl is swinging on a rope of length 5 m, which is attached to a swivel on top of a pole.

It takes her 4 seconds to complete a circle around the pole. Her mass is 40 kg.

**a** Find the tension in the rope.

**b** Find the angle between the rope and the pole.

**c** Find the radius of the circle.

**6** A particle of mass 0.3 kg is attached to one end, A, of a light inextensible string. The length of the string is 0.8 m. The particle is moving in a horizontal circle on a smooth horizontal plane. The other end, B, of the string is attached to a fixed point 0.5 m above the plane.

Find the speed of the particle when it is on the point of lifting off the plane.

**7** Particles P and Q, each of mass 0.2 kg, are attached one to each end of a light inextensible string. The string passes through a ring. P is hanging freely vertically below the ring and Q is moving in a horizontal circle with constant speed below the ring.

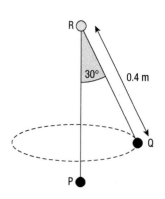

**a** Explain why the ring cannot be smooth.

The part of the string RQ makes an angle of 30° with the vertical.

**b** Find the speed of Q.

**c** Find the time it takes Q to make one revolution.

**8** A particle P is moving in a horizontal circle of radius 0.3 m with constant speed on the inside of a smooth hemisphere of radius 0.8 m.

The magnitude of the reaction between the particle and the surface of the sphere is 5 N.

**a** Find the mass of the particle.

**b** Find the speed of the particle.

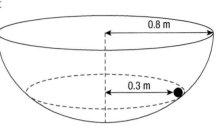

# Summary

The angular velocity of a particle describing a circle is represented by $\dfrac{d\theta}{dt}$ or $\omega$, and is measured in radians per second or revolutions per second.

A particle travelling in a circle of radius $r$ metres, with a constant angular velocity $\omega$ rad s$^{-1}$, has a speed $v$ ms$^{-1}$ round the circle where $v = r\omega$.

When a particle describes a circle of radius $r$ at a constant speed $v$ (or constant angular velocity $\omega$), the acceleration is directed towards the centre of the circle and is of magnitude $r\omega^2$ or $\dfrac{v^2}{r}$.

A force of magnitude $mr\omega^2$ or $\dfrac{mv^2}{r}$ must act towards the centre.

A string fixed at one end A and carrying at the other end a particle performing horizontal circles below A, is called a conical pendulum.

# Review

**1** Find the speed, in km h$^{-1}$, of a point on the equator of the Earth, assuming the equator to be a circle of radius 6400 km.

**2** A particle P is moving in a circle of radius 1.5 m with constant speed 5 ms$^{-1}$. Find the acceleration of P.

**3** One end A of an inelastic string AB is fixed to a point on a smooth table.

A particle P of mass 0.6 kg is attached to the other end B and moves on the table in a horizontal circle with centre A. The length of AB is 1.4 m and the tension in AB = 4 N. Find the angular speed of P.

**4** The acceleration vector of a particle P at any time $t$ is $\mathbf{a} = (3 \sin t)\mathbf{i} - (3 \cos t)\mathbf{j}$.

**a** Find the position vector of P at time $t$ given that when $t = 0$, $\mathbf{r} = 3\mathbf{j}$ and $\mathbf{v} = -3\mathbf{i}$.

**b** Hence show that P moves in a circle with constant speed.

**c** The mass of P is 1.5 kg. Find the magnitude of the force acting on P.

**5** A particle of mass 8 kg is attached to one end of a string. The other end is fixed to a point on a horizontal smooth surface. The particle is moving with speed 5ms$^{-1}$ on the surface in a circle. The tension in it is 12 N. Find the length of the string.

**6** A particle, with mass 0.1 kg, is fixed to one end of a light inextensible string. The other end is attached to a fixed point A. The particle is rotating below A in a horizontal circle with constant speed 6 ms⁻¹. The string makes an angle of 30° with the vertical.

   **a** Draw a diagram showing all the forces acting on the particle.

   **b** Find the tension in the string.

   **c** Find the radius of the horizontal circle.

## Assessment

**1** An object of mass 0.3 kg is placed on a horizontal turntable, which is rotating at a constant rate, at a distance of 0.2 m from the axis. The coefficient of friction between the object and the turntable is 0.4. Find the greatest possible value for the angular velocity if the object is not to slip outward from its position.

**2** A particle P of mass 0.3 kg is attached to one end of a light inextensible string of length 0.8 m. The other end of the string is attached to a fixed point A. P moves in a horizontal circle on a smooth horizontal plane with constant angular speed 2 rad s⁻¹.

Find the tension in the string and the force exerted by P on the plane

   **a** when A is a point on the plane

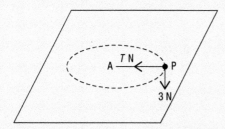

   **b** when A is a point 0.6 m above the plane.

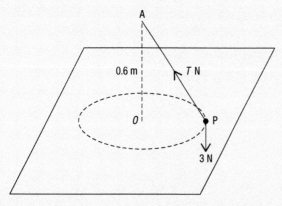

**3** Aaron, Beth, Carol and Dipak are skating. They hold hands, in this order, with their arms outstretched.

They are of similar size. Their average span from left hand to right hand is 150 cm and each of their masses is approximately 50 kg. Aaron

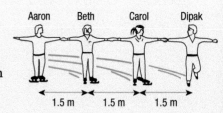

turns on a spot and the others skate round him in circles, with the same angular velocity and staying in a straight line along a radius.

**a** The angular velocity is 6 rad s⁻¹. Find the speeds of Beth, Carol and Dipak.

**b** Model the skaters as particles and assume that they are not using their skates to provide any force towards the centre.

   **i** Find the force $T_1$ which Carol exerts on Dipak.

   **ii** Find the force $T_2$ which Beth exerts on Carol.

   **iii** Find the force $T_3$ which Aaron exerts on Beth.

**4** A particle of mass 5 kg is rotating in a horizontal circle inside a smooth hollow cone whose semi-vertical angle is 60°.

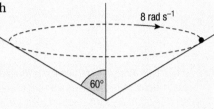

The angular speed of the particle is 8 rad s⁻¹.

**a** Draw a diagram showing the forces acting on the particle.

**b** Find the radius of the circle marked in the diagram.

**c** Find the reaction on the particle from the surface of the cone.

**5** A global positioning satellite (GPS) orbits the Earth at a height of 20 460 km above the surface of the Earth. The angular speed of the satellite is one revolution in 12 hours.

Take the radius of the Earth as 6400 km.

**a** Express the angular speed of the satellite in radians per second.

**b** Find the value of $g$ (the acceleration due to gravity) at this distance from the surface of the Earth.

**6** A particle, of mass 0.8 kg, is attached to one end of a light inextensible string. The other end of the string is attached to the fixed point $O$. The particle is set in motion, so that it moves in a horizontal circle at constant speed, with the string at an angle of 35° to the vertical. The centre of this circle is vertically below $O$, as shown in the diagram.

The particle moves in a horizontal circle and completes 20 revolutions each minute.

**a** Find the angular speed of the particle in radians per second.

**b** Find the tension in the string.

**c** Find the radius of the horizontal circle.

AQA MM2 June 2014

**7** Tom is travelling on a train which is moving at a constant speed of 15 ms⁻¹ on a horizontal track. Tom has placed his mobile phone on a rough horizontal table. The coefficient of friction between the phone and the table is 0.2.

The train moves round a bend of constant radius. The phone does not slide as the train travels round the bend.

Model the phone as a particle moving round part of a circle, with centre O and radius $r$ metres.

Find the least possible value of $r$.

AQA MM2 June 2013

# Glossary

## A

**angular speed** the rate at which an angle increases with respect to time

**angular velocity** angular speed and sense of turning (clockwise or anticlockwise)

## B

**binomial series** the series formed when $(1+x)^n$ is expanded where $n$ is any real number

## C

**centre of gravity** the point in a body through which the weight acts

**centre of mass** the point of a body about which the mass is equally distributed and for normal objects on the surface of the Earth is the same point as the centre of gravity

**chain rule** the formula $\frac{dy}{dx} = \frac{dy}{du} \times \frac{du}{dx}$ for differentiating a composite function

**coefficient of friction** the ratio of friction to normal reaction when a body is moving or is about to move over a rough surface

**common logarithms** logarithms to the base 10

**component** one of a set of vectors that are equivalent to a single (resultant) vector

**components** two or more vectors in different directions which together are equivalent to a given vector

**composite function** a function of the form $fg(x)$ where f and g are both functions of $x$.

**concurrent** going through a single point

**conical pendulum** the path traced out by a string with one end fixed and with a particle attached at the other end moving in a horizontal circle below the point of attachment

**coplanar** all in one plane

**cosecant** the reciprocal of the sine of an angle

**cotangent** the reciprocal of tangent of an angle

## D

**density** the mass per unit area of a body

**differential equation** an equation relating a function with its derivative

**displacement** the distance and direction of one point from another

**domain** the values for which a function is defined, that is the input values

## E

**equilibrium** a body is in equilibrium when it does not move or turn

**exponential decay** the rate of decrease of a quantity $x$ that is proportional to the current value of $x$

**exponential function** a function of the form $a^x$ where $a$ is a positive constant

**exponential growth** the rate of increase of a quantity $x$ that is proportional to the current value of $x$

## F

**first order differential equation** a differential equation containing the first derivative to the power 1, for example $xy + \frac{dy}{dx} = 0$

**free vector** a vector that can be in any position

**function** a function is an expression containing one variable where each value of the variable in the domain gives only one value of the function.

**function of a function** a composite function

## G

**growth factor** the constant by which a quantity grows in equal intervals of time

## H

**half-life** the time it takes for an initial mass to decay to half its original mass

## I

**implicit function** an equation containing two variables where one variable is not isolated on one side of the equation, for example $x^2 + y^2 = 1$

**improper (algebraic fraction)** a fraction where the variable in the numerator has the same or higher power than the variable in the denominator

**integrating by parts** using the formula $\int v \frac{du}{dx} dx = uv - \int u \frac{dv}{dx} dx$ to integrate a product of functions

**inverse function** the function that maps the range of a function to its domain

**iteration** a process where one value is put into a function to get another value, and that value is put into the same function to get a third value, and this is repeated

## J

**joule** the unit of work and of energy and 1 joule is the work done by a force of 1 newton moving a body 1 metre

## K

**kinematics** the study of motion without considering the forces causing the motion

**kinetic energy** the energy possessed by a moving body

## L

**lamina** a flat object modelled as having no thickness

## M

**magnitude** the size of a quantity, the magnitude of a vector is its length

**mathematical modelling** using mathematics to describe a real life situation and usually involving assumptions to simplify the situation

**mechanical energy** energy possessed by a body because of motion and position

**median** the median of a triangle is the line from one vertex to the midpoint of the opposite side

**mid-ordinate rule** a formula for finding an approximate value of the area under a curve

**modulus** the modulus of a vector is its magnitude, the modulus of a function is its size irrespective of its sign, for example, the modulus of $-2$ is 2

**moment** the moment of a force is its turning affect about an axis and is measured as the force times its perpendicular distance from the axis

## N

**Naperian logarithms** logarithms to the base e

**natural logarithms** logarithms to the base e

**newton metre** the unit of work and of energy usually called the joule

## P

**parameter** a variable that other variables can be expressed in terms of

**parametric equation** an equation expressing one variable in terms of a parameter

**partial fractions** expressing a fraction as the sum or difference of two or more simpler fractions

**position vector** a vector from a fixed point (the origin) to another point

**potential energy** the energy possessed by a body due to its position

**power** the rate at which work is done, measured in watts

**projectile** a object that is thrown or fired and then moves only under the action of its own weight

**proper (algebraic fraction)** a fraction where the variable in the numerator has a lower power than the variable in the denominator

## R

**range** the range of a function is the set of values it can take

**rational function** a fraction whose numerator and denominator are polynomials

**real numbers** numbers that can be represented as points on a number line

**resolve** to resolve a vector means to find its component in a given direction

**resultant** a single vector equivalent to a set of component vectors

**resultant vector** the single vector equivalent to the action of two or more other vectors

## S

**scalar** a quantity that is fully defined by size

**secant** the reciprocal of the cosine of an angle

**Simpson's rule** a formula for finding an approximate area under a curve

**skew (lines)** lines that do not intersect and are not parallel

**statics** the study of the forces acting on a stationary body

## T

**time of flight** the time for which a projectile is in the air

**trajectory** the path traced out by a moving object

**triangle law** the law that states when two vectors are represented by two sides of a triangle in the same sense, their resultant is represented by the third side of the triangle in the opposite sense

## U

**unit vector** a vector whose magnitude is one unit

## V

**vector** a quantity that is defined by its magnitude and direction

**volume of revolution** the volume formed when the area between a curve and one axis is rotated about that axis

## W

**watt** the unit of power where 1 watt equal to 1 joule per second

**work** work is done when a force moves an object and is measured in joules

# Answers

## 1 Functions

**Exercise 1**

**1 a** $f(x) \geq -3$    **b** $f(x) \geq -5$    **c** $f(x) \geq 0$    **d** $0 < f(x) \leq \dfrac{1}{2}$

**2 a** 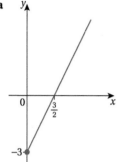    **b** 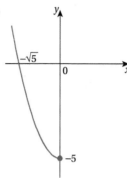    **c** 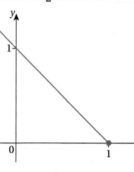    **d**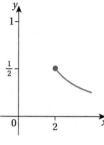

**3 a** $5, 4, 2, 0$    **b**

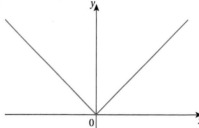

**4 a** $0, 2, 4, 5, 5$    **b** $f(x)$ 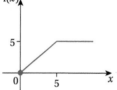    **c** $0 \leq f(x) \leq 5$

**Exercise 2**

**1 a** $\dfrac{1}{x^2}$    **b** $(1-x)^2$    **c** $1 - \dfrac{1}{x}$    **d** $1 - x^2$    **e** $\dfrac{1}{x^2}$

**2 a** $125$    **b** $15$    **c** $-1$    **d** $-1$

**3 a** $(1+x)^2$    **b** $1 + x^2$    **c** $1 + 2x$    **4 a** $\sin(3x - 4)$    **b** $3(\sin x) - 4$

**Exercise 3**

**1 a** 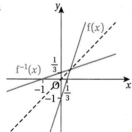    **b** 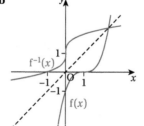    **c**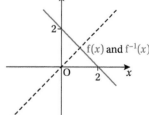

**d**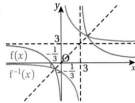

**e**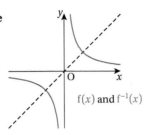

**2 a** $f^{-1}(x) = (x-1)$   **b** no   **c** $f^{-1}(x) = \sqrt[3]{x-1}$   **d** $f^{-1}(x) = \sqrt{x+4}, x \geq -4$

**e** $f^{-1}(x) = \sqrt[4]{x} - 1, x \geq 0$

**3 a** $-\dfrac{1}{3}$   **b** $\dfrac{1}{2}$

**4 a** $\dfrac{x^2+2}{3}$   **b** $\sqrt{10}$   **c** 1 or 2

**5 a** $fg(x) = 1 - x$ for $x < 1$; the inverse is also $1 - x$ and this is a function because one value of $x$ gives only one value of $1 - x$

**b** $gf(x) = \sqrt{1-x^2}$ for $-1 < x < 1$; the equation of the reflected curve is $y = \pm\sqrt{1-x^2}$ for $0 < x < 1$ and $\pm\sqrt{1-x^2}$ is not a function because one value of $x$ gives two values of $y$.

## Exercise 4

**1**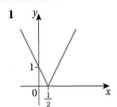

**2**

**3**

**4**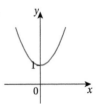

**5**

**6**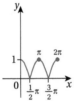

**7**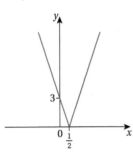

**8**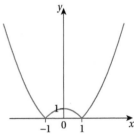

**9**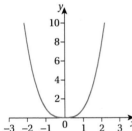

**10**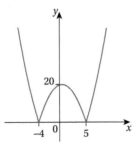

**11**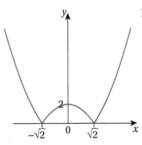

**12**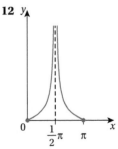

## Exercise 5

**1** $\left(\dfrac{1}{2}, \dfrac{1}{2}\right)$   **2** $(0, 0), (1, 1)$ and $(3, 3)$   **3** $(\sqrt{3}, 2\sqrt{3}), (2-\sqrt{7}, 2\sqrt{7}-4)$   **4** $(1, 1), (-1, 1)$

**5** $(1, 3), (1+\sqrt{6}, 3+2\sqrt{6})$   **6** $3$ and $\dfrac{1}{2}(\sqrt{17}-3)$   **7** $\dfrac{2}{5}$ and $1\dfrac{1}{3}$   **8** $\dfrac{2}{3}$ and $2$

**9** $-\sqrt{2}-1$ and $-1$   **10** $-1$   **11** $1-\sqrt{2}$ and $-1-\sqrt{2}$

## Exercise 6

**1** $x > -\dfrac{1}{2}$      **2** $x < \dfrac{1}{2}$      **3** $x > 1$      **4** $x < 1$      **5** $x < -\dfrac{5}{4}$ or $x > -\dfrac{1}{4}$

**6** $x < 0$ or $x > 2$      **7** $x < 0$ or $x > 1 + \sqrt{3}$      **8** $x < \dfrac{1}{4}\pi$ or $\dfrac{3}{4}\pi < x < \dfrac{5}{4}\pi$ or $\dfrac{7}{4}\pi < x$

## Exercise 7

**1**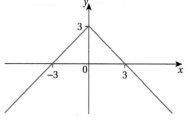

**2**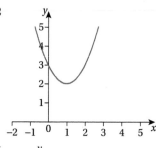

**3**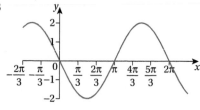

**4**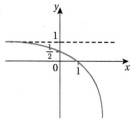

**5**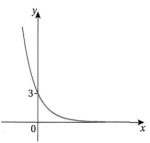

**6** one-way stretch by factor $\dfrac{1}{2}$ parallel to the $x$-axis followed by a translation by $\begin{bmatrix} \pi \\ 0 \end{bmatrix}$

**7** one-way stretch by factor $\dfrac{1}{2}$ parallel to the $x$-axis followed by a translation by $\begin{bmatrix} 1 \\ 0 \end{bmatrix}$

**8** reflection in the $x$-axis followed by a translation by $\begin{bmatrix} 0 \\ 1 \end{bmatrix}$

**9** reflection in the $x$-axis followed by a translation by $\begin{bmatrix} -\pi \\ 0 \end{bmatrix}$ or simply a translation by $\begin{bmatrix} \pi \\ 0 \end{bmatrix}$

**10** a translation by $\begin{bmatrix} 3 \\ 0 \end{bmatrix}$ followed by one-way stretch by factor 2 parallel to the $y$-axis

**11** one-way stretch by factor $\dfrac{1}{4}$ parallel to the $y$-axis followed by a one-way stretch by factor 2 parallel to the $x$-axis or a stretch by scale factor 8 parallel to the $x$-axis

**12 a** $(x-2)^2 - 10$      **b** a translation by $\begin{bmatrix} 2 \\ -10 \end{bmatrix}$ followed by one-way stretch by factor 2 parallel to the $y$-axis

## Exercise 8

**1** $\dfrac{1}{4}$      **2** $\dfrac{2x+4}{x-2}$      **3** $\dfrac{2}{3}$      **4** $\dfrac{3}{5}$      **5** $\dfrac{x}{y}$      **6** $\dfrac{1}{x-2}$

**7** $a+2$      **8** $\dfrac{5x(x+y)}{5y+2x}$      **9** $\dfrac{2a-6b}{6a+b}$      **10** $\dfrac{b-4}{3x(b+4)}$      **11** $\dfrac{x-3}{x+4}$      **12** $\dfrac{4y^2+3}{(y+3)(y-3)}$

**13** $\dfrac{1}{3(x+3)}$      **14** $\dfrac{x+2}{2x+1}$      **15** $\dfrac{x-2}{x-1}$      **16** $\dfrac{1}{2(a-5)}$      **17** $\dfrac{3}{p+3q}$      **18** $\dfrac{a^2+2a+4}{(a+5)(a+2)}$

**19** $\dfrac{x+1}{3(x+3)}$      **20** $\dfrac{4(x-3)}{(x+1)^2}$

## Exercise 9

**1** $1 - \dfrac{4}{x+2}$      **2** $2 + \dfrac{9}{x-2}$      **3** $2x+1 - \dfrac{5}{x+2}$      **4** $x-2 + \dfrac{6}{x+1}$      **5** $4x^2 + 4x + 5 + \dfrac{4}{x-1}$

**6** $2x^2 + 3x + 6 + \dfrac{14}{x-2}$

**7 a** $2 + \dfrac{4}{x-2}$  **b** $1 + \dfrac{4}{x^2-1}$  **c** $x + 2 + \dfrac{4}{x-2}$

## Exercise 10

**1** $\dfrac{3}{2(x+1)} - \dfrac{1}{2(x-1)}$

**2** $\dfrac{13}{6(x-7)} - \dfrac{1}{6(x-1)}$

**3** $\dfrac{4}{5(x-2)} - \dfrac{4}{5(x+3)}$

**4** $\dfrac{7}{9(2x-1)} + \dfrac{28}{9(x+4)}$

**5** $\dfrac{1}{x-2} - \dfrac{1}{x}$

**6** $\dfrac{3}{x-2} - \dfrac{1}{x-1}$

**7** $\dfrac{1}{2(x-3)} - \dfrac{1}{2(x+3)}$

**8** $\dfrac{7}{3x} - \dfrac{1}{3(x+1)}$

**9** $\dfrac{9}{x} - \dfrac{18}{2x+1}$

**10** $\dfrac{2}{5(x-1)} - \dfrac{1}{5(3x+2)}$

## Exercise 11

**1** $\dfrac{1}{2(x-1)} - \dfrac{1}{2(x+1)} - \dfrac{1}{(x+1)^2}$

**2** $\dfrac{11}{8(x-3)} + \dfrac{5}{8(x+1)} - \dfrac{1}{2(x+1)^2}$

**3** $\dfrac{1}{x-1} - \dfrac{1}{x-2} + \dfrac{2}{(x-2)^2}$

**4** $\dfrac{2}{x} - \dfrac{1}{x^2} - \dfrac{3}{2x+1}$

**5** $\dfrac{7}{16(x+3)} - \dfrac{1}{4(x-1)^2} + \dfrac{9}{16(x-1)}$

**6** $\dfrac{3}{(x+2)^2} + \dfrac{2}{x+2} - \dfrac{2}{x+1}$

**7** $\dfrac{2}{x-2} - \dfrac{2}{x-1}$

**8** $\dfrac{5}{9(x-1)} - \dfrac{5}{9(x+2)} - \dfrac{5}{3(x+2)^2}$

**9** $\dfrac{9}{5(x-3)} - \dfrac{3}{5(2x-1)}$

**10** $\dfrac{2}{3(x-2)^2} + \dfrac{1}{9(x-2)} - \dfrac{1}{9(x+1)}$

**11** $\dfrac{2}{3(x-4)} + \dfrac{1}{3(x+2)}$

**12** $\dfrac{10}{9(2x-1)} - \dfrac{5}{9(x-2)} + \dfrac{10}{3(x-2)^2}$

**13** $\dfrac{1}{x-3} - \dfrac{1}{x+2}$

**14** $\dfrac{1}{3(x-4)} - \dfrac{1}{3(x-1)} - \dfrac{1}{x-1}$

**15** $\dfrac{3}{2x+3} - \dfrac{1}{x+1}$

**16** $\dfrac{3}{2(x-1)} - \dfrac{9}{2(3x-1)}$

**17** $\dfrac{4}{3(2x-3)} - \dfrac{1}{x^2} - \dfrac{2}{3x}$

**18** $1 + \dfrac{1}{2(x-1)} - \dfrac{1}{2(x+1)}$

**19** $x + \dfrac{2}{x-1} - \dfrac{1}{x+1}$

**20** $1 - \dfrac{7}{4(x+3)} - \dfrac{1}{4(x-1)}$

## Review

**1 a** $f(x) > 0$  **b** $3$

**2 a** 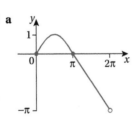  **b** $-\pi \le f(x) \le 1$

**3 a** $\sqrt{\sin x}$  **b** $0 \le x \le \pi$ ( or $2\pi \le x \le 3\pi$, and so on)

**4 a** $-\dfrac{1}{2}$

**5** reflection in the $y$-axis followed by a translation by $\begin{bmatrix} 0 \\ 3 \end{bmatrix}$

**6 a** $\dfrac{x+3}{2}$  **b** $\dfrac{2x-5}{2x+5}$

**7** $1 - \dfrac{9}{x+6}$

**8** $3x - 14 + \dfrac{43}{x+3}$

**9** $x^2 - 3x - 3 + \dfrac{2}{x-1}$

**10 a** $\dfrac{4}{7(x-3)} - \dfrac{8}{7(2x+1)}$  **b** $\dfrac{5}{7(x+1)} + \dfrac{1}{7(4x-3)}$  **c** $\dfrac{1}{t-1} + \dfrac{1}{t+1}$

**11 a** $\dfrac{9}{8(x-5)} - \dfrac{1}{8(x+3)}$  **b** $\dfrac{1}{5(x-2)} + \dfrac{6}{5(4x-3)}$  **c** $1 + \dfrac{3}{2(2x-3)} - \dfrac{3}{2x+3}$

**12 a** $\dfrac{1}{x-2} + \dfrac{1}{2(x+1)}$  **b** $\dfrac{-8}{9x} + \dfrac{1}{3x^2} + \dfrac{8}{9(x-3)}$

**1 a** $x^2 + 1, x \in \mathbb{R}, f^{-1}(x) \geq 1$   **b** 1    **2 a** $1 + \dfrac{4}{x^2 - 4}$   **b** $1 + \dfrac{1}{x-2} - \dfrac{1}{x+2}$

**3 a** reflection in the $x$-axis followed by a translation by $\begin{bmatrix} 0 \\ 1 \end{bmatrix}$

**b** 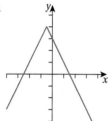   **c** $\left(-\dfrac{1}{2}, \dfrac{1}{2}\right), \left(-1\dfrac{1}{2}, \dfrac{1}{2}\right)$   **d** $x < -1\dfrac{1}{2}, x > -\dfrac{1}{2}$

**4 a** $\dfrac{4}{9(x-8)} - \dfrac{4}{9(x+1)}$   **b** $\dfrac{3}{5(x-2)^2} + \dfrac{4}{25(x-2)} - \dfrac{8}{25(2x+1)}$   **c** $\dfrac{3}{x} - \dfrac{6}{2x+1}$

**5 a** 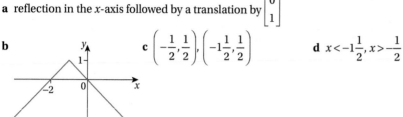   **b** $-1 \leq f(x) \leq 1$   **c** $\cos(1 - |\cos x|)$   **d** $\cos 1$

**6 a** $1 \leq f(x) \leq 21$   **b i** $f^{-1}(x) = \dfrac{1}{4}\left(\dfrac{63}{x} + 1\right)$   **ii** 21   **c i** $fg(x) = \dfrac{63}{4x^2 - 1}$   **ii** $-4$

**7** $\dfrac{3}{1-3x} + \dfrac{1}{1+x} - \dfrac{4}{(1+x)^2}$

**8 a** [graph]   **b** $x = -5$ and $x = 1$   **c** $-5 < x < 1$   **d** translation $\begin{bmatrix} 0 \\ 4 \end{bmatrix}$ followed by reflection in $y = 4$

# 2 Binomial Series

## Exercise 1

**1** $1 - x - \dfrac{x^2}{2} - \dfrac{x^3}{2}, -\dfrac{1}{2} < x < \dfrac{1}{2}$   **2** $1 - 10x + 75x^2 - 500x^3, -\dfrac{1}{5} < x < \dfrac{1}{5}$   **3** $1 + \dfrac{3}{2}x + \dfrac{3}{2}x^2 + \dfrac{5}{4}x^3, -2 < x < 2$

**4** $1 + \dfrac{3}{2}x + \dfrac{3}{8}x^2 - \dfrac{1}{16}x^3, -1 < x < 1$   **5** $\dfrac{1}{3} - \dfrac{x}{9} + \dfrac{x^2}{27} - \dfrac{x^3}{81}, -3 < x < 3$   **6** $2 - \dfrac{x}{4} + \dfrac{x^2}{64} - \dfrac{x^3}{512}, -4 < x < 4$

**7** $1 - \dfrac{x}{4} + \dfrac{3x^2}{32} - \dfrac{5x^3}{128}, -2 < x < 2$   **8** $1 + 2x + 3x^2 + 4x^3, -1 < x < 1$   **9** $\dfrac{1}{8} - \dfrac{3}{16}x + \dfrac{3}{16}x^2 - \dfrac{5}{32}x^3, -2 < x < 2$

**10** $1 - \dfrac{1}{2}x + \dfrac{3}{8}x^2 - \dfrac{5}{16}x^3, -1 < x < 1$   **11** $1 - \dfrac{1}{9}x^2, -3 < x < 3$   **12** $2 - \dfrac{3}{4}x - \dfrac{9}{64}x^2 - \dfrac{27}{512}x^3, -\dfrac{4}{3} < x < \dfrac{4}{3}$

## Exercise 2

**1** $\dfrac{1}{2(1+x)} + \dfrac{1}{2(1-x)}; 1 + x^2, -1 < x < 1$   **2** $1 + \dfrac{3}{x-1}; -2 - 3x - 3x^2 - 3x^3, -1 < x < 1$

**3** $\dfrac{2}{x} + \dfrac{2}{1-x}; \dfrac{2}{x} + 2 + 2x + 2x^2 + 2x^3, -1 < x < 1$

**4** $\dfrac{1}{5(2-x)} + \dfrac{2}{5(1+2x)}; \dfrac{1}{2} - \dfrac{3}{4}x + \dfrac{13}{8}x^2 - \dfrac{51}{16}x^3, -\dfrac{1}{2} < x < \dfrac{1}{2}$   **5** $\dfrac{1}{3(x+3)} - \dfrac{1}{3(x-3)}; \dfrac{2}{9} + \dfrac{2x^2}{81}, -3 < x < 3$

**6** $-\dfrac{1}{3(2x-1)} - \dfrac{1}{3(x+1)}, x + x^2 + 3x^3, -\dfrac{1}{2} < x < \dfrac{1}{2}$   **7** $\dfrac{1}{2(x+1)} + \dfrac{1}{x-2}; -\dfrac{x}{4} + \dfrac{3x^2}{4} - \dfrac{9x^3}{16}, -1 < x < 1$

**8** $1+\dfrac{1}{x-2}-\dfrac{1}{x+2};\dfrac{x^2}{4},-2<x<2$

**9** $\dfrac{8}{3(x-3)^2}+\dfrac{1}{9(x-3)}-\dfrac{1}{9x};-\dfrac{1}{9x}+\dfrac{7}{27}+\dfrac{5}{27}x+\dfrac{23}{243}x^2,-3<x<3$

**10** $\dfrac{1}{1-2x}-\dfrac{1}{1-x}+\dfrac{1}{(1-x)^2};1+3x+6x^2,-\dfrac{1}{2}<x<\dfrac{1}{2}$

**11** $\dfrac{9}{4(1-3x)}-\dfrac{3}{4(1-x)}-\dfrac{1}{2(1-x)^2};1+5x+18x^2,\dfrac{1}{3}<x<\dfrac{1}{3}$

### Exercise 3

**2 a** $\dfrac{1}{3(x+2)}-\dfrac{1}{3(1-x)}$    **3 a** $\dfrac{3}{2}+\dfrac{15x}{4}$    **b** $-\dfrac{1}{2}<x<\dfrac{1}{2}$    **4** $1+2x+5x^2,-\dfrac{1}{3}<x<\dfrac{1}{3}$

**5 a** $-\dfrac{121}{2}<x<\dfrac{121}{2}$    **b** 10.90872    **6 a** $-125<x<125$    **b** 4.98663

**7 a** $-\dfrac{625}{4}<x<\dfrac{625}{4}$    **b** 4.99198    **8 a** $13-\dfrac{x}{13}-\dfrac{x^2}{4394},-\dfrac{169}{2}<x<\dfrac{169}{2}$    **b** 12.92285

**9 a** $3-\dfrac{2x}{27}-\dfrac{4x^2}{2187},-\dfrac{27}{2}<x<\dfrac{27}{2}$    **b** 2.924    **10** 1.732    **11** 3.16228

### Review

**1** $1-2x+4x^2-...$    **2** $1+9x+54x^2$    **3 a** $\dfrac{2}{1-2x}-\dfrac{1}{1-x}$    **b** $1+3x+7x^2+15x^3,-\dfrac{1}{2}<x<\dfrac{1}{2}$    **4** $1+4x$

### Assessment

**1 a** $-\dfrac{1}{2}<x<\dfrac{1}{2}$    **c** $\dfrac{405}{188}$    **2 b** $a=3,b=5$    **c** 1.414

**3 a** $\dfrac{1}{1-2x}-\dfrac{1}{3-2x}$    **b** $\dfrac{2}{3}+\dfrac{16}{9}x+\dfrac{104}{27}x^2,-\dfrac{1}{2}<x<\dfrac{1}{2}$

**4 a** $\dfrac{2}{3-x}-\dfrac{1}{1+3x}$    **b i** $-\dfrac{1}{3}+\dfrac{29}{9}x-\dfrac{241}{27}x^2$    **ii** 0.4 is outside the range for which the expansion is valid.

**5 a** $1-2x+8x^2$    **b i** $\dfrac{1}{3}-\dfrac{2}{81}x+\dfrac{8}{2187}x^2$    **ii** 0.658639

# 3 Trigonometric Functions and Formulae

### Exercise 1

**1** $\dfrac{1}{3}\pi$    **2** $-\dfrac{1}{2}\pi$    **3** $\dfrac{1}{2}\pi$    **4** $-\dfrac{1}{3}\pi$    **5** $\dfrac{2}{3}\pi$    **6** $-\dfrac{1}{4}\pi$

**7** $\dfrac{3}{4}$    **8** $\dfrac{1}{8}$    **9** $2+\dfrac{1}{\sqrt{3}}$    **10** 1    **11** 0    **12** $\dfrac{1}{\sqrt{2}}$

### Exercise 2

**1 a** $60°,300°$    **b** $59.0°,239.0°$    **c** $41.8°,138.2°$    **2 a** $-140.2°,39.8°$    **b** $-131.8°,131.8°$    **c** $-150°,-30°$

**3 a** $\dfrac{\cos\theta}{\sin\theta}$    **c** $-\dfrac{1}{2}\pi,\dfrac{1}{2}\pi$    **4 a** 1    **b** $-\sqrt{2}$    **c** $-2$

**5** 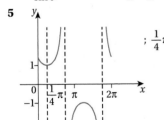 $;\dfrac{1}{4}\pi$

**6** 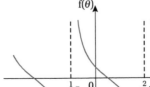  $;-\dfrac{1}{12}\pi,\dfrac{11}{12}\pi$

## Exercise 3

**1** $\tan^4 A$      **2** 1      **3** $\sec\theta\operatorname{cosec}\theta$      **4** $\sec^2\theta$      **5** $\tan\theta$      **6** $\sin^3\theta$

**7** $x^2 - y^2 = 16$      **8** $b^2x^2 - a^2y^2 = a^2b^2$      **9** $y^2(4 + x^2) = 36$      **10** $y^2(x^2 - 4x + 5) = 4$ **11** $x^2(b^2 - y^2) = a^2b^2$

**17** 57.7°, 122.3°, 237.7°, 302.3°      **18** 38.2°, 141.8°      **19** 30°, 150°      **20** 30°, 150°

**21** −0.315 rad, −2.83 rad      **22** $-\dfrac{3}{4}\pi, -0.245\,\text{rad}, \dfrac{1}{4}\pi, 2.90\,\text{rad}$      **23** $\dfrac{1}{15}\pi, \dfrac{1}{3}\pi, \dfrac{7}{15}\pi, \dfrac{11}{15}\pi, \dfrac{13}{15}\pi$

**24** none      **25** 0 and $\pi$      **26** $\dfrac{\pi}{2}$ and $\dfrac{5\pi}{6}$

## Exercise 4

**1** 0      **2** $\dfrac{1}{2}$      **3** $\dfrac{1}{4}(\sqrt{6} - \sqrt{2})$      **4** $-(2 + \sqrt{3})$      **5** $\dfrac{1}{4}(\sqrt{6} - \sqrt{2})$      **6** $\dfrac{1}{4}(\sqrt{6} + \sqrt{2})$

**7** $\sin 3\theta$      **8** 0      **9** $\tan 3A$      **10** $\tan\beta$

**11 a** 1, 115°    **b** 1, 30°    **c** 1, 310°    **d** 1, 330°

**16** 67.5°, 247.5°      **17** 7.4°, 187.4°      **18** 37.9°, 217.9°      **19** 15°, 195°

## Exercise 5

**1 a** 2, 30°      **b** $\sqrt{10}$, 71.6°      **c** 5, 36.9°      **2** $\sqrt{2}\cos\left(2\theta + \dfrac{1}{4}\pi\right)$      **3** $\sqrt{29}\sin(3\theta + 21.8°)$

**4 a** $-2\sin(\theta - 30°)$    **b** max 2 at $\theta = 300°$, min −2 at $\theta = 120°$

**5 a** $25\cos(\theta + 73.7°)$;      **b** max 28 at $\theta = 286.3°$, min −22 at $\theta = 106.3°$

**6 a** 45°      **b** 118.1°, 323.1°      **c** 0, 216.9°, 360°      **d** 0, 306.9°, 360°

## Exercise 6

**1** $\dfrac{1}{2}$      **2** $\dfrac{1}{\sqrt{2}}$      **3** $\dfrac{1}{2}\sin 2\theta$      **4** $\cos 8\theta$      **5** $-\dfrac{1}{\sqrt{3}}$      **6** $\tan 6\theta$

**7** $-\dfrac{1}{\sqrt{2}}$      **8** $\dfrac{1}{\sqrt{2}}$      **9 a** $-\dfrac{7}{25}, \dfrac{24}{25}$    **b** $\dfrac{527}{625}, \dfrac{336}{625}$    **c** $-\dfrac{119}{169}, \dfrac{120}{169}$

**10 a** $-\dfrac{336}{527}$    **b** $\dfrac{527}{625}$    **c** $-\dfrac{336}{625}$    **d** $\dfrac{164833}{390625}$

**11 a** $x(1 - y^2) = 2y$    **b** $x = 2y^2 - 1$    **c** $x = 1 - \dfrac{2}{y^2}$    **d** $2x^2y + 1 = y$

**12 a** $-\cos 2x$    **b** $3 - \cos 2x$    **c** $\dfrac{1}{2}(\cos 2x + 3)$    **d** $\dfrac{1}{2}(\cos 2x + 1)(3 + \cos 2x)$

**e** $\dfrac{1}{4}(1 + \cos 2x)^2$    **f** $\dfrac{1}{4}(1 - \cos 2x)^2$

**14 a** $\dfrac{3}{2}\pi, \dfrac{1}{6}\pi, \dfrac{5}{6}\pi$    **b** $\dfrac{1}{2}\pi, \dfrac{7}{6}\pi, \dfrac{11}{6}\pi, \dfrac{3}{2}\pi$    **c** $0, \dfrac{2}{3}\pi, \dfrac{4}{3}\pi, 2\pi$    **d** $\dfrac{1}{6}\pi, \dfrac{5}{6}\pi, \dfrac{1}{2}\pi, \dfrac{3}{2}\pi$    **e** $\dfrac{1}{3}\pi, \dfrac{5}{3}\pi$

**f** $\dfrac{1}{4}\pi, \dfrac{1}{2}\pi, \dfrac{5}{4}\pi, \dfrac{3}{2}\pi$      **g** $\dfrac{\pi}{2}$      **h** $0, \pi, 2\pi$

### Review

**1** $x^2 + \dfrac{1}{y^2} = 1$      **2** 60°, 109.5°, 250.5°, 300°      **4** $\sec^2\theta\tan^2\theta$      **5** $(x + 3)^2 - (2 - y)^2 = 1$      **6** $\dfrac{\pi}{6}$

**7** $\dfrac{2}{3}$      **8** $y = 1 - 2x^2$      **11** $\dfrac{56}{65}, -\dfrac{16}{65}$      **12** $x = 2y - 1$

**13** −155.7°, −114.3°, 24.3°, 65.7°

**14** $-\pi, 0, \pi$      **16** $\cot^2 x$      **17** 90°, 270°

**18 a** $2 - \cos 2\theta$    **b** $2 + 2\cos 4A$

**19** $5\sin(x+\alpha)$ where $\tan\alpha=\dfrac{3}{4}$; $1\min, -\dfrac{7}{3}\max$

**20** $\sqrt{2}\cos\left(x-\dfrac{1}{4}\pi\right)$; $\dfrac{1}{4}\pi$

### Assessment

**1 a** $5\cos(x+\alpha)$ where $\tan\alpha=\dfrac{4}{3}$  **b** $4+2\sec(x+\alpha)$

**c**

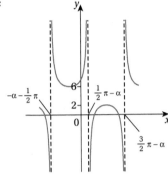

**2 a** $\sqrt{2}\sin\left(2\theta-\dfrac{1}{4}\pi\right)$  **b** $\dfrac{3}{8}\pi$

**3 a** $5\sin(x-\alpha)$ where $\tan\alpha=\dfrac{3}{4}$  **b** $\dfrac{-3}{5}$  **4 a** $0°, 233.13°, 360°$  **b** $40.2°, 252.4°$

**5 a** $18.4°, 116.6°, 198.4°, 296.6°$  **b** $8y^2=36-9x$

**6**

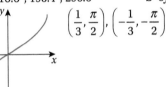

$\left(\dfrac{1}{3}, \dfrac{\pi}{2}\right), \left(-\dfrac{1}{3}, -\dfrac{\pi}{2}\right)$

**7 a** $R=\sqrt{29}$, $\alpha=1.19$  **b i** $5.09$  **ii** $0.567$ and $3.34$

## 4 Exponential and Logarithmic Functions

### Exercise 1

**1** £3.48  **2 a** £6561  **b** 7 years  **3** 4 years  **4** 7 years
**5 a** £3280.50  **b** 22 years  **6 a** 21 years  **b** 22 years  **7** 3 years  **8** 19 years

**9 a** $y=4\left(2^{\frac{x}{6}}\right)$  **b i** the area does not double exactly every 6 hours  **ii** it is growing more slowly after about 24 hours

**10 a** the initial value of the car  **b** $8876, for example, very high mileage will decrease the value more

### Exercise 2

**1 a** 7.39  **b** 0.368  **c** 4.48  **d** 0.741
**2 a**  **b**  **c**

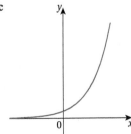

**d**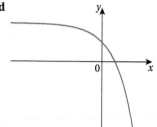

**e**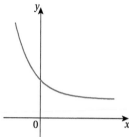

**f**

**g**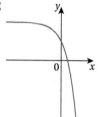

## Exercise 3

**1 a** $x = \ln 4$     **b** $\ln y = 2$     **c** $2x = \ln 3$     **d** $x - 1 = \ln 5$     **e** $\ln x = 1$

**2 a** $x = e^4$     **b** $0.5 = e^x$     **c** $x = e^y$     **d** $x = e^{\frac{3}{2}}$     **e** $(1-x)^2 = e^{1.5}$

**3 a** $3.87$     **b** $1$     **c** $0$

**4 a** $2\ln x - \ln(x+1)$     **b** $\ln(a+b) + \ln(a-b)$     **c** $\ln\cos x - \ln\sin x$     **d** $2\ln\sin x$

**5 a** $\ln\cot x$     **b** $\ln e x$     **c** $\ln(x-1)^{\frac{2}{3}}$

**6 a** $2.10$     **b** $0$     **c** $1.05$     **d** $0$

**7 a**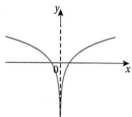

**b**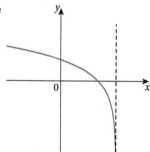

**c**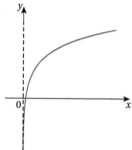

**d**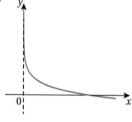

**e**

**f**

## Review

**1 a** 54.6     **b** 0.223     **c** 1.65

**2 a**

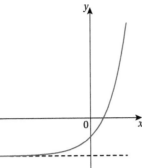

**b**

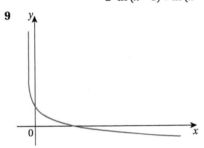

**3 a** $x = \ln 2$     **b** $\ln y = 3$     **4 a** $x = e^2$     **b** $e^x = 0.4$     **c** $x - 1 = e^y$     **5** 1.39

**6 a** $\ln x - \ln(x^2 + 1)$     **b** $\ln(x - 1) + \ln(x - 1)$     **7 a** $\ln\left(\dfrac{x-1}{x}\right)$     **b** $\ln(\tan x)$

**8** 2.30     **9**

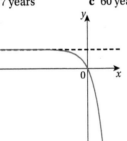

## Assessment

**1 a** £2263     **b** 17 years     **c** 60 years

**2 a** 0.717     **b**

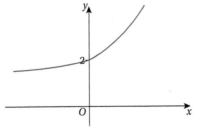

**3 a** $x - 1 = \ln 3$     **b** 1.41     **4 a** $\ln\left(\dfrac{(x+1)^2}{(x-1)}\right)$

**5 a** the purchase cost of one share     **b** $120     **c** $132.49; share prices vary unpredictably

**6 a** $f(x) < 5$     **b i** $f^{-1}(x) = \dfrac{1}{3}\ln(5 - x)$     **ii** $x = 4$

**7 a i** $f^{-1}(x) = \dfrac{1}{2}(e^x + 3)$     **ii** $f^{-1}(x) > \dfrac{3}{2}$     **iii**

**b i** $(2x - 3)^2 - 4$     **ii** $\ln(2e^{2x} - 11)$, $x = \dfrac{1}{2}\ln 8$

# 5 Differentiation

## Exercise 1

**1 a** $2e^x$
**b** $2x - e^x$
**c** $e^x$
**2** $e^2 - 2$
**3** $2 + 2e$
**4** $1$

**5** $0$
**6** $\ln 4 \ (=1.39)$
**7** there is no value of $x$ for which $e^x = -1$

## Exercise 2

**1 a** $\dfrac{3}{x}$
**b** $\dfrac{1}{x}$
**c** $-\dfrac{2}{x}$
**d** $-\dfrac{1}{2x}$
**e** $-\dfrac{5}{x}$
**f** $\dfrac{1}{2x}$

**g** $-\dfrac{3}{2x}$
**h** $\dfrac{5}{2x}$
**2 a** $(1, -1)$
**b** $(\sqrt[3]{2}, \{2 - 2\ln 2\})$
**c** $(4, \{\ln 4 - 2\})$

## Exercise 3

**1** $\dfrac{1}{8}$
**2** $\dfrac{1}{3}$
**3** $\dfrac{2}{9}$
**4** $\dfrac{1}{3}; 3y = x + 1$
**5** $\dfrac{1}{1+e}; y = \dfrac{x}{1+e}$

**6** $\dfrac{3}{2}; 2y = 3(x - 1 + \ln 3)$
**7** $1; y = x + 2 - \ln 4$
**8** $\dfrac{1}{3+4e}; y = \dfrac{x}{3+4e}$

## Exercise 4

**1 a** $\cos x + \sin x$  **b** $\cos \theta$
**c** $-3 \sin \theta$
**d** $5 \cos \theta$
**e** $3 \cos \theta - 2 \sin \theta$  **f** $4 \cos x + 6 \sin x$

**2 a** $2x - \cos x$
**b** $\dfrac{1}{x} - \sin x$
**c** $2 \cos x - e^x$
**d** $3e^x - \dfrac{1}{x} - 4 \sin x$
**e** $e\dfrac{2}{x} + 5 \cos x$
**f** $3x^2 + 3 \sin x$

**3 a** $-2$
**b** $1$
**c** $-1$
**d** $1$
**e** $2(\pi - 1)$
**f** $4$

**4 a** $\dfrac{1}{6}\pi$
**b** $\dfrac{1}{6}\pi$
**c** $\dfrac{1}{4}\pi$
**d** $\pi$

**5 a** $\left(\dfrac{\pi}{3}, \sqrt{3} - \dfrac{\pi}{3}\right)$, max; $\left(\dfrac{5\pi}{3}, -\sqrt{3} - \dfrac{5\pi}{3}\right)$, min
**b** $\left(\dfrac{1}{6}\pi, \dfrac{1}{6}\pi + \sqrt{3}\right)$, max; $\left(\dfrac{5}{6}\pi, \dfrac{5}{6}\pi - \sqrt{3}\right)$, min

**6** $y + \theta = 3 + \dfrac{1}{2}\pi$
**7** $2\pi y + x = 2\pi^3 - \pi$
**8** for example $(0, 1)$

## Exercise 5

**1** $\dfrac{\cos x}{x} - \dfrac{\sin x}{x^2}$
**2** $e^x \cos x - e^x \sin x$
**3** $\dfrac{x^3 - 2}{x} + 3x^2 \ln x$
**4** $(x+1)\cos x + \sin x$

**5** $\cos^2 x - \sin^2 x$
**6** $\dfrac{1}{x^3} - \dfrac{2\ln x}{x^3}$
**7** $\dfrac{\cos x}{x} - (\sin x)\ln x$
**8** $e^x \sin x + e^x \cos x$

**9** $2x \sin x + x^2 \cos x$
**10** $3x^2 \ln 2x + x^2$

## Exercise 6

**1** $\dfrac{e^x - xe^x}{x^2}$
**2** $\dfrac{2x(x+3) - x^2}{(x+3)^2} = \dfrac{x(x+6)}{(x+3)^2}$
**3** $\dfrac{-x^2 - 2x(4-x)}{x^4} = \dfrac{x-8}{x^3}$

**4** $\dfrac{x^2 - 3x^2 \ln x}{x^6} = \dfrac{1 - 3\ln x}{x^4}$
**5** $\dfrac{4(\sin x + \cos x) - 4x(\cos x - \sin x)}{(\sin x + \cos x)^2} = \dfrac{(4+4x)\sin x + (4-4x)\cos x}{(\sin x + \cos x)^2}$

**6** $\dfrac{(x-2)(4x) - 2x^2}{(x-2)^2} = \dfrac{2x(x-4)}{(x-2)^2}$
**7** $\dfrac{\frac{5}{3}x^{\frac{2}{3}}(3x-2) - 3x^{\frac{5}{3}}}{(3x-2)^2} = \dfrac{2x^{\frac{2}{3}}(3x-5)}{3(3x-2)^2}$
**8** $\dfrac{(3\ln x - 4)}{x^4}$

**9** $-\operatorname{cosec}^2 x$

## Exercise 7

**1** $6(3x+1)$
**2** $-4(3-x)^3$
**3** $3\cos 3x$
**4** $2e^{2x}$
**5** $\dfrac{2}{2x-1}$
**6** $-5\sin(5x)$

**7** $3x^2 \cos(x^3)$
**8** $3e^{(3x+5)}$
**9** $\dfrac{9x^2}{2\sqrt{3x^2-4}}$
**10** $\dfrac{2x-2}{x^2-2x}$
**11** $-3\sin(3x-5)$
**12** $\dfrac{3+2x}{3x+x^2}$

**13** $-10(4-2x)^4$
**14** $(2x-1)e^{(x^2-x)}$
**15** $-20\sin(5x-6)$
**16** $\dfrac{\cos x}{\sin x} = \cot x$
**17** $(1-3x^2)e^{(x-x^3)}$
**18** $-12x^2(2-x^3)^3$

**19** $\dfrac{2x-1}{3(x^2-x)^{\frac{2}{3}}}$
**20** $-\dfrac{5}{2}x^4(x^5-3)^{\frac{-3}{2}}$
**21** $2\sin x \cos x$

## Exercise 8

**1** $4\cos 4x$

**2** $2\sin(\pi - 2x)$

**3** $\dfrac{1}{2}\cos\left(\dfrac{1}{2}x + \pi\right)$

**4** $-2\cos x \sin x$

**5** $(\cos x)e^{\sin x}$

**6** $-\dfrac{\sin x}{\cos x} = -\tan x$

**7** $2x\cos x^2$

**8** $-(\sin x)e^{\cos x}$

**9** $\dfrac{\cos x}{\sin x} = \cot x$

**10** $-4\sin x\cos^3 x$

**11** $(2x-2)e^{(x^2-2x)}$

**12** $6\sec^2 6x$

**13** $\dfrac{4x-3}{2x^2-3x}$

**14** $5\cos(5x-8)$

## Exercise 9

**1** $2x + 2y\dfrac{dy}{dx} = 0$

**2** $2x + y + (x+2y)\dfrac{dy}{dx} = 0$

**3** $2x + x\dfrac{dy}{dx} + y = 2y\dfrac{dy}{dx}$

**4** $-\dfrac{1}{x^2} - \dfrac{1}{y^2}\dfrac{dy}{dx} = e^y\dfrac{dy}{dx}$

**5** $-\dfrac{2}{x^3} - \dfrac{2}{y^3}\dfrac{dy}{dx} = 0$

**6** $\dfrac{x}{2} - \dfrac{2y}{9}\dfrac{dy}{dx} = 0$

**7** $\cos x + \cos y\dfrac{dy}{dx} = 0$

**8** $\cos x \cos y - \sin x \sin y\dfrac{dy}{dx} = 0$

**9** $e^y + xe^y\dfrac{dy}{dx} = 1$

**10** $\dfrac{dy}{dx} = \pm\dfrac{1}{\sqrt{2x+1}}$

**11** $\pm\dfrac{1}{4}\sqrt{2}$

**12** $\dfrac{dy}{dx} = \dfrac{1}{1+x^2}$

**13 a** $8y + 2\sqrt{5}x + 4 = 0,\, 8y - 2\sqrt{5}x + 4 = 0$

**b** $y(2+3y_1) = xx_1 + 3y_1^2 + 2y_1 - x_1^2$

**14** $y = 2 - x$ and $y = -2$

**15** $3x + 12y - 7 = 0$

## Exercise 10

**1 a** $x = 2y^2$

**b** $x^2 + y^2 = 1$

**c** $xy = 4$

**2 a**

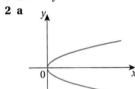

**b**

**c**

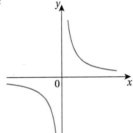

**3** $y = \dfrac{x^2}{1+x}$

**5** $y = x(x^2 - 1)$

**6** $\dfrac{x^2}{4} + \dfrac{y^2}{9} = 1$

**7 b** centre $(0, 0)$, radius 4

## Exercise 11

**1 a** $\dfrac{1}{4t}$

**b** $-\cot\theta$

**c** $-\dfrac{4}{t^2}$

**2 a** $\dfrac{dy}{dx} = 2t - t^2$

**b** $\dfrac{3}{4}$

**3** $\dfrac{3t}{2}$

**4** $\left(-\dfrac{1}{3}\sqrt{3}, \dfrac{2}{9}\sqrt{3}\right), \text{max}; \left(\dfrac{1}{3}\sqrt{3}, \dfrac{-2}{9}\sqrt{3}\right), \text{min}$

**5** $\dfrac{1}{2}\pi$

**6** $2x + y + 2 = 0$

**7 a** $6y = 4x + 5\sqrt{2}$

**b** $\left(-\dfrac{137\sqrt{2}}{97}, -\dfrac{21\sqrt{2}}{194}\right)$

**8 a** $y - t = \dfrac{1}{4t}(x - 2t^2),\, y - \dfrac{4}{t} = -\dfrac{4}{t^2}(x-t)$

**b** $y - t = -4t(x - 2t^2);\, y - \dfrac{4}{t} = \dfrac{t^2}{4}(x-t)$

**9 a** $y - \dfrac{2}{s} = s^2(x - 2s)$

**b** $\left(-\dfrac{2}{s^3}, -2s^3\right)$

**10 a** $t^2 y + x = 2t$

**b** $\left(0, \dfrac{2}{t}\right); (2t, 0)$

**11 a** $2y + tx = 8t + t^3$

**b** $\left(0, \dfrac{t}{2}(8 + t^2)\right); (8 + t^2, 0)$

## Review

**1 a** $-4\cos 4\theta$      **b** $1+\sin\theta$      **c** $3\sin^2\theta\cos\theta+3\cos 3\theta$

**2 a** $3x^2+e^x$      **b** $2e^{(2x+3)}$      **c** $e^x(\sin x+\cos x)$

**3 a** $-\dfrac{3}{x}$      **b** $-\dfrac{2}{x}$      **c** $\dfrac{1}{2x}$

**4 a** $3\cos x+e^{-x}$      **b** $\dfrac{1}{2x}+\dfrac{1}{2}\sin x$      **c** $4x^3+4e^x-\dfrac{1}{x}$      **d** $-\dfrac{1}{2}\left(e^{-x}+x^{\frac{-3}{2}}\right)-\dfrac{1}{x}$

**5** $\ln x+\dfrac{x+1}{x}$      **6** $6\sin 3x\cos 3x$      **7** $\dfrac{8}{3}(4x-1)^{\frac{-1}{3}}$      **8** $\left(3\sqrt{x}-2x\right)\left(\dfrac{3}{\sqrt{x}}-4\right)$

**9** $\dfrac{(x^4+4x^3+3)}{(x+1)^4}$      **10** $\dfrac{(x-1)\ln(x-1)-x\ln x}{x(x-1)\{\ln(x-1)\}^2}$      **11** $\dfrac{-1}{\sin x\cos x}$      **12** $2x\tan x+x^2\sec^2 x$

**13** $e^x\dfrac{(x-2)}{(x-1)^2}$      **14** $\dfrac{2\cos x}{(1-\sin x)^2}$      **15** $\dfrac{x(5x-4)}{2\sqrt{x-1}}$      **16** $-2(1-x)^2(2x+1)$

**17** $\dfrac{1}{(x+3)}-\dfrac{x}{(x^2+2)}$      **18** $\cos^2 x(4\cos^2 x-3)$      **19** $-e^{\cos^2 x}\sin 2x$

**20 a** $x=\ln 3$      **b** $x=1\ (not-1)$

**21 a** $1$      **b** $y-x=1-\dfrac{1}{2}\pi$      **c** $y+x=1+\dfrac{1}{2}\pi$

**22 a** $1+e$      **b** $y=x(1+e)$      **c** $y(1+e)+x=(1+e)^2+1$

**23 a** $2$      **b** $y=2x+1$      **c** $2y+x=2$

**24 a** $-1$      **b** $y=\dfrac{\pi}{2}-x$      **c** $y=\dfrac{\pi}{2}+x$

**25 a** for example $\left(\dfrac{1}{6}\pi,\left(\dfrac{1}{2}\pi-\sqrt{3}\right)\right)$      **b** $(1,-1)$

**26 a** $4y^3\dfrac{dy}{dx}$      **b** $y^2+2xy\dfrac{dy}{dx}$      **c** $-\dfrac{1}{y^2}\dfrac{dy}{dx}$      **d** $\ln y+\dfrac{x}{y}\dfrac{dy}{dx}$      **e** $\cos y\dfrac{dy}{dx}$

     **f** $e^y\dfrac{dy}{dx}$      **g** $\dfrac{dy}{dx}\cos x-y\sin x$      **h** $(\cos y-y\sin y)\dfrac{dy}{dx}$    **27** $\dfrac{x}{2y}$

**28** $-\dfrac{y^2}{x^2}$      **29** $-\dfrac{2y}{3x}$      **30** $-\dfrac{y(y+1)(3x+2)}{x(x+1)(y+2)}$      **31** $\dfrac{3t}{2}$

**32** $\dfrac{t}{t+1}$      **33** $-\dfrac{3}{2}\cos\theta$      **34** $-\dfrac{1}{t^2}$      **35** $-\dfrac{1}{e^t}$

**36** $2t-t^2$      **37** $y=2x+2\sqrt{2}$

## Assessment

**1 a** $\dfrac{dy}{dx}=\dfrac{1}{x}-2x$      **b** $x+y=3$      **c** $y=x+1$

**2 a** $\dfrac{dy}{dx}=\dfrac{1}{6y-2}$      **b** $x-4y+3=0$      **c** $4x+y=5$

**3 a** $\dfrac{dy}{dx}=-4x$      **b** $(0,1)$

**5 a** $2y\dfrac{dy}{dx}-2x\dfrac{dy}{dx}-2y+3\dfrac{dy}{dx}=2$      **b** $(1,1),(1,-2)$      **c** $3y-4x+1=0$ and $3y-2x+8=0$

**6 a** $y=a^2x-a^3+\dfrac{1}{a}$      **b** $\dfrac{1}{b}=a^2b-a^3+\dfrac{1}{a}\Rightarrow b=-\dfrac{1}{a^3}$

**7** $2y\sin\theta+3x\cos\theta=6$      **8** $y\cos\theta-x\sin\theta=\cos\theta$      **9** $\left(\dfrac{1}{4},1+\ln 4\right)$      **10 a** $(0,27)$

**12 a** $3e^x + \dfrac{1}{x}$  **13 a** $-\dfrac{1}{2}$  **b** $y = 2x - \dfrac{1}{2}$  **c** $-\dfrac{1}{8}$

**14** $\left(\dfrac{1}{3}, 1\right)$ and $\left(-\dfrac{1}{3}, -1\right)$

# 6 Integration

## Exercise 1

**1** $\dfrac{1}{4}e^{4x} + k$  **2** $-4e^{-x} + k$  **3** $\dfrac{1}{3}e^{(3x-2)} + k$  **4** $-\dfrac{2}{5}e^{(1-5x)} + k$  **5** $-3e^{-2x} + k$  **6** $5e^{(x-3)} + k$

**7** $2e^{\left(\frac{x}{2}+2\right)} + k$  **8** $e^{2+x} + k$  **9** $\dfrac{1}{2}e^{2x} - \dfrac{1}{2e^{2x}} + k$  **10** $\dfrac{1}{2}\{e^4 - 1\}$  **11** $2\{e^2 - 1\}$  **12** $1 - \dfrac{1}{e}$

**13** $-1 - e^{-2}$

## Exercise 2

**1** $2\ln x + k$  **2** $\dfrac{1}{4}\ln x + k$  **3** $\dfrac{3}{2}\ln x + k$  **4** $x + \ln x + k$  **5** $\dfrac{1}{2}x^2 + x - \ln x + k$

**6** $e^x + 2\ln|x| + k$  **7** $\dfrac{1}{3}\ln 2$  **8** $2 - \ln 3$  **9** $\ln 2 - 1$  **10** $e^3 - e^2 + \ln\left(\dfrac{2}{3}\right)$

**11** $\dfrac{1}{3}\left(2\ln\left(\dfrac{5}{4}\right) - 1\right)$  **12** $\ln\dfrac{3}{2} + \dfrac{38}{3}$

## Exercise 3

**1** $\dfrac{1}{8}(2x-3)^4 + k$  **2** $\dfrac{1}{15}(3x+1)^5 + k$  **3** $\dfrac{1}{25}(5x-2)^5 + k$  **4** $(2-x)^{-1} + k$  **5** $-(x+3)^{-1} + k$

**6** $\dfrac{2}{3}(1+x)^{\frac{3}{2}} + k$  **7** $\dfrac{1}{18}(3x+1)^6 + k$  **8** $\dfrac{-1}{25}(2-5x)^5 + k$

## Exercise 4

**1** $-\dfrac{1}{2}\cos 2x + k$  **2** $\dfrac{1}{7}\sin 7x + k$  **3** $\dfrac{1}{4}\tan 4x + k$  **4** $-\cos\left(\dfrac{1}{4}\pi + x\right) + k$

**5** $\dfrac{3}{4}\sin\left(4x - \dfrac{1}{2}\pi\right) + k$  **6** $\dfrac{1}{2}\tan\left(\dfrac{1}{3}\pi + 2x\right) + k$  **7** $-\dfrac{2}{3}\cos(3x - \alpha) + k$  **8** $-10\sin\left(\alpha - \dfrac{1}{2}x\right) + k$

**9** $\dfrac{1}{3}\sin 3x - \sin x + k$  **10** $\dfrac{1}{2}\tan 2x + k$  **11** $\dfrac{1}{3}$  **12** $\dfrac{1}{4}$

**13** $0$  **14** $\dfrac{1}{2}$  **15** $\dfrac{1}{2}\cos\left(\dfrac{1}{2}\pi - 2x\right) + k$  **16** $\dfrac{1}{4}e^{(4x-1)} + k$

**17** $\dfrac{1}{7}\tan 7x + k$  **18** $\dfrac{1}{2}\ln|2x-3| + k$  **19** $\sqrt{2x-3} + k$  **20** $\dfrac{-1}{3(3x-2)} + k$

**21** $\dfrac{1}{5}e^{5x} + k$  **22** $1 - \ln|x| + k$  **23** $\dfrac{1}{9}(3x-5)^3 + k$  **24** $\dfrac{1}{4}e^{(4x-5)} + k$

**25** $\dfrac{1}{6}(4x-5)^{\frac{3}{2}} + k$  **26** $-\dfrac{1}{5}\cos\left(5x - \dfrac{1}{3}\pi\right) + k$  **27** $-\dfrac{3}{2}\ln|1-x| + k$  **28** $-\dfrac{4}{3}(x-6)^{-3} + k$

**29** $\dfrac{1}{3}\sin\left(3x - \dfrac{1}{3}\pi\right) + k$  **30** $\dfrac{2}{3}x^3 - 8x + k$  **31** $\dfrac{1}{4}x^4 - 2x^3 + \dfrac{9}{2}x^2 + k$  **32** $3x - x^2 + 2x^3 + k$

**33** $-\dfrac{1}{3x^2} + k$  **34** $\dfrac{3}{2}\ln|x| + k$  **35** $\dfrac{2}{9}(x-3)^3 + k$

# Exercise 5

**1** $e^{x^4}$

**2** $-e^{\cos x}+k$

**3** $e^{\tan x}+k$

**4** $e^{x^2+x}+k$

**5** $e^{(1-\tan x)}+k$

**6** $e^{(x+\sin x)}+k$

**7** $e^{(1+x^2)}+k$

**8** $e^{(x^3-2x)}+k$

**9** $\dfrac{1}{10}(x^2-3)^5+k$

**10** $-\dfrac{1}{3}(1-x^2)^{\frac{3}{2}}+k$

**11** $\dfrac{1}{6}(\sin 2x+3)^3+k$

**12** $-\dfrac{1}{6}(1-x^3)^2+k$

**13** $\dfrac{2}{3}(1+e^x)^{\frac{3}{2}}+k$

**14** $\dfrac{1}{5}\sin^5 x+k$

**15** $\dfrac{1}{4}\tan^4 x+k$

**16** $\dfrac{1}{3(n+1)}(1+x^{n+1})^3+k$

**17** $\dfrac{3}{5}(1-\cos x)^5+k$

**18** $\dfrac{4}{9}\left(1+x^{\frac{3}{2}}\right)^{\frac{3}{2}}+k$

**19** $\dfrac{1}{12}(x^4+4)^3+k$

**20** $-\dfrac{1}{4}(1-e^x)^4+k$

**21** $\dfrac{2}{3}(1-\cos\theta)^{\frac{3}{2}}+k$

**22** $\dfrac{1}{3}(x^2+2x+3)^{\frac{3}{2}}+k$

**23** $\dfrac{1}{2}e^{(x^2+1)}+k$

**24** $\dfrac{1}{2}(1+\tan x)^2+k$

# Exercise 6

**1** $x\sin x+\cos x+k$

**2** $xe^x-e^x+k$

**3** $x^2\ln\dfrac{(3x)}{2}-\dfrac{x^2}{4}+k$

**4** $-e^{-x}(x+1)+k$

**5** $3(\sin x-x\cos x)+k$

**6** $\dfrac{1}{4}\sin 2x-\dfrac{1}{2}x\cos 2x+k$

**7** $\dfrac{1}{2}xe^{2x}-\dfrac{1}{4}e^{2x}+k$

**8** $\dfrac{1}{32}e^{4x}(8x^2-4x+1)+k$

**9** $\sin x-x\cos x+k$

**10** $x(\ln|2x|-1)+k$

**11** $xe^x+k$

**12** $\dfrac{1}{72}(8x-1)(x+1)^8+k$

**13** $\sin\left(x+\dfrac{1}{6}\pi\right)-x\cos\left(x+\dfrac{1}{6}\pi\right)+k$

**14** $\dfrac{1}{n^2}(\cos nx+nx\sin nx)+k$

**15** $\dfrac{x^2}{4}(2\ln x-1)+k$

**16** $\dfrac{3}{4}(2x\sin 2x+\cos 2x)+k$

**17** $\sin x-\dfrac{1}{3}\sin^3 x+k$

**18** $\dfrac{1}{2}e^{x^2-2x+4}+k$

**19** $(x^2+1)e^x+k$

**20** $-\dfrac{1}{4}(4+\cos x)^4+k$

**21** $e^{\sin x}+k$

**22** $\dfrac{2}{15}\sqrt{(1+x^5)^3}+k$

**23** $\dfrac{1}{5}(e^x+2)^5+k$

**24** $\dfrac{1}{4}e^{2x-1}(2x-3)+k$

**25** $-\dfrac{1}{20}(1-x^2)^{10}+k$

**26** $\dfrac{1}{6}\sin^6 x+k$

# Exercise 7

**1** $\ln(4+\sin x)+k$

**2** $\dfrac{1}{3}\ln|3e^x-1|+k$

**3** $\dfrac{1}{4(1-x^2)^2}+k$

**4** $\dfrac{1}{2\cos^2 x}+k$

**5** $\dfrac{1}{4}\ln(1+x^4)+k$

**6** $\ln|x^2+3x-4|+k$

**7** $\dfrac{2}{3}\sqrt{2+x^3}+k$

**8** $\dfrac{1}{2-\sin x}+k$

**9** $\ln|\ln x|+k$

**10** $\dfrac{-1}{5\sin^5 x}+k$

**11** $-\ln|1-x^2|+k$

**12** $-2\sqrt{1-e^x}+k$

**13** $\dfrac{1}{6}\ln|3x^2-6x+1|+k$

**14** $\dfrac{-1}{(n-1)\sin^{n-1} x}+k\,(n\neq 1)$

**15** $\dfrac{1}{(n-1)\cos^{(n-1)} x}+k\,(n\neq 1)$

**16** $-\ln(4+\cos x)+k$

**17** $\dfrac{1}{2}\ln|x(x-2)|+k$

**18** $-\dfrac{1}{e^x-x}+k$

**19** $\ln 3$

**20** $\ln\sqrt{2}$

**21** $\dfrac{55}{1152}$

**22** $\dfrac{e-1}{2(e+1)}$

**23** $0$

**24** $\ln 2$

## Exercise 8

**1** $2\ln\left|\dfrac{x}{x+1}\right|+k$

**2** $\ln\left|\dfrac{x-2}{x+2}\right|+k$

**3** $\dfrac{1}{2}\ln|x^2-1|+k$

**4** $\dfrac{1}{2}\ln\left|\dfrac{(x+2)^3}{x}\right|+k$

**5** $\ln\dfrac{(x-3)^2}{|x-2|}+k$

**6** $\dfrac{1}{2}\ln\dfrac{|x^2-1|}{x^2}+k$

**7** $x-\ln|x+1|+k$

**8** $x+4\ln|x|+k$

**9** $x-4\ln|x+4|+k$

**10** $\ln\dfrac{|1-x|}{x^4}+k$

**11** $x+\dfrac{1}{2}\ln\dfrac{|x+1|}{|x-1|}+k$

**12** $x+\ln\dfrac{|x+1|}{(x+2)^4}+k$

**13** $\dfrac{1}{2}\ln|x^2-1|+k$

**14** $\dfrac{-1}{x^2-1}+k$

**15** $\ln\left|\dfrac{x-1}{x+1}\right|+k$

**16** $\ln|x^2-5x+6|+k$

**17** $\ln\dfrac{(x-3)^6}{(x-2)^4}+k$

**18** $\ln\left|\dfrac{(x-3)^3}{x-2}\right|+k$

**19** $4+\ln 5$

**20** $\ln\dfrac{1}{6}$

**21** $\dfrac{1}{2}\ln\dfrac{12}{5}$

**22** $\ln\dfrac{5}{3}$

**23** $\dfrac{5}{36}$

**24** $1-\dfrac{3}{2}\ln\dfrac{7}{5}$

## Exercise 9

**1** $\dfrac{1}{4}(2x+\sin 2x)+k$

**2** $\sin x-\dfrac{1}{3}\sin^3 x+k$

**3** $-\dfrac{1}{15}\cos x(15-10\cos^2 x+3\cos^4 x)+k$

**4** $\tan x-x+k$

**5** $\dfrac{1}{32}\{12x-8\sin 2x+\sin 4x\}+k$

**6** $\dfrac{1}{2}\sec^2 x-\ln|\sec x|+k$

**7** $\dfrac{1}{32}\{12x+8\sin 2x+\sin 4x\}+k$

**8** $\dfrac{1}{3}\cos x(\cos^2 x-3)+k$

## Exercise 10

**1** $\dfrac{512}{15}\pi$

**2** $\dfrac{1}{2}\pi(e^6-1)$

**3** $\dfrac{1}{2}\pi$

**4** $\dfrac{64}{5}\pi$

**5** $\left(\dfrac{\pi}{2}\right)^2$

**6** $8\pi$

**7** $8\pi$

**8** $\dfrac{3}{5}\pi\left(\sqrt[3]{32}-1\right)$

**9** $\dfrac{1}{2}\pi(e^2-1)$

## Review

**1** $-3e^{-x}+k$

**2** $e^{x^2}+k$

**3** $\dfrac{1}{6}(3\tan x-4)^2+k$

**4** $\dfrac{4}{3}\ln|x|+k$

**5** $\dfrac{1}{4}\tan^4 x+k$

**6** $\dfrac{1}{18}(3x+4)^6+k$

**7** $-e^{\cos x}+k$

**8** $\dfrac{1}{4}\sin\left(4x+\dfrac{\pi}{7}\right)+k$

**9** $\dfrac{1}{2}e^{(x^2-2x+3)}+k$

**10** $\dfrac{-1}{30}(1-x^3)^{10}+k$

**11** $\dfrac{1}{3}(2x+1)^{\frac{3}{2}}+k$

**12** $\dfrac{1}{3}(x^2+1)^{\frac{3}{2}}+k$

**13** $\dfrac{-1}{3}(2-\sin x)^3+k$

**14** $\dfrac{1}{2}e^{2x}(x^2-x+1)+k$

**15** $\dfrac{1}{2}(x+1)^2\ln(x+1)-\dfrac{1}{4}(x+1)^2+k$

**16** $x^2\sin x+2x\cos x-2\sin x+k$

**17** $\dfrac{1}{2}e^{2x+3}+k$

**18** $\dfrac{1}{6}(2x^2-5)^{\frac{3}{2}}+k$

**19** $xe^x-e^x+k$

**20** $x\ln x-x+k$

**21** $\dfrac{1}{2}x-\dfrac{1}{12}\sin 6x+k$

**22** $-\dfrac{1}{2}e^{-x^2}+k$

**23** $\dfrac{1}{3}\sin^3 x+k$

**24** $\dfrac{10u-7}{110}(u+7)^{10}+k$

**25** $-\dfrac{1}{12}(x^3+9)^{-4}+k$

**26** $\dfrac{1}{2}\ln(1-\cos 2y)+k$

**27** $\frac{1}{2}\ln|2x+7|+k$

**28** $-\frac{2}{9}(1+\cos 3x)^{\frac{3}{2}}+k$

**29** $\frac{1}{2}\ln|x^2+4x-5|+k$

**30** $\frac{2}{3}\ln|x+5|+\frac{1}{3}\ln|x-1|+k$

**31** $\frac{1}{3}x\sin 3x+\frac{1}{9}\cos 3x+k$

**32** $x\ln|5x|-x+k$

**33** $\frac{3}{8}\pi^2$

### Assessment

**1 a** $-\frac{1}{2}e^{(1-x)^2}+k$

**b** $\ln(x^2+1)+k$

**c** $\frac{1}{5}\sin^5 x+k$

**2 a** $-\frac{1}{4}(4+\cos x)^4+k$

**b** $\frac{1}{16}(2x+3)^8\left(\frac{8}{9}x-\frac{1}{6}\right)+k$

**c** $-\frac{1}{9}(3x+1)e^{(2-3x)}+k$

**3** $3(x^2-3)^{\frac{1}{2}}+k$

**4 a** $\frac{1}{2(x-3)}-\frac{1}{2(x+1)}$

**b** $\frac{1}{2}\ln\left(\frac{x-3}{x+1}\right)+k$

**5 a** $-(9-y^2)^{\frac{3}{2}}+k$

**b** $-\frac{\pi}{8}$

**6** $\frac{33\pi}{5}$

**7** $\left(\frac{\pi}{2}\right)(e^2-4e+5)$

**8 a** $A=2, B=3$

**b** $y=x^2+3\ln(2x^2-x+2)+1-3\ln 5$

**9 a** $-\frac{1}{2}x^2\cos 2x+\frac{1}{2}x\sin 2x+\frac{1}{4}\cos 2x+c$

**b** $\pi\left(\frac{\pi^2}{8}-\frac{1}{2}\right)$

# 7 Differential Equations

### Exercise 1

**1** $y^2=A-2\cos x$

**2** $\frac{1}{y}-\frac{1}{x}=A$

**3** $2y^3=3(x^2+4y+A)$

**4** $x=A\sec y$

**5** $(A-x)y=1$

**6** $y=\ln\dfrac{A}{\sqrt{1-x^2}}$

**7** $y=A(x-3)$

**8** $x+A=4\ln|\sin y|$

**9** $u^2=v^2+4v+A$

**10** $16y^3=12x^4\ln|x|-3x^4+A$

**11** $y^2+2(x+1)e^{-x}=A$

**12** $\sin x=A-e^{-y}$

**13** $2r^2=2\theta-\sin 2\theta+A$

**14** $u+2=A(v+1)$

**15** $y^2=A+(\ln|x|)^2$

**16** $y^2=Ax(x+2)$

**17** $4v^3=3(2+t)^4+A$

**18** $1+y^2=Ax^2$

**19** $Ar=e^{\tan\theta}$

**20** $y^2=A-\text{cosec}^2x$

**21** $v^2+A=2u-2\ln|u|$

**22** $e^{-x}=e^{1-y}+A$

**23** $A-\dfrac{1}{y}=2\ln|\tan x|$

**24** $y-1=A(y+1)(x^2+1)$

### Exercise 2

**1** $y^3=x^3+3x-13$

**2** $e^t(5-2\sqrt{s})=1$

**3** $3(y^2-1)=8(x^2-1)$

**4** $y=e^x-2$

**5** $y^2=-e^{-3x}+e^{-3}+4$

**6** $(y+1)^2(x+1)=2(x-1)$

**7** $4y^2=(y+1)^2(x^2+1)$

### Exercise 3

**1** $s\dfrac{ds}{dt}=k$

**2** $\dfrac{dh}{dt}=k\ln(H-h)$

**3** $\dfrac{dh}{dt}=kV$

**4** $\dfrac{dm}{dt}=-km$

**5** $\dfrac{dh}{dt}=-k(H-h)$

**6** $\dfrac{dd}{dt}=\dfrac{k}{d}$

**7 a** $\dfrac{dn}{dt}=0$

**b** $\dfrac{dn}{dt}=k\sqrt{n}$

**8 a** $\dfrac{dn}{dt}=k_1 n$

**b** $\dfrac{dn}{dt}=\dfrac{k_2}{n}$

**c** $\dfrac{dn}{dt}=-k_3$

### Exercise 4

**1 a** $\dfrac{dh}{dt}=\dfrac{k}{h^3}$

**b** $h^4=5t+1$

**c** 16 minutes

**2 a** $\dfrac{dy}{dx}=k\sqrt{x}$

**b** $y=0.4x^{\frac{3}{2}}+1.6$

**c** 1.2

**3 a** $\dfrac{dn}{dt}=kn$

**b** 17.1 hours (3 sf)

**4 a** $\dfrac{dm}{dt}=-km$

**b** $m=1000e^{-\frac{t\ln 2}{500}}$

**c** 1000 years

**5 a** $-\dfrac{dm}{dt}=km$; $m=50e^{-kt}$ where $k=0.002\,554\ldots$

**b** 26.8 g (3 sf)

**1** $\dfrac{dr}{dt} = kr^2$      **2** $x^3y = y - 1$      **3** $y = \tan\left\{\dfrac{1}{2}(x^2 - 4)\right\}$      **4 a** $y^2 = 2x$

**5 a** $\dfrac{dN}{dt} = kN$      **b** 33      **c** E.g. there were not 200 rabbits on the island

**6 a** $y = 100e^{-0.105x}$      **b** E.g. the rate of inflation is very unlikely to remain constant.

## Assessment

**1** $y^2 = 4x^2 + 5$      **2** $\sin y = -1 - \cos x$      **3 a** $\dfrac{dy}{dx} = \dfrac{k}{xy}$      **b** $y^2 = 100 - \dfrac{48 \ln|x|}{\ln 2}$

**4 c** $k = 0.0357$ (3.s.f.)      **d** 44°C (nearest degree)

**5** About 4 to $4\dfrac{1}{2}$ hours before discovery assuming Newton's Law of Cooling

**6** $y = \dfrac{9}{3x\cos x - \sin 3x + 10}$

**7 a** $x = \dfrac{1}{20}\left(\dfrac{4+5t}{1+t}\right)^2 - \dfrac{4}{5}$      **b i** $\dfrac{dr}{dt} = \dfrac{k}{r^2}$      **ii** 3 m

# 8 Numerical Methods

## Exercise 1

**9** 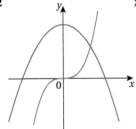 ; 3

**10**

**12** 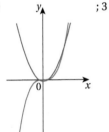 ; 1 and 2

**13**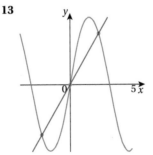

**14**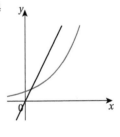

## Exercise 2

**1 b** $x = \dfrac{1}{10}(2 + x^2 - x^3)$      **c** yes         **2 b** $x = \dfrac{1}{9}(3x^3 - 2x^2 + 2)$      **c** yes

**3 b** $x = \dfrac{1}{6}(1 - 2x^3 - x^2)$      **c** yes         **4 a** $x = \dfrac{1}{8}(8 - x^2)$      **c** yes

**5 b** 0.71, 0.61      **6 b** 0.97, 0.92      **7 b** 0.198, 0.194
     **c** 2.6786, 2.608..., 2.389..., 2.389..., 1.787... The values are moving away from the larger root.
     (They converge to the smaller root.)

## Exercise 3

**1 a** 21      **b** 21.3      **2 a** 0.648      **b** 0.671      **3 a** 1.79      **b** 1.62
**4 a** 1.32      **b** 1.29      **5 a** 5.58      **b** 5.54

## Review

**2 c** 2.20 (3 sf)    **3** 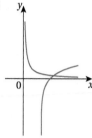 ; 3 and 4    **4** 9360 (3 sf)    **5** −2.60 (3 sf)    **6** 1.81 (3 sf)

## Assessment

**1 a**

**d** 1.24, 1.26    **2 b** 0.682 (3 sf)    **3 a** 79.8 (3 sf)    **b** 80.5 (3 sf)

**4 a** 2.00    **b** $\dfrac{1}{2}(x^2+1)\ln(x^2+1)-\dfrac{x^2}{2}+k$    **c** $\dfrac{1}{2}(5\ln 5-4)=2.02$

**5** 2.449    **6** 2.541    **7 c** $x_2=3.578,\ x_3=3.568$

# 9 Vectors

### Exercise 1

**1 a** $\dfrac{1}{2}(\mathbf{a}+\mathbf{c})$    **b** $\dfrac{1}{2}(\mathbf{b}+\mathbf{d})$    **2 a** $\dfrac{1}{4}(\mathbf{a}+2\mathbf{b}+\mathbf{c})$    **b** $\dfrac{1}{4}(\mathbf{b}+2\mathbf{c}+\mathbf{d})$

### Exercise 2

**1 a** $\begin{bmatrix} 3 \\ 6 \\ 4 \end{bmatrix}$    **b** $\begin{bmatrix} 1 \\ -2 \\ -7 \end{bmatrix}$    **c** $\begin{bmatrix} 1 \\ 0 \\ -3 \end{bmatrix}$

**2 a** $(5, -7, 2)$    **b** $(1, 4, 0)$    **c** $(0, 1, -1)$

**3 a** $\sqrt{21}$    **b** 5    **c** 3

**4 a** 6    **b** 7    **c** $\sqrt{206}$

**5 a** $\begin{bmatrix} 3 \\ 0 \\ 4 \end{bmatrix}$    **b** $\begin{bmatrix} 2 \\ -2 \\ 2 \end{bmatrix}$    **c** $\begin{bmatrix} 2 \\ 3 \\ 3 \end{bmatrix}$    **d** $\begin{bmatrix} -6 \\ 12 \\ -8 \end{bmatrix}$

**6** $\lambda = \dfrac{1}{2}$    **7 b, e** and **f**

**8 a** neither    **b** parallel    **c** equal

**9 a** $\begin{bmatrix} -2 \\ -3 \\ 7 \end{bmatrix}$    **b** $\begin{bmatrix} 0 \\ -1 \\ 3 \end{bmatrix}$    **c** $\begin{bmatrix} -2 \\ -2 \\ 4 \end{bmatrix}$

**10** $\sqrt{62}, \sqrt{10}, 2\sqrt{6}$　　　　**11** $\sqrt{5}$　　　　**12** $\overrightarrow{AB} = \begin{bmatrix} -1 \\ 3 \\ 0 \end{bmatrix}, \overrightarrow{BD} = \begin{bmatrix} 0 \\ -1 \\ -4 \end{bmatrix} \overrightarrow{CD} = \begin{bmatrix} -2 \\ 2 \\ -6 \end{bmatrix} \overrightarrow{AD} = \begin{bmatrix} -1 \\ 2 \\ -4 \end{bmatrix}$

## Exercise 3

**1** $\sqrt{13}; 3$　　　　**2 a** no　　　　**b** no　　　　**c** no

## Exercise 4

**1 a** $\begin{bmatrix} 2 \\ -1 \\ -5 \end{bmatrix}$　　　**b** $\begin{bmatrix} 0 \\ 3 \\ -5 \end{bmatrix}$　　　**c** $\begin{bmatrix} -2 \\ 4 \\ -1 \end{bmatrix}$

**2 a** $\mathbf{r} = \begin{bmatrix} 1 \\ -3 \\ 2 \end{bmatrix} + t \begin{bmatrix} 5 \\ 4 \\ -1 \end{bmatrix}$　　**b** $\mathbf{r} = \begin{bmatrix} 2 \\ 1 \\ 0 \end{bmatrix} + t \begin{bmatrix} 0 \\ 3 \\ -1 \end{bmatrix}$　　**c** $\mathbf{r} = t \begin{bmatrix} 1 \\ -1 \\ -1 \end{bmatrix}$

**3** $\mathbf{r} = \begin{bmatrix} 4 \\ 5 \\ 10 \end{bmatrix} + s \begin{bmatrix} 1 \\ 1 \\ 3 \end{bmatrix}$　　$\mathbf{r} = \begin{bmatrix} 2 \\ 3 \\ 4 \end{bmatrix} + s \begin{bmatrix} 1 \\ 1 \\ 5 \end{bmatrix}$　　$\mathbf{r} = \begin{bmatrix} 4 \\ 5 \\ 10 \end{bmatrix} + s \begin{bmatrix} 1 \\ 1 \\ 5 \end{bmatrix}$

**4 a i** $\mathbf{r} = \begin{bmatrix} 1 \\ 7 \\ 8 \end{bmatrix} + s \begin{bmatrix} 3 \\ 1 \\ -4 \end{bmatrix}$; **ii** $\mathbf{r} = \begin{bmatrix} 1 \\ 1 \\ 7 \end{bmatrix} + s \begin{bmatrix} 2 \\ 3 \\ -6 \end{bmatrix}$ $(7, 9, 0), \left(0, \frac{20}{3}, \frac{28}{3}\right), (-20, 0, 36)$ **b ii** $\left(\frac{10}{3}, \frac{9}{2}, 0\right), \left(0, \frac{-1}{2}, 10\right), \left(\frac{1}{3}, 0, 9\right)$

**b** $\mathbf{r} = \begin{bmatrix} 1 \\ 1 \\ 7 \end{bmatrix} + s \begin{bmatrix} 2 \\ 3 \\ -6 \end{bmatrix}; \left(\frac{10}{3}, \frac{9}{2}, 0\right), \left(0, -\frac{1}{2}, 10\right), \left(\frac{1}{3}, 0, 9\right)$

## Exercise 5

**1 a** parallel　　**b** intersect at $\begin{bmatrix} 1 \\ 2 \\ 0 \end{bmatrix}$　　**c** skew　　**2** $-3, \begin{bmatrix} -1 \\ 3 \\ 4 \end{bmatrix}$

## Exercise 6

**1 a** 30　　　　**b** 0　　　　**c** $-1$

**2 a** $7, \frac{1}{3}\sqrt{7}$　　**b** $14, \sqrt{\frac{7}{19}}$　　**c** $3, \frac{3}{58}\sqrt{58}$　　**d** $1, \frac{1}{5}$

**3** 4　　　　**4 a** $-\sqrt{\frac{7}{34}}$　　**b** $\begin{bmatrix} 0 \\ 6 \\ 3 \end{bmatrix}, 33.2°$

**7 a** $10\sqrt{3}$　　**b** $41 - 20\sqrt{3}$　　**8** $112.2°$

**9 a i** $\begin{bmatrix} 0 \\ 4 \\ -4 \end{bmatrix}$　　**ii** $\begin{bmatrix} 4 \\ 4 \\ 0 \end{bmatrix}$　　**b** $\hat{A} = 90°, \hat{E} = 19° \hat{H} = 71°$

**10** no; **a** and **b** may be perpendicular　　　**11 b** $122°$　　**c** 17.3 sq units

**12 a** $0°$　　**b** $27.9°$　　**c** $90°$

**13 b** $\left(\frac{8}{3}, \frac{5}{6}, \frac{23}{6}\right)$　　**c** $\frac{\sqrt{66}}{6} = 1.35$ (3 sf)

**14 b** $\left(-\frac{14}{13}, \frac{23}{13}, \frac{4}{13}\right)$　　**c** $\frac{\sqrt{(260)}}{13} = 1.24$

## Review

**1 a** $\overrightarrow{AC}$　　**b** $\overrightarrow{AD}$

**2 a** $\mathbf{b} - \mathbf{a}$　　**b** $\mathbf{c} - \mathbf{b}$　　**c** $\mathbf{c} - \mathbf{a}$

**3** $(2, 1, 0)$

**4 a** 3
**b** $\begin{bmatrix} -2 \\ 2 \\ -1 \end{bmatrix}$
**c** $\begin{bmatrix} 4 \\ 0 \\ \dfrac{-1}{2} \end{bmatrix}$

**5** $\mathbf{r} = \begin{bmatrix} 2 \\ 1 \\ 0 \end{bmatrix} + t \begin{bmatrix} -1 \\ 1 \\ -2 \end{bmatrix}$
**6 a** $\mathbf{r} = \begin{bmatrix} 3 \\ 2 \\ -1 \end{bmatrix} + t \begin{bmatrix} 3 \\ 4 \\ -5 \end{bmatrix}$
**b** $\left( \dfrac{3}{2}, 0, \dfrac{3}{2} \right)$

**7 b** $(11, -5, -12)$
**8** 0
**9** 86.5°
**10** 27.2°

**11 a** $(0, 0, 3)$
**b** $\sqrt{3}$

**Assessment**

**1 a i** $\begin{bmatrix} 2 \\ 2 \\ -2 \end{bmatrix}$
**ii** $\begin{bmatrix} -2 \\ 2 \\ 0 \end{bmatrix}$
**b** 90°

**2 a** $\mathbf{r} = \begin{bmatrix} 2 \\ 1 \\ 0 \end{bmatrix} + t \begin{bmatrix} 2 \\ -1 \\ 3 \end{bmatrix}$
**b** $\mathbf{r} = \begin{bmatrix} 2 \\ 1 \\ 0 \end{bmatrix} + s \begin{bmatrix} 1 \\ 0 \\ -5 \end{bmatrix}$
**c** 47.0°
**3 a** $\mathbf{r} = \begin{bmatrix} 13 \\ -4 \\ 2 \end{bmatrix} + t \begin{bmatrix} 5 \\ 0 \\ 1 \end{bmatrix}$

**4** $\dfrac{\sqrt{315}}{7} = 2.54$
**5** $a + b - 3c = 6$

**6 a** 60°
**b** $(15, 6, 2)$
**c** $(11, 0, 0)$ and $(23, 4, -8)$

**7 a** $\mathbf{r} = \begin{bmatrix} 5 \\ 1 \\ -2 \end{bmatrix} + \lambda \begin{bmatrix} -1 \\ -2 \\ 5 \end{bmatrix}$ or $\mathbf{r} = \begin{bmatrix} 4 \\ -1 \\ 3 \end{bmatrix} + \lambda \begin{bmatrix} -1 \\ -2 \\ 5 \end{bmatrix}$
**b i** $(7, 5, -12)$   **ii** $(-23, 5, 0)$

# 10 Mathematical Modelling and Kinematics

Answers are given to 3 significant figures unless otherwise stated in the question.
The value of $g$ is taken as 9.8.

## Exercise 1

Model each object as a particle, ignore any resistance unless specifically mentioned.
Assume cables, tow-bars etc. to be straight, light and inextensible.
Any sensible answer for the reasonableness of assumptions is acceptable.

**1 a** 6550 N, 5680 N, 5630 N
**b** 904 N, 784 N, 776 N
**2 a** 1480 N
**b** 3330 N

**3 a** 3.08 ms$^{-2}$
**b** −0.25 ms$^{-2}$
**c** 80 s

**4** $a = 2$ m/s$^2$, $T = 600$ N
**5 a** $35(2g + 1)$ N; a particle or a small block
**b** $435(2g + 1)$ N

**6** $600g$
**7 a** 2 ms$^{-2}$
**b** 600 N

## Exercise 2

**1** $3\mathbf{i} + 6\mathbf{j} + 4\mathbf{k}$
**2** $(5, -7, 2)$
**3** $\sqrt{21}$

**4 a** $3\mathbf{i} + 4\mathbf{k}$   **b** $2\mathbf{i} - 2\mathbf{j} + 2\mathbf{k}$
**5 a** $-2\mathbf{i} - 3\mathbf{j} + 7\mathbf{k}$
**b** $-\mathbf{j} + 3\mathbf{k}$
**c** $-2\mathbf{i} - 2\mathbf{j} + 4\mathbf{k}$

## Exercise 3

**1 a** $\mathbf{v} = 6t^2\mathbf{i} + 6t\mathbf{j}$, $\mathbf{a} = 12t\mathbf{i} + 6\mathbf{j}$
**b** $\mathbf{v} = 24\mathbf{i} + 12\mathbf{j}$, $\mathbf{a} = 24\mathbf{i} + 6\mathbf{j}$
**2 a** $4y = 3x^2$
**b** $(y + 4)^2 = 16x$
**c** $y = \dfrac{2}{3}$

**3** $v = t\mathbf{i} - 2t\mathbf{j}$; $\mathbf{r} = \left( \dfrac{1}{2}t^2 + 3 \right)\mathbf{i} + (1 - t^2)\mathbf{j}$

**4 a** $-\mathbf{j}$
**b** $18\mathbf{i} + \mathbf{j}$
**c** $8\mathbf{i}$

**5 a** $(-\sin t)\mathbf{i} + 3t^2\mathbf{j} - (\cos t)\mathbf{k}$
**b** $-\mathbf{i}$
**6 a** $\mathbf{r} = 16\mathbf{i} + 8\mathbf{j}$
**b** 17.9 m

**7 a** $v = t(3\mathbf{i} - 2\mathbf{j})$        **b** $10.8 \text{ ms}^{-1}$        **8 a** $= 2\mathbf{i} + 6\mathbf{j}$; $6.32 \text{ ms}^{-2}$

**9** $v = 9\mathbf{i} + 14\mathbf{j}$, $\mathbf{a} = 9\mathbf{i} + \dfrac{57}{2}\mathbf{j}$     **10 a** $v = 9\mathbf{i} - \dfrac{16}{t^2}\mathbf{j} + \mathbf{k}$     **b** $\mathbf{a} = 9t\mathbf{i} + \dfrac{16}{t}\mathbf{j} + (t+2)\mathbf{k}$

**11 a i** $v = 4\cos\alpha\mathbf{i} + (4\sin\alpha - gt)\mathbf{j}$    **ii** $\mathbf{r} = 4t\cos\alpha\mathbf{i} + \left(4t\sin\alpha - \dfrac{1}{2}gt^2\right)\mathbf{j}$    **b** $y = x\tan\alpha - \dfrac{gx^2}{32\cos^2\alpha}$

### Review

**1** Model the balls as particles, the string as light and inextensible and the peg as smooth.
The peg is unlikely to be perfectly smooth, but it may be smooth enough to make the calculation
reasonably accurate.

**2** $9\mathbf{i} + 5\mathbf{j}$              **3 a**           **b** $0.364\mathbf{i} + 7.71\mathbf{j}$     **c** $2.7$ degrees     **d** $7.72\,(3\text{sf})$

**4 a** $v = 6t^2\mathbf{i} + 6t\mathbf{j}$, $\mathbf{a} = 12t\mathbf{i} + 6\mathbf{j}$
  **b** $v = 24\mathbf{i} + 12\mathbf{j}$, $\mathbf{a} = 24\mathbf{i} + 6\mathbf{j}$

**5 a** $v = (2t+1)\mathbf{i} - 2t\mathbf{j}$, $\mathbf{a} = 2\mathbf{i} - 2\mathbf{j}$
  **b** $v = 3\mathbf{i} - 2\mathbf{j}$, $\mathbf{a} = 2\mathbf{i} - 2\mathbf{j}$

**6 a** $v = e^t\mathbf{i} - 2t\mathbf{j} + 3t^2\mathbf{k}$, $\mathbf{a} = e^t\mathbf{i} - 2\mathbf{j} + 6t\mathbf{k}$
  **b** $v = e^6\mathbf{i} - 12\mathbf{j} + 108\mathbf{k}$, $\mathbf{a} = e^6\mathbf{i} - 2\mathbf{j} + 36\mathbf{k}$

**7** $v = t\mathbf{i} - 2t\mathbf{j}$, $\mathbf{r} = (\dfrac{1}{2}t^2 + 3)\mathbf{i} - (1 - t^2)\mathbf{j}$     **8** $\mathbf{a} = \mathbf{i} - \mathbf{k}$     **9** $25.6 \text{ m}$

### Assessment

**1 a** $v = 8\mathbf{i} + \left(23 - \dfrac{27}{t^2}\right)\mathbf{j} + 2\mathbf{k}$    **b** $\mathbf{r} = 8t\mathbf{i} + \left(23t + \dfrac{27}{t} - 10\right)\mathbf{j} + (2t - 6)\mathbf{k}$     **c** $52.1 \text{ m}$     **d** $7.72 \text{ ms}^{-1}$

**2 a** $12\mathbf{i} + 6\mathbf{j}$           **b** $14.3$

**3 a** $v = \mathbf{i} - \dfrac{\pi}{2}\mathbf{j}$          **b** $\mathbf{a} = 0$        **c** $y = -\cos\left(\dfrac{\pi x}{2}\right)$

**4 a** $v = (3\cos 3t)\mathbf{i} - (3\sin 3t)\mathbf{j}$    **b** $3$

**5 a** $(0.3\mathbf{i} + 0.4\mathbf{j}) \text{ ms}^{-1}$        **b** $68.0 \text{ m}$       **c** $(5.5\mathbf{i} + 4\mathbf{j}) \text{ ms}^{-1}$

**6 a** $(500 - 17.5t - 0.25t^2)\mathbf{i} + (200 - 27t + 0.3t^2)\mathbf{j}$

  **b** $t = 40$            **c** $-7.5\mathbf{i} - 15\mathbf{j}$

  **d** no – the helicopter will follow a curved path, not move in a straight line.

# 11 Forces in Equilibrium, Friction and Moments

### Exercise 1

**1**      **2**      **3**      **4**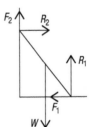

**5 a**    **b**   **c**      **6**

**7**      **8**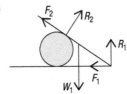

## Exercise 2

1 7.17 N, 33°

2 1.43 N, 170°

3 22.9 N, 234°

4 12.1 N, 38°

5 8.28 N, 51°

6 17.2 N, 63°

7 8.54, 69°

8 13 m/s, 23°

9 2 m, 90°

10 5 N, 127°

11 8.22 N, 49°

12 261 m, bearing 305°

13 292 N, 6°

14 1.01 N, 97°

## Exercise 3

1 a $P = 30\sqrt{3}, Q = 30$  b $P = 12, Q = 12\sqrt{3}$  c $P = 100 \sin 70° = 34.2, Q = 100 \cos 20° = 94.0$

2 a $P = 24, Q = 22\sqrt{3}$  b $P = 24, Q = 15\sqrt{3}$  c $P = 8, \theta = 60°$

3 a $P = 13\sqrt{3}, Q = 0$  b $\theta = 60°, P = 4\sqrt{3}$  c $P = 10, Q = 8$

4 $5g$ N

5 90°; $T_1 = 12$ N, $T_2 = 16$ N

6 17.3 N

7 50 N

8 (anticlockwise from P) $Pi, Qj, \dfrac{5\sqrt{3}}{2}i - \dfrac{5}{2}j, \dfrac{-3\sqrt{3}}{2}i - \dfrac{3}{2}j, 2\sqrt{3}i - 6j; P = -3\sqrt{3}, Q = 10$

9 a $q = -16, p = 0$  b $p = 4, q = -7$

10 a $a = 7, b = -11$  b $-11i + 2j$

11 0.577

12 $\dfrac{1}{3}$

13 22.4 N

14 0.789

15 16.2 N

16 $\dfrac{1}{2}$

17 a 17.4 N  b 9.18 N

## Exercise 4

1 a 12 N m ⟲  b 19.2 N m ⟳  c 10.5 N m ⟳

2 a 11 N m ⟳  b 1.5 N m ⟳  c 3 N m ⟳  d 0

  e 2.5 N m ⟳  f 0  g 32 N m ⟳

3 a 1 N m ⟲  b 2.54 N m ⟲  4 a 1.73 N m  b 7.20 N m

## Exercise 5

1 A: $11\dfrac{2}{3}$ N, B: $13\dfrac{1}{3}$ N

2 A: $13\dfrac{1}{4}$ N, B: $21\dfrac{1}{4}$ N

3 A: 26 N, B: 6 N

4 A: 36 N, B: 0

5 $T_1 = 10$ N, $T_2 = 5$ N

6 $T_1 = 30$ N, $T_2 = 0$

7 $T_1 = 35$ N, $T_2 = 70$ N

8 $T_1 = 0, T_2 = 45$ N

9 About 1 m

10 a 150 N, 570 N  b 312 N, 408 N (3 sf)  c each 360 N

11 a 50 kg  b 12 kg

12 1.375 m

## Exercise 6

1 34°

2 a 200 N  b 100 N  c $\dfrac{1}{2}$

3 28°

4 513 N; $\mu = 0.521$

5 1.22 m

6 20.6°

7 12 N; 12 N ↑

8 a 12 N  b 20.8 N

9 $12\sqrt{3}$ N (20.8)

10 a $53\dfrac{1}{3}$ N  b 44.3 N at 16° to PQ  12 a 11.5 N  b 5.77 N

13 1.13, 26.7 N

## Review

1 a 10 N, no  b 40 N, no  c 40 N, yes

2 a $\dfrac{3}{7}$ m  b 500 N  3 685 N  4 a 40 N

## Assessment

1 b 9.56 N  c 139 N

2 0.309

   Assume no other resistance to motion; model the trolley as a particle.

3 a 5.5 N m  b 2.75  4 a 0.825 m  b 0.8 m

5 a 81.3 N  b 49.4 N at 81° to the downward vertical.

**6 a** $0.775\,\text{ms}^{-1}$

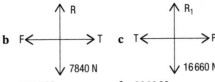

**b** F$\longleftarrow$T with R up, 7840 N down

**c** T$\longleftarrow$P with $R_1$ up, 16 660 N down

**d** 3140 N     **e** 3180 N     **f** 3260 N

**7** 0.268

# 12 Centres of Mass

## Exercise 1

**1** $4.44a$      **2** $2.5a$      **3** $2.2a$      **4** $2.6a$

**5** $1.75a$      **6** $3.09a$      **7** $\left(\dfrac{13}{5}, \dfrac{16}{5}\right)$      **8** $\left(\dfrac{6}{7}, \dfrac{16}{7}\right)$

**9** $\left(\dfrac{1}{5}, \dfrac{11}{5}\right)$      **10** $\left(-\dfrac{4}{5}, \dfrac{6}{5}\right)$      **11** $5\mathbf{i} + 4.6\mathbf{j}$

## Exercise 2

**1 a** $(2, 2)$      **b** $(4, 3)$      **c** $(2, 1)$

**2 a** $(3.5, 5.5), (7, 4)$      **b** $(1.5, 3.5), (5, 1.5)$

**3**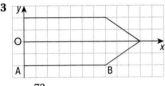

$\bar{X} = \dfrac{73}{17}, \bar{Y} = 0$

**4**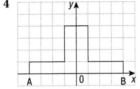

$\bar{X} = 0, \bar{Y} = \dfrac{19}{14}$

**5**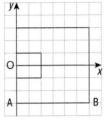

$\bar{X} = 3.25, \bar{Y} = 0$

**6** $(5.6, 4.3)$      **7** $\left(\dfrac{16}{3}, \dfrac{14}{3}\right)$      **8** $(12, 7.2)$      **9** $\left(\dfrac{10}{7}, \dfrac{6}{7}\right)$

**10 a** 5 cm      **b** 2 cm      **11 a** 2 m      **b** 7 m

**12** $\dfrac{18}{7}$ cm      **13** 4330 km

Model the Earth and Moon as particles.

## Exercise 3

**1** 25°      **2** 19°      **3** 43°      **4** 38°

**5** 74°      **6** 88°      **7 a** 4.90 cm      **b** 8.79 cm

## Review

**1** $3.9a$      **2** $3.5a$      **3** $\left(\dfrac{1}{3}, 1\right)$      **4** $(2, 1)$

**5** $(3, 3), (6, 4)$      **6 a** $\dfrac{13}{3}$ cm      **b** $\dfrac{25}{3}$ cm      **c** 20.4°

**7** 45°

## Assessment

**1 a** $\left(\dfrac{63}{14}, \dfrac{57}{14}\right)$      **b** 30.6°

**2 a** The lamina is symmetrical about the line.      **b** 4.07 cm

  **c** 53.6°      **d** at G: 247 N, at H: 204 N

**3 a** 12.2 cm      **b** 77.1°

**4 a i** 2 cm    **ii** 3.5 cm      **b** 29.7°

**5** 5.5 m

**6 a** Symmetry      **b** 13.6 m      **c** 42°

# 13 Motion in One, Two and Three Dimensions

## Exercise 1

**1 a** $(g+1)\,\text{N}$    **b** $(g+5)\,\text{N}$    **2 a** $583\,\text{kg}$    **b** $5070\,\text{N}$

**3 a**    **b** $96\,\text{N}$    **4** $41\,\text{m}$

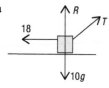

**5 a** $2.39\,\text{ms}^{-2}$    **b** $24.4\,\text{N}$    **c** $47.1\,\text{N}$    **6** $0.342$

## Exercise 2

**1** $v = 9\mathbf{i} + 14\mathbf{j};\ r = 9\mathbf{i} + \dfrac{57}{2}\mathbf{j}$

**2 a i** $6\mathbf{i}$    **ii** $4\mathbf{j}$    **iii** $v = 6\mathbf{i} + 4t\mathbf{j},\ r = 6t\mathbf{i} + 2t^2\mathbf{j}$    **b** $18y = x^2$

**3 a** $8\mathbf{i} + 4\mathbf{j};\ 4\mathbf{i} + 2\mathbf{j}$    **b** $r = 2t^2\mathbf{i} + (t^3 + 3)\mathbf{j}$

**4 a** $14\mathbf{i} + (6 + 2t)\mathbf{j}$    **b** $a = 7\mathbf{i} + (3 + t)\mathbf{j},\ v = 7t\mathbf{i} + (3t + \tfrac{1}{2}t^2)\mathbf{j}$    **c** $14\mathbf{i} + \dfrac{22}{3}\mathbf{j}$

**5** $2\mathbf{i} + \dfrac{3}{2}\mathbf{j} + \dfrac{1}{2}\mathbf{k}$    **6 a** $v = 6\mathbf{i} + (\cos t)\mathbf{k},\ a = -(\sin t)\mathbf{k}$    **b** $F = -4(\sin t)\mathbf{k}$

**7 a** $a = 2\mathbf{i} - 3\mathbf{j} - \dfrac{1}{2}e^t\mathbf{k}$    **b** $v = 2t\mathbf{i} - 3t\mathbf{j} - \dfrac{1}{2}e^t\mathbf{k}$

## Exercise 3

**1 a** $21.4\,\text{ms}^{-1}\nearrow$    **b** $22.5\,\text{ms}^{-1}\searrow$

**2 a** $12.4\,\text{m}$    **b** $15.0\,\text{m}$    **c** $7.86\,\text{m}$

**3 a** $25.4\,\text{ms}^{-1},\ (24, 13.1)$    **b** $24.0\,\text{ms}^{-1},\ (48, 16.4)$    **4** $1\,\text{second},\ 3.76\,\text{m}$

**5 a** $27.7\,\text{ms}^{-1}$    **b** $78.4\,\text{m}$

**6** $x = 4.33,\ y = 1.28;\ 4.51\,\text{m}$    **7 a** between $0.8\,\text{s}$ and $4\,\text{s}$    **b** $32\,\text{m}$

**8 a** $15.2\,\text{ms}^{-1},\ 10.2°$ below the horizontal    **b** $10.4\,\text{ms}^{-1},\ 15.1°$ below the horizontal

**9** $2.04\,\text{seconds},\ 20.4\,\text{m}$    **10 a** $2.64\,\text{seconds}$    **b** No, the ball lands $24.4\,\text{m}$ from the wall

**11** $2.70\,\text{seconds}$    **12 a** $32.5\,\text{m}$    **b** $212.5\,\text{m}$

## Exercise 4

**1** $7.35\,\text{m},\ 50.9\,\text{m}$    **2** $21.7°$ or $68.3°$    **3** $19.8\,\text{ms}^{-1},\ 40\,\text{m}$    **4** $26.1\,\text{ms}^{-1},\ 26\,\text{m}$

**5** $26.3°,\ 129\,\text{m}$    **6 a** $49.0\,\text{m}$    **b** $53.4\,\text{m}$    **7** $y = 0.839x - 0.00334x^2$ (3 s.f.), $15.4\,\text{m}$

**8 a** $11.8°$    **b** $5.87\,\text{m}$    **9 a** $y = \dfrac{4}{3}x - \dfrac{5}{9}x^2$    **b** $y = -x - \dfrac{1}{90}x^2$

**10 a** $1.96\,\text{seconds}$    **b** $25.1\,\text{m}$    **11** $30.7°$    **12** $15.0°$

### Review

**1 a**    **b** $R = 3g - \dfrac{T}{2}$    **c** $9.98\,\text{ms}^{-2}$

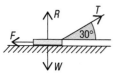

**2 a** $v = 2\mathbf{i} - 3e^{-t}\mathbf{j} + 3t^2\mathbf{k}$    **b** $a = 3e^{-t}\mathbf{j} + 6t\mathbf{k}$    **c** $F = 9e^{-t}\mathbf{j} + 18t\mathbf{k}$

**3** $3.58\,\text{seconds}$    **4** $40.1\,\text{m}$    **5 a** $y = x\tan\theta - 0.00378x^2\sec^2\theta$    **b** $73.9°$ or $17.5°$

### Assessment

**1 a**    **b** $373\,\text{N}$

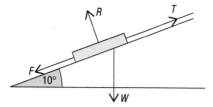

**2 a** $a = 3i + \sin t\,j$    **b** $F = 18i + 6\sin t\,j$    **c** 18 N    **d** $0, \pi, 2\pi$    **e** $r = 2 + \dfrac{3}{2}t^2 i - \sin t\,j$

**3 a** 1.72 seconds    **b** 13.3 ms$^{-1}$    **c** $y = 2x - 0.0362x^2$    **d** yes, it passes 3.5 m above the tree

**4 a** 0.269 seconds    **b** 11.3° below the horizontal    **c** No; hits net 0.14 m below cord

**5 a** 4.45 ms$^{-2}$; 26.7 N    **b** 2.64 ms$^{-2}$; 35.8 N

**6 a** $a = (6t - 1.2t^2)i + 2e^{-2t}j$    **b** $v = (7 + 3t^2 - 0.4t^3)i - (3 + e^{-2t})j$    **c** 10.1 ms$^{-1}$

**7 a** 1.28 s    **b** 15.7 ms$^{-1}$    **c** 20.0 ms$^{-1}$

# 14 Work, Power and Energy

### Exercise 1

**1** (anticlockwise from 20 N)
$20\sqrt{3}$ J, 0, −24 J, 0

**2** (anticlockwise from 8 N)
−24 J, 0, 48 J

**3** (anticlockwise from 7 N)
28 J, 0, −8 J, −16 J

**4** 61.7 J    **5** 5970 J

**6** 3390 J; assume all crates raised to exactly shelf floor level    **7** 24 kJ    **8** 108 kJ

**9 a** 180 J    **b** 282 J Assume constant speed    **10** 2800 kJ    **11 a** 8110 J    **b** 8110 J

**12** 1440 J; assume steady speed and rope doesn't stretch

**13** 25 N; assume steady speed and rope doesn't stretch

**14** 100 J; assume steady speed and rope doesn't stretch

**15** $\dfrac{1}{4}$; assume steady speed and rope doesn't stretch    **16** 11 kJ; 14 kJ; assume constant speed

### Exercise 2

**1** 261 W    **2** 1.96 kW    **3** 36 W    **4** 4.8 W

**5** 8.41 W    **6 a** 24 kW    **b** 1800 N    **c** 20 ms$^{-1}$

**7** 1000 kW    **8 a** 17.2 ms$^{-1}$    **b** 39.5 ms$^{-1}$

**9 a** 94.1 kW    **b** 5880 N    **c** 16 ms$^{-1}$    **10** 0.65 ms$^{-2}$    **11** 1.78 ms$^{-2}$    **12** 0.585 ms$^{-2}$

### Exercise 3

**1 a** 431    **b** 0.816

**2 a** 324 J    **b** 60 000 J    **c** 2000 J    **d** 65 J

**3 a i** 0    **ii** 4230 J    **b** 4760 J    **c** 1910 J    **d** 1380 J    **e** 4230 J

**4** Treat woman as a particle    **a** 144 J    **b** 598 J

**5 a** 79 kJ    **b** 25 m/s (90 k/h)    **6 a** 10 m/s    **b** 25 m

### Exercise 4

**1** 22.9 N; assume string light and inextensible    **2** 576 J

**3** 1.64 N; 6.41 ms$^{-1}$; assume resistance constant

**4** 11 J; 6.2 N (assume zero resistance)    **5** 12.4 ms$^{-1}$

**6 a** 6.20 ms$^{-1}$    **b** 4.77 ms$^{-1}$    **7 a** 137 J    **b** 148 J

**8 a** 16.1 m    **b** 2.99 ms$^{-1}$    **9** 3.4 N; assume resistance constant

**10 a** 16 J    **b** 36 J

### Exercise 5

**1** 1.84 m    **2** 9.86 ms$^{-1}$

**3 a** 4.3 m    **b** 6.1 ms$^{-1}$    **c** 9.1 ms$^{-1}$

**4** 1.63 m    **5** 4.85 ms$^{-1}$    **6** 3.93 ms$^{-1}$    **7** 3.13 ms$^{-1}$

**8** 0.3 m    **9** 5.6 ms$^{-1}$    **10** 4.8 ms$^{-1}$

**11 a** 6.3 ms$^{-1}$    **b** 3.1 ms$^{-1}$

### Review

**1** 4350 J    **2** 1390 N    **3** 50 J    **4** 28.6 ms$^{-1}$

**5** 4000 N    **6** 294 J    **7** 87.5 J    **8** 21600 N    **9** 18.0 ms$^{-1}$

1  **a**  265 N             **b**  164 N             **c**  245 J             **d**  913 J
2  **a**  31.4 ms⁻¹         **b**  22.7 ms⁻¹         **c**  25.6 ms⁻¹         3  30 kW
4  **a**  203 J             **b**  34 J              **c**  0.077            5  1.9 m
    Model girl plus seat as a particle – a very rough approximation.
    Model rope as light and inextensible, assume no air resistance – both reasonable.
6  3.6 kg
7  **a**  29 800 J          **b**  23 100 J          **c i**  52 900 J        **ii**  37.3 ms⁻¹

# 15 Uniform Circular Motion

## Exercise 1

1  **a**  1.91              **b**  10.5             2  0.00175            3  **a**  $6.94 \times 10^{-4}$   **b**  $7.27 \times 10^{-5}$
4  **a**  1.2               **b**  12.5             **c**  4
5  **a**  1.6 rad s⁻¹       **b**  1.6 ms⁻¹         6  51.2               7  432
8  **i**  $\overrightarrow{PO}$   **ii**  $\overrightarrow{QO}$   9  30              10  13.9            11  $\mathbf{a} = -(\cos t)\mathbf{i} - (\sin t)\mathbf{j}$
12 **a**  0.0338 ms⁻² towards the centre of earth   **b**  0.0169 ms⁻² towards, and at right angles to, the earth's axis

13  $\dfrac{48\,200}{g}$ m          14  4.2 rad s⁻¹

## Exercise 2

1  10 N                    2  16.7 m              3  **a**  80 ms⁻²          **b**  8 N
4  **a**  42.7 N           **b**  14.4 rad s⁻¹
5  Model passenger as a particle and assume plane's angular velocity constant.
    **a**  horizontal: 1130 N; vertical: 588 N      **b**  1280 N at 27° to the horizontal
6  Model satellite as particle, assume earth spherical, assume satellite orbiting steadily.
    **a**  4750 N          **b**  7860 ms⁻¹        **c**  0.00121 rad/s    **d**  86.5 minutes
7  **a**  29.4 N           **b**  12.1 rad s⁻¹     **c**  2.42 ms⁻¹         8  **b**  $-(6 \sin t)\mathbf{i} - (6 \cos t)\mathbf{j}$

## Exercise 3

1  **a**  288 N            **b**  84°              2  **a**  21.6 N          **b**  3.90 rad s⁻¹
3  0.392 m                 4  **a**  5.88 N        **b**  2.8 ms⁻¹
5  **a**  493 N            **b**  37°              **c**  3.0 m             6  2.76 ms⁻¹
7  **a**  tension in the two parts of the string cannot be equal for Q to move in a circle      **b**  1.06 ms⁻¹      **c**  1.18 s
8  **a**  0.4 kg           **b**  1.5 ms⁻¹

## Review

1  1675 kmh⁻¹              2  16.7 ms⁻²            3  2.18 rad s⁻¹          4  **a**  $-(3 \sin t)\mathbf{i} + (3 \cos t)\mathbf{j}$     **c**  4.5
5  16.7 m                  6  **a**                **b**  1.13 N            **c**  63.6 cm

1  4.4 rad s⁻¹             2  **a**  0.96 N, 3 N    **b**  1.2 N, 2.28 N     **3a** 9 ms⁻¹, 18 ms⁻¹, 27 ms⁻¹
    **b  i**  $225\omega^2$ N       **ii**  $375\omega^2$ N    **iii**  $450\omega^2$ N
4  **a**                                           **b**  0.0884 m          **c**  56.6 N

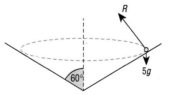

5  **a**  0.000145 rad s⁻¹     **b**  0.565 ms⁻²

6  **a**  $\dfrac{2\pi}{3}$ radians per second      **b**  9.57 N           **c**  1.56 m            7  115

# Index